ETHICS ON T

ZOO AND AQUARIUM

BIOLOGY AND CONSERVATION SERIES

SERIES EDITORS

Michael Hutchins, *American Zoo and Aquarium Association*

Terry L. Maple, *Zoo Atlanta*

Chris Andrews, *National Aquarium at Baltimore*

Published in cooperation with the American Zoo and Aquarium Association

This series aims to publish innovative studies in the field of zoo and aquarium biology and conservation. Priority will be given to those contributions that focus on the interface between captive and field conservation and seek to merge theory with practice. Topics appropriate for this series include but are not limited to zoo- and aquarium-based captive breeding and reintroduction programs, animal management science, philosophy and ethics, public education, professional training and technology transfer, and field conservation efforts.

ETHICS

ON THE ARK

ZOOS, ANIMAL WELFARE, AND WILDLIFE CONSERVATION

EDITED BY BRYAN G. NORTON, MICHAEL HUTCHINS, ELIZABETH F. STEVENS, AND TERRY L. MAPLE

with assistance from John Wuichet

SMITHSONIAN INSTITUTION PRESS • Washington and London

Excerpt from *Slaughterhouse-Five,* by Kurt Vonnegut, Jr., copyright © 1968, 1969 by Kurt Vonnegut, Jr., is used by permission of Delacorte Press/Seymour Lawrence, a division of Bantam Doubleday Dell Publishing Group, Inc.

Copy Editor: Lorraine Atherton
Production Editor: Deborah L. Sanders
Designer: Janice Wheeler

Library of Congress Cataloging-in-Publication Data
    Ethics on the ark : zoos, animal welfare, and wildlife conservation / edited
        by Bryan G. Norton . . . [et al.].
        p.   cm.
    Papers from a workshop held in Atlanta, Georgia, in March 1992.
    Includes bibliographical references.
        ISBN 1-56098-515-1   ISBN 1-56098-689-1
        1. Captive wild animals—Breeding—Moral and ethical aspects—
    Congresses. 2. Wildlife conservation—Congresses. 3. Zoos—Philosophy—
    Congresses. 4. Zoo animals—Congresses. 5. Animal welfare—Congresses.
    I. Norton, Bryan G.
    SF408.3.E84 1995
    639.9'3'01—dc20   94-37139

British Library Cataloguing-in-Publication Data available

Manufactured in the United States of America
02 01 00 99 98 97   5 4 3

∞ The paper used in this publication meets the minimum requirements of the American National Standard for Permanence of Paper for Printed Library Materials Z39.48-1984.

Dedicated to the memory of Robert W. Loftin

1938–1993

philosopher, ornithologist, conservationist, and teacher

CONTENTS

CONTRIBUTORS

Karen Allen is director of both public affairs and the Alliance for Conservation at Conservation International. She is a former public relations director for the AZA and the Audubon Park and Zoological Garden in New Orleans. She has worked for both the CBS Television Network and the CBS affiliate in New Orleans. As a journalist, she covered natural history and conservation issues in Central and South America. She has delivered a number of papers on animal rights and animal welfare issues and is currently researching the topic of cultural sensitivity and conservation.

Benjamin Beck is associate director of biological programs at the National Zoological Park, chair of the AZA's Reintroduction Advisory Group, deputy chair of the IUCN Species Survival Commission Reintroduction Specialist Group, and a member of the propagation groups of the AZA gorilla and orangutan Species Survival Plans, the Great Ape Taxon Advisory Group, and the Scientific Advisory Board of the Dian Fossey Gorilla Fund. A primatologist by training, he is author of the books *Animal Tool Behavior* and *The Use and Manufacture of Tools by Animals*. Dr. Beck has coordinated the preparation, reintroduction, and postrelease monitoring of 135 golden lion tamarins in Brazil and maintains a database on reintroduction programs for captive-born animals.

Roger Caras is president of the American Society for the Prevention of Cruelty to Animals (ASPCA), vice president and vice chairman of the board for the Humane Society of the United States, and author of more than sixty books on animals. He has served as special correspondent for animals and the environment on such television programs as *ABC World News Tonight, 20/20,* and *Good Morn-*

ing America and currently produces a daily feature for CBS Radio. Roger Caras holds professorships at a number of universities, sits on the boards of humane and conservation organizations, and has been awarded three honorary doctorates and many international awards.

William Conway, president of the Wildlife Conservation Society (formerly New York Zoological Society), directs the Bronx Zoo and oversees a wildlife conservation program with projects in forty-six nations. He led development of the AZA's Accreditation and Species Survival Plan programs and conducts a major conservation program in Patagonia. He has written extensively about wildlife and received several honorary doctorates.

Betsy Dresser is research director of the Center for Reproduction of Endangered Wildlife (CREW) at the Cincinnati Zoo and Botanical Garden. Dr. Dresser is a reproductive physiologist who has developed innovative technological approaches to animal propagation, including cryopreservation, artificial insemination, embryo transfer, and in vitro fertilization. Dr. Dresser is also an associate professor of obstetrics and gynecology at the University of Cincinnati College of Medicine, and she is a leading specialist in the field of reproductive technologies for animals and humans.

David Ehrenfeld is professor of biology at Rutgers University, New Brunswick, New Jersey, where he teaches courses in ecology. He received his medical degree from Harvard Medical School and his doctorate in zoology from the University of Florida. He is the author of *The Arrogance of Humanism* (1978, 1981) and the founding editor of *Conservation Biology.* His latest book is *Beginning Again: People, Nature, and the New Millennium.*

Ardith Eudey is Asia vice chair of the Primate Specialist Group (PSG), IUCN Species Survival Commission, and editor of *Asian Primates,* a newsletter of the PSG. For more than twenty years she has studied the distribution and habitat preference of sympatric species of macaque monkeys in west-central Thailand. Her professional interests also include reconciling conservation objectives with the socioeconomic needs of hill tribes and other minorities. She has taught biological anthropology at California State University, Hayward; University of California, Davis; and University of Nevada, Reno. Presently, she is updating the *Action Plan for Asian Primate Conservation: 1987–91,* which she compiled for the PSG.

Roger Fouts is professor of psychology and director of the Chimpanzee and Human Communication Institute at Central Washington University in Ellensburg, Washington. He is best known for his research of the past twenty-seven years concerning the acquisition of American Sign Language by chimpanzees. He is also a member of the Board of Trustees of the Psychologists for the Ethical Treatment of Animals and active in efforts to improve laboratory conditions and the psychological well-being of captive chimpanzees.

Valerius Geist is professor and program director for environmental science in the Faculty of Environmental Design at the University of Calgary. An ethologist, ecologist, and evolutionist, Dr. Geist has authored several books, including his classic work in behavioral ecology, *Mountain Sheep,* and in the biology of health, *Life Strategies, Human Evolution, Environmental Design.* His research has been applied to the transformation of strip mines into custom bighorn sheep habitats. His recent work includes a major monograph on the deer family and the editing of a multiauthor book on wildlife conservation policies. Three trade books published with Michael Francis on mule deer, elk, and mountain sheep combine good science with good photography for the general public. Dr. Geist has been honored for his work by the Wildlife Society and the American Association for the Advancement of Science.

David Hancocks is executive director of the Arizona-Sonora Desert Museum in Tucson, Arizona. He is a registered architect and has carried out design and planning work for zoos in England, France, Singapore, Australia, and the United States and champions the notion that zoo architects should recognize the animals as their principal clients. From 1974 to 1985 he was director of the Woodland Park Zoological Gardens in Seattle, Washington, where the concepts of bioclimatic zoning and landscape immersion were established. He was one of the earliest practitioners of what has come to be known as environmental enrichment for animals in zoos. David Hancocks is a fellow of the Discovery Institute, Seattle.

Eugene Hargrove is associate professor of philosophy and chair of the Department of Philosophy and Religion Studies at the University of North Texas, where he is founder and editor of the journal *Environmental Ethics*. Dr. Hargrove received his degree in philosophy from the University of Missouri. He is the author and editor of several books, including *Foundations of Environmental Ethics,* and numerous journal articles.

Michael Hutchins is director of conservation and science for the American Zoo and Aquarium Association (AZA), consulting editor for *Zoo Biology,* and adjunct professor, University of Maryland Graduate Program in Sustainable Development and Conservation Biology. He and his staff administer the AZA Species Survival Plan, an internationally recognized captive breeding and field conservation program for more than one hundred species of endangered or threatened animals. Dr. Hutchins received his doctorate in 1984 from the University of Washington, Seattle. His dissertation research focused on the behavioral ecology of Rocky Mountain goats in Olympic National Park.

Dale Jamieson is professor of philosophy at the University of Colorado, Boulder, adjunct scientist in the Environmental and Societal Impacts Group at the National Center for Atmospheric Research, and former director of the Center for Values and Social Policy. He is author of "Against Zoos," which originally appeared in *In Defense of Animals,* edited by Peter Singer, and has since been widely anthologized. In addition to publishing many articles, he is the coeditor of *Interpretation and Explanation in the Study of Animal Behavior* and *Reflecting on Nature: Readings in Environmental Philosophy.* He is currently at work on a book on global environmental change.

Fred Koontz is director of the Science Resource Center for the Wildlife Conservation Society and cochair of the AZA Old World Monkey Advisory Group. Dr. Koontz received his doctorate in zoology from the University of Maryland. Dr. Koontz is a member of the IUCN Species Survival Commission's Reintroduction Specialist Group and the Captive Breeding Specialist Group. He is also a member of the AZA's Small Population Management Advisory Group. In addition, Dr. Koontz is the author or coauthor of many publications in the field of zoology.

Robert Lacy is a population geneticist in the Department of Conservation Biology of the Brookfield Zoo, Chicago. He conducts research on the effects of inbreeding and hybridization in endangered species, develops techniques for the analysis of pedigrees and the management of wildlife breeding programs, and assists wildlife agencies in the use of computer models of population viability to help guide the recovery of endangered species. He is a member of the IUCN SSC Captive Breeding Specialist Group and the AZA Small Population Management Advisory Group. He has served as a consultant on population biology and genetic aspects of conservation to wildlife agencies and zoos around the world.

Donald Lindburg is head of the Behavior Division for the Center for Reproduction of Endangered Species, which is operated by the Zoological Society of San Diego. He is also editor of the book *The Macaques* and editor of the journal *Zoo Biology.*

Linda Lindburg is managing editor of the journal *Zoo Biology.* She is former president of the Markham Park Zoological Society in Broward County, Florida.

Robert Loftin, professor of philosophy and lecturer in ornithology, University of North Florida, died on 13 August 1993, at the age of 54, after a second bout with cancer; the editors dedicate this book to his memory. He was a distinguished philosopher, an ardent conservationist, a beloved teacher, and an enthusiastic ornithologist. Dr. Loftin received his degree in philosophy from Florida State University and published extensively on the treatment of animals, including the articles "The Morality of Hunting" and "The Medical Treatment of Wild Animals," both published in *Environmental Ethics,* for which Dr. Loftin was the book review editor.

Terry Maple is president and chief executive officer of Zoo Atlanta, professor of psychology at the Georgia Institute of Technology, affiliate scientist at the Yerkes Primate Research Center of Emory University, and an elected fellow of the American Psychological Association. For seven years Dr. Maple was editor-in-chief of the journal *Zoo Biology.* He is author of more than eighty scientific publications, including *Captivity and Behavior* (1979), *Orangutan Behavior* (1980), and *Gorilla Behavior* (1982). In 1993, his new book *Zoo Man* was published by Longstreet Press.

Rita McManamon is senior vice president for veterinary services at Zoo Atlanta and serves as the veterinary advisor for the AZA Orangutan Species Survival Plan. She has extensive experience in wildlife rehabilitation and emergency veterinary medicine. Dr. McManamon teaches veterinary students at Tuskegee University, the University of Georgia, and Auburn University. Through travel and professional contacts, she is frequently consulted on international animal health and welfare issues.

Bryan Norton is professor of environmental public policy at the Georgia Institute of Technology. Much of Dr. Norton's past work has focused on questions of biological diversity. He is the author or editor of several books, including *Why Preserve Natural Variety?, Toward Unity among Environmentalists,* and *Ecosystem Health: New Goals for Environmental Management.* His current work focuses on defining ecosystem health and determining parameters for sustainability. Dr. Norton's degree is in philosophy, from the University of Michigan. He is a member of the Environmental Economics Advisory Committee of the Environmental Protection Agency's Science Advisory Board and a fellow of the Hastings Center.

Tom Regan is professor of philosophy at North Carolina State University, where he has twice been elected outstanding teacher. He has been named Alumni Distinguished Visiting Scholar at the University of Calgary and has served as visiting distinguished professor of philosophy at Brooklyn College. His books include *The Case for Animal Rights* (1983), which was forwarded for the Pulitzer prize, *Bloomsbury's Prophet: G. E. Moore and the Development of His Moral Philosophy* (1986), and *The Thee Generation: Reflections on the Coming Revolution* (1991). In addition to his scholarly activities, his work as a video author and director has earned major international awards.

Elizabeth Stevens is director of Zoo Atlanta's Conservation Action Resource Center, where she coordinates the zoo's conservation, research, and education programs. After completing her B.S. in zoology at Duke University, she began her graduate studies at the Institute for Behavioral Physiology at the University of Tübingen in West Germany and then obtained her doctorate in biology from the University of North Carolina at Chapel Hill. Her graduate studies focused on the evolution of vertebrate social systems, and she conducted field research in Africa, South America, and on the North Carolina coast. Dr. Stevens first became involved in wildlife conservation at zoos during her curatorial internship at the National Zoological Park, Smithsonian Institution, in Washington, D.C.

Robert Vrijenhoek is professor of marine and coastal sciences and director of the Center for Theoretical and Applied Genetics at Rutgers University. He is a former editor of *Evolution* and currently an associate editor for *Conservation Biology* and the *Journal of Heredity*. Dr. Vrijenhoek received his degree in 1972 from the University of Connecticut, Storrs. His recent research has focused on gene flow, molecular systematics, and biodiversity in deep-sea hydrothermal vent organisms, requiring many dives in deep submergence vehicles such as *Alvin* and *Shinkai 6500*.

Frederic Wagner is professor in the Department of Fisheries and Wildlife and associate dean of the College of Natural Resources, Utah State University. He is an expert on natural resources management policy, having worked extensively on issues related to the control of predators and exotic species. Dr. Wagner received his degree in zoology and wildlife management from the University of Wisconsin and is currently director of the Ecology Center at Utah State University. Interested mainly in animal ecology, particularly wildlife ecology, Dr. Wagner has focused his recent work on the role of science in natural resources policy.

Chris Wemmer is associate director of conservation at the National Zoological Park. A mammalogist who has worked extensively both in zoos and in the field, he is coauthor of *Pere David Deer: The Biology and Management of an Extinct Species*. Dr. Wemmer received his degree from the University of Maryland and is currently director of the National Zoo's Conservation and Research Center in Front

Royal, Virginia. His special interest is in facilitating wildlife conservation and research in developing countries. Dr. Wemmer has administered and advised field projects on the Bengal tiger, the greater one-horned Asian rhino, and domesticated Asian elephants.

John Wuichet is associate fellow for the U.S. Army Environmental Policy Institute in Atlanta, Georgia. He has contributed to the Army-wide *Threatened and Endangered Species Policy Development Plan.* He holds a master's degree in environmental public policy from the Georgia Institute of Technology and a bachelor's degree with honors in philosophy from Oglethorpe University.

FOREWORD

The people of every period of human history have had prevailing ways of thinking about their relationship to the world around them. From time to time, after intervals of centuries, great events—a change of dynasty, the emergence of a new religion, the invention of a new technology—herald a change in the dominant worldview. Which comes first, the change in worldview or the great event, is not important here. What concerns me in this writing is that the years of transition between worldviews are always difficult and dangerous. Entrenched ideas and entrenched power structures struggle to maintain dominance or merely to stay alive: the struggles are often vicious and desperate, like those of a mighty predator cornered by ruthless hunters. Millions of people are affected. Such a period culminated in the nineteenth century, when the industrial revolution, backed by a persuasive misinterpretation of the new Darwinian theory, finally swept away the old way of living. But the Age of Control—as we might call the worldview that accompanied and supported the industrial revolution—has had a surprisingly short reign. Now, as it begins to give way to what Wendell Berry calls the Age of Modesty, the transition period is especially turbulent, confusing, and bloody.

Institutions have a tough time during these transition periods. Many do not make it; of the few that do, most are changed out of all recognition. Zoos are among the institutions of antiquity that have survived a number of transitions, but the contemporary zoo is largely a product of the nineteenth century and the first two-thirds of the twentieth. In many ways, the zoo has come to typify the themes of the Age of Control: exploration, domination, machismo, exhibitionism, assertion of superiority, manipulation. Indeed, even as the century and millennium come to an end, the poorer-quality zoos still offer little more than gratification of these sordid and discredited ideals. The good zoos are something else again.

The good zoos—guided by their professional organization, the American Zoo and Aquarium Association—have made a remarkable decision, embodied in this book. The decision, almost unique among contemporary institutions, was to undergo a searching examination of every aspect of their institutional character, practice, and goals, as a preliminary step toward considering deep, fundamental changes that would align zoos with the new worldview. This decision could be called a product of enlightened self-preservation, yet I think there is a less cynical way of explaining it. Zoos, aquariums, and botanical gardens appeal to some of the most profound parts of our human nature—perhaps they are a pleasant and comforting reminder to us of our origins in and myriad ties to nature. This appeal has transcended a succession of worldviews, although the zoo has always adapted itself to the culture of the moment. Now, once again, our views of ourselves and our place in nature are changing, and it is up to those who know and care for zoos, if they wish to save this ancient institution, to see that they, like the animals in them, evolve and survive.

The only way for self-examination to work is if it is completely honest, and uncompromising honesty is the most striking feature of this book. If you are a reader accustomed to a televised diet of lies and deception, you many find yourself looking for the missing information, the stacked arguments, and the clever simplifications. You won't find any. You will find passionate and scientifically convincing essays describing the critical role zoos are playing in conserving the vanishing species of the world's ecosystems, and in educating children and adults about the value and the plight of nature. And you will find passionate and carefully reasoned essays that prove to you that zoos are conserving little or nothing, that they are teaching and reinforcing destructive ways of relating to nature, that in fact society would be better off if zoos became extinct.

There are many issues and many sides in the great zoo debate. Do the rights and welfare of individual animals—this badger, that tortoise—take precedence over the rights and welfare of whole species or even ecosystems; can a hybrid orangutan be sacrificed to make zoo space for a presumably more important orangutan with a genotype closer to that of her wild ancestors? Are zoos playing God (the old worldview) when they manipulate and even risk killing individual wild or captive animals to learn or achieve something that will preserve their endangered populations? Can zoos predict the outcomes of their actions well enough to justify such interference? Is it more important to try to save the charismatic megavertebrates that the public likes to look at—the rhino and the macaw—or should zoos give equal effort to inconspicuous little animals—the mouse and the topminnow? Or should they concentrate on keystone species, large or small, that sustain the integrity of whole ecosystems? Or should zoos abandon altogether their possibly ineffectual pretensions to wise stewardship and endorse the unzoolike conservation method of simple protection of wild land, with active management replaced by benign neglect? Should zoos stop trying to conserve and display exotic species, focusing instead on local animals and habitats? Should they attempt to make displays indicate, to the observers if not to the occupants of the enclosures, the appearance of the animals' usual ecosystems? Should they forget about the unobstructed display of animals, letting them hide in burrows or bushes if that is what the creatures prefer? Or should zoos abandon traditional animal displays completely, providing only a parklike series of habitat types to look at, in which animals occasionally appear as they go about their business? Or can zoos display the habitats alone, showing the animals on conveniently placed video monitors?

I agree with Dale Jamieson (Part Two) that "conflicts of value are intrinsic to wildlife management" and that "no philosophy can make all these conflicts disappear." If you read this book and embrace one of its two main points of view (pro-zoo or pro–animal rights) to the total exclusion of the other, perhaps you have missed the point. The point is that, at their best, both sides are right, and in such cases the only thing to do is to keep searching for common ground, moving ahead on a path that is mutually acceptable if not always entirely satisfactory. Appropriately, because the symposium upon which this volume was based was the beginning of a dialogue, the path is not yet fully mapped, but it can be dimly perceived. It takes us past at least three patches of common ground: the area of the need for significant improvement of the day-to-day welfare of animals in zoos, the area of teaching the public the importance of saving wild habitats, and the area of redesigning zoos away from the display of caged animals and toward the display of natural areas in a parklike setting.

The new worldview that emerges as the dust of battle settles—whenever that may be—will inevitably contain parts of the old. We can no more prevent this than we can abolish history or reject deep-seated parts of our nature. Nor should we try—the Age of Control contributed or preserved many human inventions of enormous value. The zoo may be one such invention, and this book shows how it might be handed on in peace and cooperation from one warring camp to another, even in a world of upheaval.

DAVID EHRENFELD

PREFACE

One effect of the expansion of human dominance across the globe has been a corresponding expansion of human responsibility. Today, we face new moral dilemmas that never presented themselves to our forefathers because our activities are so much more pervasive. More and more features of the earth and the physical and biological systems that form it are intentional or unintentional effects of human activities. All species modify their environment, but only modern humans have so successfully melded information processing with technological control to consciously remake their habitat in a few generations. It is no exaggeration to say that the success of the human species has, at least so far, been an unmitigated disaster for most other species and for nearly all ecological communities. The human ability to control nature through more and more powerful technologies raises more and more dilemmas, because with conscious control comes moral responsibility; and as humans dominate more and more natural systems with their conscious plans and technological strategies, they cannot avoid responsibility for the devastation of natural habitats and for the resulting death of individuals and the endangerment of populations and species.

The excruciating dilemmas and paradoxes addressed in this book are new in the sense that they arise when our new and expanded responsibilities apparently demand conflicting actions that are advocated by sincere, well-meaning, and passionately committed moral individuals such as animal rights activists, conservationists, and scientific resource managers. In particular, we are concerned with the conflict between our increasing awareness of the often negative impacts of our expanding activities on the health and welfare of individual animals and the equally urgent recognition that human activities profoundly affect ecological and evolutionary processes. It is inevitable that in our attempts to save processes, species, and ecosystems, we will sometimes be faced with choices that negatively impact individual members of sentient species. Becoming, as Aldo Leopold said, not the "conqueror" of the land community but a "plain member and citizen of it" will require of humans that we accept responsibility for all of our actions (or inactions), reduce conflict among our varied obligations to protect individuals and ecosystem processes whenever possible, and ultimately choose wisely which principles should be given priority when unavoidable conflicts arise.

This book results from a workshop, held in Atlanta, Georgia, from 19 to 22 March 1992, at which participants engaged in more than two days of presentations and committee work. The upshot of that conference is the essays presented here and the committee recommendations presented in the appendix. This project, from the start, has been both plagued and, in a more profound sense, blessed by an inherent tension between two charges we imposed upon ourselves: on the one hand, to present the wide variety of opinion that exists among experts and the public concerning our moral obligation to protect wild species and animals, and on the other hand, to achieve as much consensus as possible and prepare recommendations that might help guide future actions of the board of directors of the American Zoo and Aquarium Association (AZA, formerly the American Association of Zoological Parks and Aquariums, AAZPA). This book includes both attempts by publishing, in the main body, the full range of opinions that were expressed at the conference and by providing, in the appendix, an accounting of how much agreement was reached and a guide to further discussions and research. In the appendix, we have reproduced the recommendations of six working groups, which included as balanced as possible a representation of varied viewpoints. The groups were instructed to reach as much consensus as possible but to recognize and characterize differences that could not be resolved.

The purpose of this book is to focus on ethical issues associated with captive breeding programs, especially Species Survival Plans, but to do so while recognizing that these ethical issues can be fully understood only in their broader conservation and social context. While we need to explore concerns regarding the welfare of animals maintained in captivity for breeding programs, we must also recognize that a comprehensive understanding of those ethical dilemmas will require a broader exploration of the role of captive breeding programs in conservation, and also the entire role of zoos and aquariums in modern society.

So we have chosen to explore three broad themes and six issue areas. The six issue areas can be grouped according to whether they address the impact of captive breeding programs on conservation more broadly or issues regarding the care and treatment of animals in captive breeding programs. The book therefore has four main sections addressing the following questions: Part One, What is the role of zoos in conservation and society today and what should their role be in the twenty-first century? Part Two, What should be the main target of our attempts to save wild animals (genes, individuals, populations, species, or ecosystem processes)? Part Three, How do, and should, captive breeding programs interact with in situ conservation efforts? And Part Four, What types and levels of care must be afforded animals that are included in captive breeding programs? Parts Three and Four are broken into three subsections, each of which presents a range of opinion on prominent issues corresponding to areas of discussion of the six working groups. These subsections are presented here as point-counterpoint exchanges in which the first writer states, in general terms, a position representative of the zoo community. Those statements are then followed by one or more reactions from representatives of differing viewpoints. We feel it is important that readers recognize the comprehensiveness of the areas of agreement and how few concrete disagreements exist regarding what policies are required by responsible zoos *if* it is granted that animals should be held in captivity for the purpose of protecting their species from extinction. This implies that there was strong consensus favoring the recommendations—everyone thought that we should at least institute these changes. The moral qualms some participants felt regarding the justification for holding animals captive at all imply that it would be much harder to attain agreement on exactly what more should be done to honor our moral obligations to individual animals who are now captive. Thus, while a few areas remain at issue—especially the perplexing surplus-animal problem—we believe a great deal of consensus was achieved in favor of important policies in all areas, despite the divergent values and worldviews of our participants. Readers are thereby urged to balance the apparent message of diversity of opinion expressed in the chapters with the message of the appendix that, while this dialogue must continue, there are many consensually acceptable steps that can be taken right now to protect individual animals, reduce suffering, and ensure that captive propagation has a fully positive impact on efforts to protect wild species in their natural habitats.

BRYAN NORTON

ACKNOWLEDGMENTS

A project such as this is possible only with the help, cooperation, and support of many people and organizations. The idea for this project was initially discussed in the early eighties by Pamela Parker and Eugene Hargrove, and although that time was not ripe for a full-blown project, we have benefited from those early discussions. We are indebted to the Ethics and Values Studies program of the National Science Foundation for funding the conference in Atlanta (National Science Foundation grant NSF 90-77) and especially to Rachelle Hollander and Vivian Weil, who facilitated the project from their positions in the EVS office.

The AZA board of directors deserves recognition for their support of this project. It was planned and implemented during the tenure of three board presidents: Charlie Hoessle, director, Saint Louis Zoological Park; Steve Wylie, director, Oklahoma City Zoological Park; and Dr. Dennis Merritt, director of collections, Lincoln Park Zoological Gardens, Chicago. The organizers and editors thank these individuals and other board members for their willingness to engage in constructive dialogue and their vision for the future of professionally managed zoos and aquariums. Thanks are also due to former AZA deputy director David Jenkins, current AZA executive director Syd Butler, and AZA chief administrative officer Robert Wagner for their encouragement and support.

We are especially indebted to the staff of Zoo Atlanta, who coordinated the local arrangements for the conference. Celia Grams, budget officer for the School of Public Policy, Georgia Institute of Technology, administered with remarkable good humor the complex process of coordinating the grant budget. George Rabb provided opening remarks to set the stage for the conference. Dennis Merritt, Eugene Hargrove, David Hancocks, Bruce Read, Peter Jaszi, and Douglas Myers served as moderators of the working-group sessions. We also wish to thank members of the zoo staff and graduate students in the School of Public Policy who served as recorders for the working-group sessions. These include Carolyn Harings, Lori Perkins, Craig Piper, Sene Sorrow, and Gail Bruner Lash. Carolyn Harings also assisted in arranging the conference and began the process of assembling the chapters for the book. John Wuichet carried the complex process forward to completion, dealing with the countless details required to turn diverse products of contributors into a final manuscript. In general, the editors wish to thank the many people who participated and helped out in many ways—the project could not have been a success without their help.

ETHICS ON THE ARK

ZOO CONSERVATION AND

ETHICAL PARADOXES

William Conway

I t is a paradox that so many human beings agonize over the well-being of an individual animal yet ignore the millions daily brutalized by the destruction of their environments. But it is not exceptional. We are inspired by wild-ness, equating it with freedom and beauty, and each day we destroy more wildness. We endow animals with traits we find selfless in ourselves, knowing in our hearts that their concerns may have more to do with the next meal than the next world. We are touched with sadness at the plight of vanishing species but much more readily brought to tears by the difficulties of E.T., Dumbo, or Mickey Mouse. And we are just as easily concerned with Star Wars Wookies as with Australian wallabies, with Ninja Turtles as with giant tortoises. We are moved to indignation by the animal researcher's scalpel but thrilled with admi-ration for cheetahs gobbling the guts of agonized gazelles. Poorly equipped to discern data from deceit, we populate our concepts with caricatures that differ not only from culture to culture but also from time to time. We contend forever with short-term perceptions in a long-term world.

The consideration of ethical paradoxes in zoos, aquariums, and also wildlife conservation is timely. As the twentieth century concludes we confront a rela-tionship between humans and animals that has fundamentally and profoundly changed. We have long eaten them, skinned them, studied them, and displaced them, and now we surround them. For most, the only hope of survival lies in our care for them.

To a population biologist, the number of animals in zoos is so small as to be inconsequential. Its significance rests in the fact that zoo collections now provide the most important contact with living wild animals that many people will ever have. Nevertheless, there are ethical paradoxes in the very nature of zoos and their use of wild animals. We must ask ourselves, In what contexts? Against which background?

SOME DEFINITIONS

In the first place, what do we mean by "zoos and aquariums"? The U.S. Department of Agriculture accords exhibitor's permits to more than 1,700 animal collections. They range from shopping mall petting zoos to not-for-profit public institutions with large scientific and teaching staffs and several thousand animals. Among them are 164 institutions accredited by the American Zoo and Aquarium Association (AZA), an association that has made conservation its highest priority. These AZA zoos and aquariums are the professional zoo community, and they are what is meant when I speak of zoos. While we do not doubt that other institutions and their treatment of animals raise important issues, these are beyond the concern of this book.

And what do we mean by "an ethical paradox in zoos"? Zoos seek to inspire public interest in wild creatures and nature, to provide ecological education, and to help save wild species from extinction, but in doing so they confine wild animals away from nature and manage their lives.

For that matter, what do we mean by "animals"? The cynical animal rightist's syllogism "a rat is a pig, is a dog, is a boy" is not sufficiently discerning. Whatever the taxonomy of domestic animals, a rhino is not a roach, a gorilla not a goose, a tiger not a turtle. Wild creatures differ endlessly. These differences are reflected in their forms, needs, behavior, limitations, and abilities. When we use the word "animals" generically, we must do so with great care. We must remember the disparities between the needs of long-nosed bats and leopard cats and consider that preservation of biological diversity is key not only to the objectives of zoos and conservation but ultimately to the well-being of civilization.

Most particularly, in what special context does the modern zoo operate? It is one of extinction. The current rate of species extinction is about 10,000 times the natural background rate that existed before the appearance of human beings (Wilson 1989). This context is also one in which human individuals and institutions address short-term concerns that display ignorance and lack of rational priorities required to preserve life and its ongoing evolution. No powerful government anywhere on earth has placed environmental conservation among its top priorities. These realities are central to the aspirations of zoos and conservationists—and to the paradoxes they face.

UTILIZATION OF WILDLIFE BY HUMANS

Animal collecting for zoos is regularly listed among the major threats to wild animals—a paradox (World Resources Institute 1992, 288). Indeed, comparative wildlife utilization by humans deserves special consideration because zoos are often pictured as cavernous sinkholes of wildlife—another paradox. Both are mythical. I can find no instance where zoo collecting has been among the significant causes of endangerment in the decline of any species (Conway 1968). Far from wildlife sinkholes, zoos have comparatively few animals.

For example, the number of whitetail deer annually shot by New York state hunters alone is about the same as the total of mammals, birds, reptiles, and amphibians now living in all the nation's zoos combined. Only some 600,000 such vertebrates live in all the world's zoos listed in the *International Zoo Yearbook* (Olney et al. 1989). Consider the following figures.

The annual take of oceanic fishes is nearly 90.7 metric tons (100 million tons). Terrestrial takes are equally remarkable, especially of uncommon species. A 1982 study revealed that the 574,000 people of the rural population of the Brazilian state of Amazonas were annually killing 2,824,662 mammals and 530,884 birds. Adding reptiles raises the count to 3.5 million vertebrates killed by these people each year for food. At the same time, another study showed that the inhabitants of three little Waorani villages in Ecuador killed 3,165 mammals, birds, and reptiles, including 562 wooly monkeys, in one year (Robinson and Redford 1991). A 1975–1976 study found that the Peruvians of the Department of Loreto annually kill 370,000 monkeys for consumption and sale—more than acquired by zoos in the last 300 years. Buenos Aires records list the export of the skins of 21,534,299 foxes, pumas, guanacos, nutrias, lizards, and the like between 1976 and 1979 (Robinson and Redford 1991) and even larger kills take place each year in Africa and Asia. U.S. wildlife imports from 1980 through 1985 averaged 12,000 to 14,000 primates, 6 million raw fur skins, 800,000 live birds, 300,000 to 500,000 live reptiles, 2 million to 4 million reptile skins, and 125 million live ornamental fish (Fitzgerald 1989). In contrast to such extractions from nature, about 93 percent of all the mammals and 75 percent of all the birds added to AZA zoo collections in recent years were zoo-bred (Conway 1986).

Of course, the vast majority of captive animal breeding is domestic and takes place outside zoos. After all, the U.S. cattle herd is 98 million strong, the pigs 56 million, the sheep 10.7 million, while the number of chickens is astronomical. More than 5 billion animals are killed for food each year in the United States, 90 percent poultry (Donnelley and Nolan 1990). Besides, we have nearly 118 million dogs and cats, of which we annually kill more than 13 million whose care we are unwilling to pay for, perhaps 2,000 a week in New York City. A further 17 million to 22 million animals are consumed each year in research, testing, and education (Donnelley and Nolan 1990). Just keeping all these domestic animals has an enormous effect upon wild animals.

Worldwide, between 10 billion and 15 billion domestic animals are utilizing former wildlife habitat. More than a billion of these are ruminants, and they dominate at least 3 billion hectares of land (Myers 1984), a piece nearly the size of Africa. These great populations of introduced domestics impose untold suffering upon wild animals while preempting their natural habitats. Another 1.5 billion hectares are even more thoroughly compromised for crops. Besides, in the United States, there are far more exotic ungulates outside than inside zoos. Approximately 300,000 to 500,000 deer, antelope, and wild sheep are being kept and bred on ranches, nearly 200 times the number of their conspecifics at zoos.

HUMAN POPULATION GROWTH, HABITAT LOSS, AND EXTINCTION

E. O. Wilson and Paul Ehrlich calculate that one-fourth of all species on earth will be lost in fifty years if tropical forests continue to be felled at the current rate (Stevens 1991). But the pace is accelerating. At present logging levels, less than 10 percent of the tropical forests will remain standing in twenty years (Steadman 1991). In 1989, Wilson estimated that the loss from deforestation could be as much as 4,000 to 6,000 species per year (Wilson 1989). He now calculates that 50,000 species of plants and animals are being "doomed to extinction" each year, about 6 an hour, based upon species/area models of tropical forests. While such high figures for mostly unseen little species are arguable, what we can see happening to the large ones is not.

All of the big storks and cranes, most of the big cats and crocodilians, all of the great sea turtles, most of the antelope and wild cattle and equines, and even the vast majority of small American, Asian, and European migratory birds are in decline. Only one of the five species of rhinos now has a population as high as 5,000.

The conversion of wild lands to farmlands and developments condemns uncounted millions of wild animals each year to starvation, suffering, and death. Yet, while almost every U.S. state makes it illegal to withhold food or care from captive animals, not one prosecutes those who act the same way toward animals not captive (see New York State Agriculture and Markets Law Section 26).

Sadly, the people who "own" wildlife by virtue of having colonized and dominated its habitats often consider it competitive or a nuisance and actively seek to extinguish it. It is telling that zoo scientists have been accused of playing God (Mann and Plummer 1992) when they try to save a species by captive propagation but would simply be called investors if they contributed to its destruction—perhaps because this destruction is so directly tied to human increase.

Every 2.7 years, humanity adds to its numbers the equivalent of another USA in people to be fed, housed, clothed, and educated. By 2050, human population is expected to be 9 billion to 12 billion. Wild creatures are becoming captives in people-surrounded islands of their former habitats.

Cotton fields border one side of Tanzania's great Serengeti National Park while agriculturists and cattle grazers steadily encroach upon Kenya's Mara and Aberdares preserves. Africa's last 500 mountain gorillas dwell in less than 600 square kilometers (231.6 square miles) of forest encircled by some of the most intensely cultivated farms in the world. In Venezuela, cracids, monkeys, and tinamous are besieged in Henri Pittier and San Esteban national parks. Elephants, gaur, deer, and gibbons are surrounded in Thailand's poorly protected Khao Yai National Park.

But the majority of wild creatures have no reserves at all. Even where parks exist and overhunting, agricultural encroachment, and political catastrophe are avoided, the survival of wild animals is not guaranteed.

THE SHORTCOMINGS OF SIMPLE PROTECTION AND BENIGN NEGLECT

In 1947, Florida's Everglades National Park was established to protect the finest assemblage of large wading birds on the continent. Hunting and encroachment were stopped. Nevertheless, by 1989, the populations of these birds had declined by 90 percent. We manned the borders but failed to care for its contents. In 1987, a sobering study of seven of the largest national parks in western North America found that twenty-seven species of mammals had gone extinct locally in one or more of these sacrosanct reserves (Newmark 1987). Such tragedies are being replicated around the world.

Whether there is global climate change or not, wildlife communities and their habitats are not static. They constantly change in makeup and form. Although Kenya's 184 square kilometers (71 square miles) of Lake Nakuru National Park have been protectively fenced in, much of its wildlife is dependent upon rains so inconstant that the lake dries out every few years. Now, only the birds will be able to move to traditional emergency areas. Management of prey populations cannot be delegated to predators in undersized reserves, nor can recovery from floods, nutritional deficiencies, and epidemic disease be left to the healing powers of natural cycles. What is happening over most of the world is the progressive loss of wild land, the insularization of parks and refuges and their gradual alteration. Setting land aside does not assure species preservation if managed by benign neglect.

In this shifting climate, the World Conservation Union (IUCN) is now considering basing its efforts upon a philosophy of utilization. It hopes to demonstrate that wildlife preservation will generate employment and revenue. Hunting, fishing, the pet trade, ecotourism, and commercial harvesting for food are all endorsed as keys to "sustainable development" (IUCN 1992). It argues that animals must earn their keep.

DIRECTIONS AND DILEMMAS

Inexorably, many parks and reserves are becoming megazoos—hence subject to zoo paradoxes. In the constricted wild lands of the future, only continuous monitoring and management can deal with the ways animals affect each other and their habitats. Will Nakuru's reintroduced hyenas control the burgeoning impala population? Will they drive off the flamingos?

Except in very big reserves, large species will have small populations that may not be demographically or genetically viable. Recent zoo mammal studies have found a 10 percent decrease in survival where there was a 10 percent increase in inbreeding (Ralls et al. 1988). Animal by animal, management will be needed to effect translocations for genetic interchange between isolated reserves or to deal with the threat of damaging exotics and disease just as in zoos.

Surpluses and dominance pose special problems in unnaturally confined parks and zoos. In parks, one abundant species cannot be permitted to so multiply that it destroys the homes and foods of others. In zoos, it must not be given space or resources essential for the survival of others. An aggressive bull rhino or cock pheasant cannot be allowed to so monopolize females that no males of other family lines can pass on their genes. Moreover, even the rarest species—bred on a sustaining basis in a population of finite size—will produce animals unavoidably surplus to carrying capacity for their particular sex or genetic line.

On one hand, zoos and parks must inspire public sympathy for the goals of sustaining species and protecting the welfare of individuals, creating a reverence for wildlife. On the other, they must teach that the obligations of care include managing the natality and even the mortality of individuals in relation to the well-being of their species and ecosystems (Seal 1991). As we care for animals in ever smaller fragments of nature in less natural situations or in zoos, they require ever more intensive management if they are to survive.

THE ZOO ROLE

Except for zoos and zoolike institutions, no other conservation or animal welfare organizations actually provide ongoing animal-by-animal care for wild creatures, sustaining them generation after generation. Increasingly, modern zoos are defined by that commitment. Except for AZA institutions, it may be that no other U.S. animal welfare or conservation organizations subject themselves to regular inspection and accreditation, insist upon a common code of ethics, and provide ongoing scientific instruction and training for their members.

It seems to me that zoos are destined to become ever more important centers for conservation research, conservation action, and education. In 1990–91 AZA institutions conducted nearly 390 conservation and science projects in sixty-three countries and produced more than 400 technical publications on wildlife biology and conservation (Hutchins et al. 1991).

Zoo science is contributing basic biological information and technological know-how to the increasingly demanding tasks of wildlife care in constricted habitats. Increasingly also, zoos are conducting cutting-edge conservation science in nature, utilizing their specialized institutional bases and expertise.

Uniquely, zoo conservation action bends recreational dollars to conservation purposes. Money otherwise used to go to the ball game, the movies, or a symphony is converted to endangered-species propagation and conservation education programs (Conway 1989). A coordinated propagation program, the AZA Species Survival Plan now seeks to sustain a heritage of at least a few of the world's most endangered creatures, thus buying time for conservation and restoration efforts in nature for creatures that would otherwise be lost. Providing animals or their genes for reintroduction may become a specialized zoo discipline on behalf of all society, and inherent in the zoo biologists' view is the conviction that without their efforts, many species will become extinct that otherwise would not.

Perhaps the most serious threat to the well-being of many wild creatures is that they will be ignored, condemned by the growing masses of humanity to the same closets of irrelevance and curiosity as silent movies and trilobites. But those masses are concentrated in cities where the zoos are. The most profound and moving lessons zoo education has to offer are simply well-cared-for, well-exhibited, living animals. They live with us, daily creating news and arousing interest, acting as ambassadors for their kind. They do not permit us to ignore the fact that their kind exists. In the United States, well over 100 million people took advantage of that lesson in 1991, and nearly 14 million participated in formal zoo education programs, including 10 million schoolchildren and 35,000 teachers.

Today, the collapse and extinction of wildlife communities is far advanced. Although nature is humanity's home, wildlife is becoming its ward. A species' management, essential for its own survival, must not be threatened by ignorance and insensitivity. Animal welfare must not become irrelevant. Our task is to establish a workable, morally and scientifically acceptable way of dealing with the substance and the perception of paradox in our relations with wild creatures. Whatever this way, we must not lose sight of the fact that the ethical dilemmas in zoos and conservation most urgently requiring resolution are those that threaten biological diversity and the continued renewal of life. Ecosystems and wildlife in the twenty-first century will be a nature that we recreate or care for—a final paradox.

ACKNOWLEDGMENTS

I thank Robert Cook, Michael Hutchins, and Dan Wharton for helpful suggestions and comments.

REFERENCES

Conway, W. 1968. The consumption of wildlife by man. *Animal Kingdom* 7(3): 18–23.
———. 1986. The practical difficulties and financial implications of endangered species breeding programs. *International Zoo Yearbook* 24/25:210–219.
———. 1989. The prospects for sustaining species and their evolution. In *Conservation for the Twenty-First Century,* ed. D. Western and M. Pearl, 199–209. New York: Oxford University Press.
Donnelley, S., and K. Nolan, eds. 1990. Animals, science, and ethics. *Hastings Center Report* 20, No. 3 (May–June): Special supplement, 1–32.
Fitzgerald, S. 1989. *International Wildlife Trade: Whose Business Is It?* Washington, D.C.: World Wildlife Fund.
Hutchins, M., R. Wiese, K. Willis, and S. Becker, eds. 1991. *AAZPA Annual Report on Conservation and Science.* Bethesda, Md.: AAZPA.
IUCN. 1992. CITES Conference of the Parties, held in Kyoto, Japan, March 2–13.
Mann, C. C., and M. L. Plummer. 1992. The butterfly problem. *Atlantic,* January, 47–59.

Myers, N., ed. 1984. *Gaia: An Atlas of Planet Management.* Garden City, N.Y.: Anchor Press, Doubleday & Co.

Newmark, W. 1987. A land-bridge island perspective on mammalian extinction in western North American parks. *Nature* 325:430–432.

Olney, P., P. Ellis, and F. Fisken, eds. 1989. Zoos and aquaria of the world. *International Zoo Yearbook* 28:283–366.

Ralls, K., J. D. Ballou, and A. Templeton. 1988. Estimates of lethal equivalents and the cost of inbreeding in mammals. *Conservation Biology* 2(2): 185–193.

Robinson, J., and K. Redford. 1991. *Neotropical Wildlife Use and Conservation.* Chicago: University of Chicago Press.

Seal, U. S. 1991. Fertility control as a tool for regulatory, captive, and free-ranging wildlife populations. *Journal of Zoo and Wildlife Medicine* 22(1): 1–5.

Steadman, D. 1991. Extinction of species: Past, present, and future. In *Global Climate Change and Life on Earth,* ed. R. L. Wyman, 156–169. New York: Routledge, Chapman & Hall.

Stevens, W. K. 1991. Species loss: Crisis or false alarm. *New York Times,* 20 August, sec. C.

Wilson, E. O. 1989. Threats to biodiversity. *Scientific American* 261(3): 108–116.

World Resources Institute. 1992. *The 1992 Information Please Environmental Almanac.* Compiled by World Resources Institute. Boston: Houghton Mifflin.

PART ONE

THE FUTURE OF ZOOS

hat will be the role of zoos in the twenty-first century? Zoos are already engaged in changing both their missions and their messages as we near the end of the twentieth century, and it is almost certain that zoos and their employees will have more and more responsibilities offered them, even thrust upon them, as human expansion rolls unchecked into the next century. Since zoos have become the most common home of Species Survival Plans and many zoo biologists are active in forming and operating captive breeding programs, this section is designed to examine captive breeding programs in their larger, institutional setting. But this section must be both hardheadedly realistic and at the same time speculative because, as noted, the role and tasks of zoos are constantly changing in response to the rapidly worsening situation in the wild for many species, not to mention changing public attitudes. To understand our obligations connected with captive breeding programs we must understand the programs in this broader context of zoo missions, messages, and social milieu—and we must understand them even more broadly as a part of obligations regarding the protection of natural variety. We present chapters by experts from several fields to portray and examine the cultural and conservation context in which zoos will chart their future course.

Eugene Hargrove, an environmental ethicist who has explored the history of attitudes toward environmental protection, discusses the history of attitudes toward wild animals in captivity and speculates about the possible future of those attitudes in a postmodern worldview. Terry Maple, a behavioral primatologist and director of Zoo Atlanta, describes zoos as increasingly important in the struggle to save biodiversity. He presents a vision of zoos as leaders in conservation activities both in situ and ex situ, and he advocates strong commitments by zoos to both education and conservation zoology. While Maple's vision represents an expansion of the zoo role along the current trajectory of change—doing better and better what the largest and most prominent zoos have already

11

been doing—David Hancocks, an architect by training but now the director of the Arizona-Sonora Desert Museum, articulates a different role for zoos of the future. Hancocks expresses concern about overemphasizing a few large-scale, high-profile conservation efforts likely to affect only a few of the larger, charismatic and exotic species. He favors a more localized mission of zoos that specialize in protecting and teaching about endangered local fauna and habitats and that generate support for conservation in local communities. Tom Regan questions whether zoos can be justified according to the principles of several prominent theories of moral obligation. He argues that neither standard utilitarian ethics nor the extant forms of holistic moral theory—while each might justify keeping animals in zoos—is an adequate theory of morality. On a rights theory, which he thinks is the correct and applicable moral theory, the activities of zoos are not morally defensible. Dale Jamieson, an ethicist who has long been a critic of zoos, recognizes that there will be conflicts between advocates for animals and advocates for conservation, recounts some of the reasons to be wary of actions that harm individual animals for more abstract causes like species preservation, and argues that there is a moral presumption against keeping wild animals in captivity. The opinions of these authors are very diverse, suggesting that the debate regarding permissible treatment of animals in captivity must proceed without a clear consensus regarding the current value or future justification for the institutional frameworks represented by zoos and their conservation programs.

THE ROLE OF ZOOS IN THE

TWENTY-FIRST CENTURY

Eugene Hargrove

etermining the role that zoos will play in the twenty-first century is extremely difficult because we have now reached the end of the third major period in Western civilization—the modern period. What the new period will be like remains, at this time, a matter of speculation. Although, in this century, across many fields of learning and inquiry, the basic presuppositions of the modern period have systematically been rejected, we still do not have a very clear picture of the new suppositions that are supposed to take their place. It is for this reason that the new period is tentatively named "postmodern," simply the period after the modern period. If the changes in perspective in the new period turn out to be as dramatic as those that occurred between the medieval and modern periods, virtually any scenario may be possible, from the end of the zoo as an outmoded modern institution to a continuation of its current role or even a dynamic new role in terms of a postmodern vision. Because zoological gardens are characteristic and distinctive products of the modern period, the first two possibilities are most likely. Nevertheless, because the history of ideas is full of strange twists and turns, and given that the postmodern period is still undefined, a new role for zoos remains a practical possibility.

Zoological gardens, like botanical gardens, began to appear during the transition from the late medieval to early modern periods, when natural-history scientists began classifying new-world biota. Botanical gardens were created to display the plants discovered worldwide during the period of European exploration that began with the voyages of the Spanish and the Portuguese in the fifteenth century. The Catholic Church was at first opposed to gardens in general on the grounds that they inappropriately promoted the earthly and the sensual,

13

drawing attention away from God and the spiritual. In its first concession the church reluctantly accepted gardens as symbols of the Garden of Eden and the promise of a pleasant afterlife (Clark 1976, chs. 1–2). The new gardens, however, quickly degenerated into celebrations of human domination over nature, in which plants were arranged and reshaped into more perfect geometrical patterns and forms. In its second concession the church approved importation of foreign plants, provided they were mentioned in the Bible. Soon after plants from the Holy Land became commonplace, the church lost control and foreign plants flooded into Europe from temperate zones all over the planet. The introduction of foreign plants produced comparative interest in indigenous plants and speeded the transition from the formal to the informal garden, in which plants were displayed in a natural state without reshaping and without placement in geometrical patterns (Hargrove 1989, 83). In these new gardens the main foci were the actual, natural scientific and aesthetic properties of plants, reflecting the close connections among gardeners, natural-history scientists, poets, and painters. This shift in focus also reflected a significant difference between the physical sciences and natural history. Whereas physicists and chemists attended to primary properties of natural objects such as length, width, and depth (measuring rather than observing), biologists, botanists, and geologists on the other hand attended to secondary properties such as colors, textures, tastes, sounds, and smells—all observable properties of equal interest to humanists (Hargrove 1989, 78).

These developments were facilitated by a basic change in thought patterns from a form characteristic of the medieval period, symbolic thinking, to the form now characteristic of the modern period, representational thinking (Huizinga 1924, 182–194). When someone in the Middle Ages looked at an image of a plant or animal (for example, a lily or a lamb), that person normally began mentally spinning through a host of passages in the Bible in search of the most appropriate line or parable. A lily would trigger a thought about the "lilies of the field," the lamb a passage from John, "Behold the lamb of God, who taketh away our sins." Once the modern period began, however, any person shown similar images automatically thought instead of a real lily or a real lamb. In philosophy this new form of thought had a predictable but unproductive effect, sending philosophers on a hopeless quest to prove that they knew the world existed (Hargrove 1989, 35–37). In art, however, its effect was more positive. A three-hundred-year experiment in representational painting developed an aesthetic taste first for composed landscape paintings and finally for natural landscapes viewed from a perspective that mimicked the composition of a painting—what is today called the "roadside scenic view." This representational aesthetic perspective was essential to the establishment of such nature preserves as Yosemite and Yellowstone in the nineteenth century and continues to support preservationist concerns through orgies of representational thinking in various forms—from Ansel Adams's photographs on Sierra Club calendars to public television's nature documentaries. Without representational thinking it probably would not be possible to stir public concern about the preservation of places ordinary people will most likely never visit—for example, the Arctic National Wildlife Refuge and the various wildlife preserves in Africa.

Preservationist concern for the animals represented in zoological parks is a special variant of representational thinking in which the focus of attention is not on the particular but on the universal—that is, not on the individual animal but on the species. When a visitor to a zoo sees an animal for the first time, he or she tries to construct a mental image of the essential properties of the animal. Once this image is complete, the visitor then goes on to determine whether the animal on display is a good example or representation of that species (see Hargrove 1989, 122–124). This kind of thinking originated in Plato's theory of forms, according to which natural objects participate with varying degrees of success in perfect ideas called forms. That theory of participation found its way into the minds of zoo visitors via Aristotle, who classified about five hundred plants and animals in terms of genus and species, a system of classification later elaborated upon by his modern counterparts, including Linnaeus.

Because of this focus on the species rather than the individual, wildlife preservation has special features that are not shared by other forms of nature preservation. The preservation of geological forms and ecological systems is essentially the preservation of places. These places are not conceptualized in terms of a set of universalizable essential properties. Rather they are viewed as the unique and individual creations of natural history that cannot be usefully reduced to a set of universal characteristics. Our preservationist concern for natural areas translates into an attempt to protect the unique features of specific places. Our preservationist concern for species, however, translates, quite to the contrary, into an attempt to preserve the universal features of species as exemplified in a large pool of individuals that are participating in those characteristics. This concern does not require the protection of each individual animal, only the protection of enough individual animals to keep the species classification active. Concern for the protection of each individual animal does not arise until the pool of animals representing a species becomes so small that the death of any individual may lead to the extinction of the species. At that time individual wild animals may begin receiving better medical attention than many humans receive. Should the population increase to acceptable levels, however, concern for the individual animal once again disappears. It is the focus on the species over the individual that produces the conflict between animal rights activists and environmentalists. Environmental ethicists debated the relationship of environmental ethics and animal welfare ethics in the early eighties, concluding at first that they were incompatible (Callicott 1980) and finally that they are complementary (Warren 1983), although each side of the debate dealt with different animals (domesticated animals and wild animals) and with different histories of ideas (Hargrove 1992, xix–xx). The history of ideas behind animal liberation or animal rights is completely different from the history of ideas behind species preservation (Hargrove 1989, ch. 4). Animal liberation arose out of concern about the treatment of domestic animals in the nineteenth century. The key phrase was "unnecessary suffering" expressed in terms of classical utilitarianism in which good and evil are defined in terms of pleasure and pain. In contrast, species preservation in the nineteenth century arose out of a preevolutionary and pre-ecological concern that extinctions would destroy irreplaceable links in the Great Chain of Being. The key phrase was "wanton destruction," which permitted (nonwanton) use without regard to suffering as long as there was no chance that the species would become extinct. George Catlin, for example, though calling for protection of the buffalo against wanton destruction, on one occasion studied the suffering of a buffalo he had shot on a sporting hunt, putting the animal out of its misery only at the end of the day when it was time to go back to the fort (Catlin n.d., 256–257).

The conceptual background of species preservation is actually more contorted and confusing than any other element in environmental or animal welfare ethics. As already noted, the general model for representational thinking in nature presentation involves particular images representing particular places on this planet. The model for such thought in species preservation, in contrast, involves two peculiarities. First, the individual living animal represents a biological classification with the epistemological status of a Platonic form. In other words, the real represents a mental construction, the opposite of the primary model. Second, because this conception of the species–individual relationship developed before evolutionary and ecological theory and has not fully taken them into account, it is not entirely compatible with current biological theory (see Hargrove 1989, 129–132). Evolution, once it became accepted, should have weakened the case for species preservation, since it converts the biological classification from a group of essential, eternal, universal properties into a group of contingent properties that a population of interbreeding animals happens to have at a given moment in natural history. In terms of evolution, moreover, extinction is natural, both in terms of evolutionary failure and evolutionary success. Indeed there can be no long-term failure unless life is completely extinguished on this planet. Barring that circumstance, speciation will begin again, and sooner or later massive biodiversity will once more appear.

Similarly, ecology teaches us that the extinction of particular species within natural systems simply produces minor adjustments in which other species fill the vacated niches—for example, the expansion of the coyote following the extermination of the wolf in this country. Species preservation was not adversely affected by evolution and ecology simply because the older conception of species was carried forward uncritically. Instead of weakening the species concept, they enhanced it by providing a dynamic narrative dimension—natural history—which could not be derived from the static conception of species as immutable and unchanging, which was derived from Plato and Aristotle.

Judging by the current focus of zoos on species preservation, in the twenty-first century these institutions will try to be biological arks—indeed, a flotilla of arks—for individual animals representing biological classifications that are in danger of becoming inactive. Though commendable, this mission may be a difficult, perhaps even an impossible one, because it relies (1) on a form of thinking—representational thinking—that is expected to go out of fashion and (2) on a conception of species that is preevolutionary and in conflict with concern for the individual. With regard to representational thinking, it is possible that this form of thought will continue to play the same role that it has throughout the modern period. However, it should be noted that art, which more than any other human activity promoted the environmental dimension of representational thinking, abandoned that perspective more than a century ago. As one art professor recently put it to me, there are only so many leaves backlit by the sun that can appear in a Sierra Club calendar. When the last leaf is shot, she concluded, people may finally catch up with artists and embrace a conception of beauty that is essentially antinature. Should this scenario occur, the relevance of zoos and the environmental movement in general will depend on the degree to which they can utilize new forms of thought.

Happily, there appear at this time to be good prospects for a successful conversion. Most scholars who speculate about the postmodern period predict the emergence of narrative thinking as the dominant mode of thought (Cheney 1989). In accordance with narrative thinking, which has considerable affinities with the symbolic thinking of the medieval period, disputes will be settled by telling stories rather than by presenting reasoned arguments. Unfortunately, however, the stories about animals—the kind presented in nature documentaries—will largely be narratives of animals in natural systems rather than stories about the living representatives of species displayed in the zoos of the twentieth century, who usually do not have interesting adventures that can be the subjects of a narrative. If narrative thinking displaces representational thinking, ultrarealistic replicas of natural habitats may not be enough to hold the attention of park visitors. The ultimate example of the narrative experience is the Disney theme park, which is little concerned with representational realism. The plastic crocodile rising out of the water on the Safari Ride in Disney World or Disneyland is connected with fictional crocodiles, for example, those found in such cartoon feature films as *Peter Pan* and *The Jungle Book,* not the crocodile of the natural world. In competing with Disney-style narrative, the display of living representatives of species may very well soon become the exception rather than the rule. While it may still be appropriate to display animals selectively, to provide some direct contact between visitors and living specimens, postmodern, narrative-oriented visitors of the twenty-first century may not need, or *desire,* to see a complete set of megafauna. If so, the triumph of narrative thinking could encourage the development of partially zooless zoos in which the animals live in remote locations, roaming under nearly natural conditions, away from the direct observation of zoo visitors, who would observe them indirectly through narrative documentaries (for example, videotapes).

Such zoos would have the advantage of resolving the conflict between environmental and animal welfare ethics. In the current zoo setting, the animals remain not only representatives of their species but also individual animals in their own right, triggering, as noted above, conflicting moral intuitions from two complementary histories of ideas. This conflict may be justifiable if zoos are arks that will eventually land and discharge their passengers into reconstructed habitats. It is less justifiable if they are on a boat ride that will never end. If it becomes clear in the twenty-first century that zoos are on an endless journey, the continued conflict between the two ethical perspectives may require us to dock at mini natural habitats or, when that is not feasible, simply to bring the journey of some species to a natural conclusion—extinction. Docking the arks, however, should it become necessary, need not mean the end of zoological societies as institutions. It could simply mean the abandonment of the representational display of animals in order to remain current with new forms of thought in the postmodern period (for example, conservation education and bioregional interpretation via narrative modes of thought).

REFERENCES

Callicott, J. B. 1980. Animal liberation: A triangular affair. *Environmental Ethics* 2:311–324.

Catlin, G. n.d. *Letters and Notes on the Manners, Customs, and Conditions of the North American Indians.* Vol. 1. New York: Dover Publications.

Cheney, J. 1989. Postmodern environmental ethics: Ethics as bioregional narrative. *Environmental Ethics* 11:117–134.

Clark, K. 1976. *Landscape into Art*. New York: Harper & Row.

Hargrove, E. C. 1989. *Foundations of Environmental Ethics*. Englewood Cliffs, N.J.: Prentice-Hall.

———. 1992. *The Animal Rights/Environmental Ethics Debate: The Environmental Perspective*. Albany: State University of New York Press.

Huizinga, J. 1924. *The Waning of the Middle Ages: A Study of the Forms of Life, Thought, and Art in France and the Netherlands in the Fourteenth and Fifteenth Centuries*. New York: St. Martin's Press.

Warren, M. A. 1983. The rights of the nonhuman world. In *Environmental Philosophy: A Collection of Readings,* ed. R. Elliot and A. Gare, 109–134. St. Lucia: University of Queensland Press.

TOWARD A RESPONSIBLE ZOO AGENDA

Terry Maple

However paradoxical it may sound, the truth is actually this: the free animal does not live in freedom; neither in space nor as regards its behaviour towards other animals.—Heini Hediger (1964)

Nine years ago I was sitting on the sidelines as the Atlanta Zoo crashed and burned. Like most citizens in our community, I was a helpless witness to a series of stupid (and often mysterious) managerial blunders. The organizational demise of Atlanta's neglected zoo became a national scandal and a media circus, moving the *New York Times* to render the following opinion in June 1984: "Atlanta has given neither money nor thought to its disintegrating zoo. How a community treats animals says something about the human beings who run it. Unless Atlanta wants to commit itself to a professionally operated zoological park, would it not be better to forget about having one at all?" These events were a terrible embarrassment to our ambitious city. Aspiring to greatness, Atlanta in 1984 was a city humbled by the image of its hapless zoo. After the zoo crisis, the city slipped from first in 1983 to eleventh in Rand McNally's 1984 *Most Liveable Cities* index, wherein the zoo crisis was identified as a factor contributing to its degraded image. As the local and national media shed light on the zoo's problems, I began to appreciate the value of awareness and introspection. The people of Atlanta were beginning to see their zoo for what it really was—an emperor in his new clothes. For its lack of self-criticism, the zoo was now exposed for all to see.

NO ZOO IS AN ISLAND

Fortunately, the citizens of Atlanta elected to confront and resolve the zoo's longstanding problems. In the summer of 1984, I was recruited as the "reform director" who would implement sweeping changes. It was a considerable challenge, as I was trained for a career in the academy, not the zoo. As a professor of psychology with specialized expertise in nonhuman primate behavior, I was a theoretician, who gave advice to zoo professionals but didn't have to make hard decisions. My books, *Orangutan Behavior* (Maple 1980), and *Gorilla Behavior* (Maple and Hoff 1982), were required reading for serious curators, but my entire management experience was a mere ten months at the Audubon Zoo on leave from Georgia Tech. Although a frequent consultant to zoo directors, my own operating credentials were modest by comparison.

Those limitations notwithstanding, I was the recipient of an unusual opportunity. Accepting the responsibility of running a dysfunctional zoo, I would have to apply strong medicine of my own. At this moment in the zoo's long history, the quality of my ideas and the fitness of my leadership would be severely tested. I was inexperienced, but I was not unprepared. Since the Humane Society of the United States had labeled Atlanta's one of the ten worst zoos in the nation, it was an honorable assignment. Even then I was confident that Atlanta's zoo could be a role model for other neglected zoos. Rising from its ashes, it would become Atlanta's most spectacular phoenix story (Maple and Archibald 1993).

Enjoying the advantage of hitting rock bottom, I never intended simply to render first aid to a wounded institution. I wasn't hired merely to nudge the zoo from crisis to mediocrity. In fact, the citizens of Atlanta issued nothing less than a mandate for excellence. This mandate reflected the caliber of our community, and it reminded me that our zoo would be only as great as the people whom we recruited to our cause. The growing quality of the world's zoos is testimony to the knowledge and commitment of our talented employees, experienced volunteers, and well-connected governing boards. In communities throughout the land, the people have spoken, loud and clear—they will support only high-quality zoological parks and aquariums.

The Atlanta Zoo was so far gone in 1984 that it needed an entirely new identity. We had to infuse it with new ideas, and new blood. I reasoned that we had to recruit experienced staff from successful institutions, so we engaged in a hiring blitz and raided an assortment of talent from America's best zoos. From the beginning we sought to recruit *ideological diversity*. We wanted bright, creative, autonomous people who could build entire departments from scratch. Innovative members of my staff built a bird department, a human resources system, financial networks, security units, an education division, and links to universities. Each key hire has left a legacy within the organizational structure that is now Zoo Atlanta. In the grip of this spirit of innovation, we envisioned a forum where zoo biologists and their critics might dare to engage in meaningful dialogue and debate. This concept has been finally expressed in the Atlanta conference Animal Welfare and Conservation: Ethical Paradoxes in Modern Zoos and Aquariums. Zoo Atlanta is a small zoo—a collection of fewer than one thousand animals on only 15.18 hectares (37.5 acres)—but we think big. Our benchmark conference, conceived in partnership with AZA and the Georgia Institute of Technology, was the product of a compelling idea—all zoos and aquariums, regardless of size, must confront the salient issues of their time.

As a reform director in Atlanta, I was able to reconstruct the organizational system to reflect contemporary challenges and industrywide priorities. Because I had spent my entire scientific career writing about captive environments and their effects on behavior, I recognized the importance and potential of innovations in habitat design and behavioral enrichment. As we formulated a collective vision for the emerging Zoo Atlanta, we deliberately considered the roles of individual-animal welfare and conservation at the species level. We found comfort in both camps, and our experience clearly demonstrates that a win-win strategy is not only possible but also reasonably easy to achieve.

During a decade of reform we have willingly participated in several parallel revolutions in the zoo world. Surely, the move from cages to naturalistic habitats has been revolutionary. Zoos have been transformed, almost overnight, from prisons into bioparks, becoming lush greenswards to enhance the well-being of animals and people. In this way, zoos have become a part of the natural world and thoroughly therapeutic in their effects on visitors. As Heini Hediger phrased it, zoos have become "the biotope of modern man, a secondary nature, a place for recreation, teaching and research, badly needed in our time" (1969).

IN PRAISE OF ZOO STANDARDS

Since 1980 the AZA has operated a zoo accreditation system. Accreditation in the zoo and aquarium world requires an on-site inspection at least once every five years conducted by at least two experienced zoo professionals. The accredited zoo must meet high standards in animal management, facility cleanliness, veterinary care, financial stability, conservation, education, and stewardship. These standards are growing tougher and more comprehensive with each passing year, but I have heard no complaints from within the ranks. In fact, much of the pressure for higher standards is generated from within. We all want better zoological parks.

When I first became zoo director in 1984 I was given some good advice by experienced accreditation inspectors. They suggested that we should not be in a hurry for approval but instead work diligently and carefully to meet each accreditation criterion. There was much to do, and we wanted to succeed, so we advanced carefully toward our goal, first achieving accreditation in 1987 and again in 1992. The first step to meeting the standards is to admit your deficiencies. Inexplicably, zoos in trouble often founder in an altered state of denial. It should be recognized that you cannot argue, beg, or force your way into accreditation— you have to earn it. In my opinion, the accreditation program of AZA ranks high among the zoo world's most important accomplishments. The process makes every zoo better, and it keeps the profession focused on issues of quality. Because our product is better (and more trustworthy), public confidence is strengthened.

As tough as they are today, future zoo standards are surely going to be tougher. I expect that they will be advanced by more quantitative methods of assessment, and they will progress in keeping with our own commitments to excellence. These standards are not imposed by our communities, although they are consistent with the general public's interest in zoological parks of quality. In fact, zoo professionals themselves are eager to improve the zoo workplace and the systems that support it. The zoo industry's enthusiasm for excellence is our greatest source of strength, and it is manifest from top to bottom in zoos today. The toughest zoo critics have recognized that zoo professionals, from keepers to directors, are truly dedicated to the welfare of each and every animal in the zoo's collection. Within the network of accredited zoos and aquariums, the animals' quality of life will surely continue to improve.

DEBATING THE VALUE OF ZOOS

In his 1985 critique of zoos, Dale Jamieson asserted that zoos generally do not live up to their own goals—that zoo animals are deprived of freedom for little social or scientific good, and that zoos cause suffering without producing compensatory benefits for animals or people.

In a concise response, Chiszar, Murphy, and Iliff offered six salient points or rebuttals (1990), paraphrased herein:

1. Zoos and aquariums host from 300 million to 400 million visitors each year, providing the benefits of meaningful education and recreation.

2. Zoos have been traditionally connected with public education, and the favorable response of school systems indicates that they clearly value the association.
3. Organized Species Survival Plans (SSPs) focus attention and resources on the problems of captive propagation for threatened and endangered species.
4. Zoo exhibits now reflect expanding ecological consciousness and public advocacy.
5. There is an admirable history of success in clinical work, basic science, and large-scale projects that have required the combined efforts and cooperation of many zoos.
6. At least a dozen taxa have been saved from extinction by the combined efforts of the world's zoos.

Clearly, as Chiszar and his collaborators illustrated, the world's zoos are beginning to live up to their lofty goals, and the confinement of a small population of wildlife is serving useful purposes for people and animals alike. In fact, the goals of zoos are comparable to those of other essential institutions such as museums, libraries, and universities. Today's zoos do much more than entertain the public—they are education, science, and conservation centers, with unique standing in their communities (Conway 1969). In fact, zoos may be regarded as pathways to other valued urban programs, and an appropriate "global perspective on animals, environments and people" (Chiszar et al. 1990).

Zoo environments also succeed in promoting variability, health, and longevity, and zoo professionals take seriously their obligation to minimize animal stress, boredom, trauma, and disease. Furthermore, zoos willingly submit to inspections by outside agencies that evaluate the quality of facilities, procedures, staffing, supervision, and records. Regarding the charge of cruelty, the authors forcefully concluded, "There is no reason to believe that zoo animals necessarily live under cruel or damaging conditions in order to provide visitors with entertainment and education or by enabling zoos to engage in conservation and research."

ON THE MEANING OF "ZOO"

Webster's Ninth New Collegiate Dictionary offers a secondary definition of "zoo" as "a place, situation, or group marked by crowding, confusion, or unrestrained behavior." (Aquariums, our sister institutions, are not burdened by this problem.) There must be some zoos like this, but *high-quality* zoos are well-organized, disciplined, and highly professional institutions. People are prone to refer to a raucous party, a disorderly high school, or a chaotic convention as a zoo, as in "this place is a zoo!" However, the modern zoo is the antithesis of chaos. In fact, zoos today are more cooperative, organized, and communicative than at any time in their history. This state of readiness is dictated by the gravity of our situation, as wildlife and wilderness disappear at alarming rates. Given these circumstances, the modern zoo cannot afford to be out of control.

Recently the New York Zoological Park (Bronx Zoo) changed its name to become the world's first wildlife conservation park. The leaders of the venerable New York Zoological Society (now the Wildlife Conservation Society) made this change to emphasize their commitment to conservation, a nuance that had gone unrecognized by the public. "Zoo" had become a loaded term, with lots of negative baggage—or so it appeared in New York. If any zoo is qualified to be labeled a conservation park, it is the Bronx Zoo. No world zoo has done more to save wildlife and wild places throughout the world. The limitations inherent in the word "zoo" lead us to substitute terms with greater breadth. This, in part, is why zoos have chosen to call themselves zoological gardens or parks. Many American zoos have added "Botanical Gardens" to their titles, recognizing the importance of landscape. M. H. Robinson is the strongest proponent of the biopark concept (1990), which combines the strengths of natural history and cultural museums, botanical gardens, and zoos to achieve an integrated whole.

However, another way to approach the problem of finding our identity is to change the meaning of "zoo" itself. This is accomplished by continuing to focus our values to reflect our commitments to conservation, education, science, and animal welfare. These are compatible directives enabling us to create zoos that enlighten people and enrich the lives of animals.

The modern zoo is a place where quality is much more important than quantity. The best zoos no longer brag about the size of their collections, as specialization is more cost-effective and provides better environments for a collection that has been "rightsized" for good reason. Animals presented well, in complex and naturalistic enclosures, fulfill their ambassadorial function—they remind us to act on behalf of that vulnerable, wild constituency that they represent. Having fewer animals is better when they live in appropriate groupings within high-quality habitats. It is no longer how many animals but how they live that really matters.

One sure indicator of the shift in how we characterize the zoo is revealed in our vision or mission statements. The one that I regard as the most inspiring is Minnesota Zoo's catchphrase "Strengthening the bond between people and the living earth." At Kentucky's Louisville Zoo they are committed to "bettering the bond between people and the planet," while Maryland's Baltimore Aquarium seeks to "create an understanding of the environment and the ecological balance of life." In 1989, our centennial year, we adopted the motto "Conservation leadership for our second century." Of course, it is very difficult to capture the full mission of a zoo in one sentence, but these examples illustrate the high ideals of zoos and aquariums today.

ROADSIDE IMPOSTORS

At our Conservation and Animal Welfare conference in Atlanta, it was agreed that zoos of quality are fundamentally different from roadside attractions that dare to call themselves zoos. The assembled participants also recognized the opportunity to shut down those roadside attractions that were clearly cruel to animals. Zoos, it was proposed, need to get out front on this issue and partner with animal welfare groups to close the offending entities. An example of this type of partnership has recently occurred in the case of Ivan the gorilla.

Ivan lived in the B&I Department Store in Tacoma, Washington, where he occupied a small, concrete enclosure. He led a solitary life and had not seen a conspecific since he was captured as an infant in the wilds of Africa twenty-six years ago. His plight was brought to the attention of the public by Allison Argo, a filmmaker who produced a National Geographic special entitled *The Urban Gorilla*. Argo and others carried on a public campaign to translocate Ivan to a reputable zoo where he could be socialized.

For more than two years, zoo directors spoke out on this issue at the request of the Progressive Animal Welfare Society (PAWS), an animal welfare organization in the state of Washington. I was among the most active of the critics, along with Woodland Park Zoo director David Towne. When I visited Tacoma and Seattle, I appeared at a press conference to promote Ivan's release to a zoo. As I said to the assembled press, "Gorillas should be managed by professional staff in qualified zoos. No gorilla should live in a department store." This cooperative relationship with PAWS is the precise scenario envisioned by our conference participants. It has since resulted in the successful acquisition of this gorilla by the zoo community and has led to his eventual acclimation to group life in a naturalistic habitat at Zoo Atlanta (Boss 1994).

There are other situations where zoos have taken a position shared by animal welfare groups. For example, the SSP committee for orangutans, during my tenure as coordinator, took the position that orangutans should not be surplused to the entertainment industry. The policy was a result of commercial exploitation of the animals with no redeeming educational features. We collectively cringed at the sight of orangutans in demeaning costume, dancing and gesticulating to patrons of nightclubs and casinos. In our judgment it is morally wrong for a zoo to make orangutans available for that purpose, and we have condemned and discouraged the practice.

One reason that apes have been so easily acquired by entertainers is, plain and simple, poor planning on the part of zoos. Great apes are long-lived, and zoos must be prepared to deal with their longevity and its by-products, multiple offspring. A *strategic collection plan* helps to avoid unnecessary or excessive breeding and the surplus problems that result. Such planning prepares us to retire animals if they can no longer serve the public on exhibit. Superior zoos carefully plan their collections, animal by animal, taxon by taxon, and thoughtfully evaluate the present and future role of each individual. Such detailed planning is new to many zoos, but the approach will surely be endorsed by responsible animal welfare organizations and embraced by all zoos.

BRAVE ZOO WORLD IN THE TWENTY-FIRST CENTURY

Some zoo critics have suggested that zoos save species at the expense of individuals. While this may have occurred in some instances, it is certainly not a necessary outcome of species survival programs. Careful planning can result in win–win strategies whereby individuals and the species as a whole can benefit, but it is not always easy to orchestrate such conclusions. I am familiar with one case in which the gorilla Timmy was moved from the Cleveland Zoo to the Bronx Zoo. The case stimulated considerable debate in the local and national press, resulting in litigation brought by a Cleveland animal rights group. The case was unusual in that behaviorists hired by the animal rights group took the position that it would be an unfortunate disruption of Timmy's successful social relationship if he was separated from his cagemate, a female who could not conceive. Our side argued that in this case, the good of the species was a higher priority, but that Timmy's opportunity to meet other females was likely to produce equally successful social relationships. Timmy adjusted quite well to life in the Bronx (a move approved by the courts), and he eventually copulated. Timmy became a father on 18 July 1993. It is important that translocations of gorillas and other animals be conducted so that the individual as well as the species benefits from the move, but such harmony cannot always be achieved immediately. The commitment of zoos to the welfare of individual animals is worthy of emphasis here.

I have considerable confidence that the new direction of zoos will ensure widespread public confidence in our institutions. From this position we should become more effective advocates of both animal welfare and global conservation. We will lead by our examples, and communities will support these efforts with increasing pride of ownership and participation. If we continue along this path, our achievements will be praiseworthy indeed. I would project the following six items as achievable outcomes early in the next century:

1. Collectively, zoos and aquariums will be the most active, well-funded, and effective conservation organizations in the world.
2. Zoos and aquariums will be routinely allied with research and teaching institutions throughout the world, providing a multitude of opportunities to advance basic and applied zoo biology.
3. Zoos and aquariums, experts in managing ecotourists, will be routinely sought by developing countries to help plan, operate, monitor, and evaluate parks and protected reserves.
4. While actively promoting the conservation of wildlife and ecosystems, zoos and aquariums will still practice selected rescues of highly endangered species and will continue to operate as the last (but hopefully temporary) refuge for ecosystems that are no longer safe for animals.
5. Zoos and aquariums will be the acknowledged leaders in the successful marketing of a universal conservation ethic and a cooperative network for environmental problem-solving.
6. As mainstream conservationists, zoos and aquariums will inspire individuals, groups, and corporations to work actively on behalf of wildlife and wildlife habitat locally and globally.

A highlight of this new century will be the expected rapprochement between good zoos and responsible animal welfare organizations. Putting our differences aside and finding common ground, we can focus public and private resources on the problem of upgrading facilities and operating standards. It is highly likely that within the next twenty-five years, some organization, working with zoo professionals, is going to develop a comprehensive and objective rating system for zoological parks and aquariums. This system may work something like the star or diamond systems work for restaurants, or it may offer the detail of a livable-city or best-college index. If the system is developed correctly, it will help communities to assess the national standing of their local zoo. Other than AZA accreditation, there is no system of comparison. A good rating system would stimulate healthy competition and give every zoo a relevant target. It seems to me that if major universities are willing to subject themselves to departmental rankings (e.g., the top twenty schools in microbiology), then we might consider ranking the top twenty zoos in education, conservation, and science. However, such a system must be developed and endorsed by experienced zoo professionals.

One man's crystal ball suggests that there may be fewer zoos in the future, but they will surely be superior in every way. They are likely to be the product of effective public-private partnerships, with zoo managers exercising greater business autonomy and local governments providing the capital to build state-of-the-art facilities. By the turn of the century, all good zoos will be naturalistic, but they will also be struggling to bring people into closer (and safer) contact with animals. This trend will be a result of the continuing estrangement of humankind from wild animals. In the natural world, things will probably get worse before they get better, but zoos will play an important role in rallying the public to strengthen environmental policies.

ISSUES AND ANSWERS

I opened this chapter with a quotation from Heini Hediger on freedom. This and other issues have tended to divide the zoo community and its critics from animal welfare organizations. However, it is possible to agree with Hediger and still endeavor to provide animals with optimum environments in the zoo. Modern zoo professionals, working in zoos and in diminished national parks encircled by human habitation, are attempting to preserve wildlife autonomy by enrichment and education. In the zoo, we must expand opportunities for animals and simulate the way of nature; in national parks we must teach local people to preserve environmental complexity so that animals can still make a living in their altered habitat. Freedom may be a relative term, but animals can fulfill their species potential only in meaningful and productive niches. We can continue to debate issues such as freedom, or we can work to improve the lives of both captive and wild animals. One thing is certain, the vast numbers of animals adapted to zoo life are not going back to the wild. In many cases, the wild no longer exists for them. Zoo animals serve the useful purpose of public education, and for this reason we owe them the best possible quality of life.

The debate among zoo professionals and responsible critics has already borne fruit. We must continue this dialogue as a challenge to old methods, a bridge to self-awareness, and a pathway to new ideas. Constructive criticism should always be welcomed, and we should continue to seek dialogue with any organizations that can apply their resources to help us achieve better results. While zoos and aquariums are clearly changing for the better, our most dynamic epoch is just ahead in the challenge and opportunity of a new century.

REFERENCES

Boss, K. 1994. Gorilla welfare: Ivan's 30-year odyssey from jungle to shopping center to zoo. *Pacific (Seattle Times)*, 24 July, 12–18.

Chiszar, D., J. B. Murphy, and W. Iliff. 1990. For zoos. *Psychological Record* 40:3–13.

Conway, W. G. 1969. Zoos: Their changing roles. *Science* 161:48–52.

Hediger, H. 1964. *Wild Animals in Captivity*. New York: Dover Publications.

———. 1969. *Man and Animal in the Zoo*. London: Routledge & Kegan Paul.

Jamieson, D. 1985. Against zoos. In *In Defense of Animals,* ed. P. Singer, 109–117. Oxford: Basil Blackwell.

Maple, T. L. 1980. *Orangutan Behavior*. New York: Van Nostrand Reinhold.

Maple, T. L., and E. Archibald. 1993. *Zoo Man: Inside the Zoo Revolution*. Atlanta, Ga.: Longstreet Press.

Maple, T. L., and M. P. Hoff. 1982. *Gorilla Behavior*. New York: Van Nostrand Reinhold.

Robinson, M. H. 1990. Afterword: The once and future zoo. In *Smithsonian's New Zoo,* by Jake Page, 198–205. Washington, D.C.: Smithsonian Institution Press.

LIONS AND TIGERS AND BEARS, OH NO!

David Hancocks

The public zoo as we know it today is essentially a nineteenth-century concept. It emerged in 1828, when the Zoological Society of London opened the doors to its Zoological Garden in Regent's Park. The idea of presenting wild animals to the public in a parklike setting immediately gained enormous appeal, and public zoos began appearing in the major cities of Europe, Australia, and North America (Hancocks 1971). The zoo concept was perfect for the Victorian era. There was a rapidly emerging middle class, the novelty of leisure time for family entertainment, a new focus and belief in things edifying and educational, the opening up of new lands, the growth of empires, exploration of dark continents, discoveries of strange animals. It all combined perfectly to make the zoo a place of tremendous popularity.

For the greater part of their history public zoos remained basically unchanged from that original concept, although, in recent years, we have seen areas of significant change: most zoos have become increasingly active in breeding programs, especially for some endangered species; some have been giving emphasis to the development of naturalistic exhibits, though most improvements have been only cosmetic and restricted to areas viewed by the public; and there have been valiant but scattered attempts at behavioral enrichment.

My concerns are that these areas of progress are being superimposed on a concept that is insufficient for the coming century, and that this nineteenth-century concept is inherently and fundamentally flawed.

The Victorian zoogoers wanted only to satisfy their curiosity and merely look at specimens of exotic wild animals. That reason for going to the zoo was sufficient for the day. The challenges that face us now are for our increasingly

31

urban populations to develop understandings of the complexities of nature, to develop a new respect for nature, and to comprehend and support the values of biological diversity.

The Western mind seems to cope with diversity by analytical thinking. We made sense of the apparent chaos of nature by using a systematic approach, dividing everything into kingdoms and classes and subdivisions (Berlin 1973). Thus, back in the nineteenth century, while some people were developing zoological gardens, others were laying out botanical gardens, and aquariums, and establishing museums of other disciplines: geology, entomology, archaeology, and anthropology. And so, on Sunday afternoons, the industrious Victorian family could choose some particular place to be exposed to some particular subdivision of the curiosities of the natural world.

We continue to perpetuate that subdivided view of nature. In almost every city you find the zoological garden on one side of town, the botanical garden on the other, and the natural history museum somewhere in between. This tidy-minded approach stems from an outdated perspective. It goes against the grain of nature. Meanwhile, the amount of destruction and stress that is being placed on our ecosystems is occurring at a rate far in excess of anything similar that has happened in human history (Ehrlich and Ehrlich 1981). If that story is to be told, and understood, we cannot hope to do it in the piecemeal way that is required by the present disparate natural history institutions that remain from a past era.

Moreover, despite the presence of all our zoological, botanical, and other natural history facilities, we know that, sadly, our culture has a gigantic problem of scientific illiteracy. If we are to make progress in this area and help people to develop a holistic view and a wider perspective, then we need to develop new tools for the job. Merely presenting bits of the picture in different places hasn't worked, and won't work. What can zoos do about this? Are zoos trapped, by nomenclature, into focusing only on zoology? Must they try to tell the story of nature using only one chapter of the book? If so, they will fail.

Might it be too radical, as a start, for zoos to consider changing their name, so that they could escape the tyranny of one scientific discipline? The Bronx Zoo has renamed itself the Wildlife Conservation Park, jettisoning the term "zoo." Its new title, however, is not likely to be easily adopted. "Biological park," proposed by the director of the National Zoo, Michael Robinson, doesn't roll easily off the tongue either, certainly not as easily as the word "zoo." We face the same problem with the name Arizona-Sonora Desert Museum. But the greater breadth of meaning in "biopark" or "museum" means that institutions with such a title can develop exhibits that focus on all aspects of the wilderness. They have the potential to reveal and explain the connections and interdependencies between all components of nature, and that means not just the obvious interrelationships between plants and animals but all the complexities of our ecosystems.

There are, for example, fascinating, important, and distinct connections that can be told between radiolarians and clouds, between snails and the greenhouse effect, between granite and trees, between grasses and human civilization. Perhaps the greatest challenge facing zoological parks is to devise ways in which such stories can be told, to develop exhibits that explain the relationships between microbes and flowers and climate and snakes and minerals and elephants. And evolution. And extinction.

Such a change would mean a close reexamination of the zoo concept. It would require some long and concentrated rethinking. In practical terms it would demand a shift away from that distorted, minimalist, and fragmented view of nature that typically, if unwittingly, zoos present, with their emphasis on the charismatic and especially the diurnal, social, large, colorful, cute, typically mammalian and usually African species.

There are maybe as many as 30 million species of animals on this planet (Wilson 1992). Yet, ask people to start naming wild animal species and you will likely find most will start to falter and stumble after a dozen, and likely run dry at two dozen. The few that they name will almost certainly be those charismatic species, and especially the mammals, that you find in virtually every zoo collection—especially the elephants, giraffes, zebras, hippos, rhinos, chimpanzees, lions and tigers and bears.

Recently, at the Desert Museum, we have been planning a desert grasslands exhibit. There had been a suggestion for a pronghorn exhibit in the grasslands. For reasons of animal health care and limited resources I decided this species was not appropriate. I was then chastised for removing "one of the most interesting animals in the grasslands." Admittedly pronghorns are the largest and most visible species in this habitat, but we should not confuse mere size with interest.

In further research about desert grasslands animals, I learned about a curious little mouse that kills and eats other small animals. This little creature, called the grasshopper mouse, even stands on its hind legs and bays at the full moon, in tiny high-pitched howls, like a wolf (Bailey 1929). I feel compelled to ask whether this species is not an equally good candidate for one of the most interesting animals in the grasslands. I also must ask whether any zoo visitor would even be aware of the existence of this animal. More pertinent, maybe, is some recent research that reveals that the presence of certain species of rodents in the desert grasslands determines whether the habitat remains as open grassland or becomes desert scrub (Brown and Heske 1990). So perhaps mice are not only as interesting an animal as pronghorns, but they also might even be more critical to the habitat. Something similar is almost certainly true in most other habitats: it's often the small life forms that have the greatest biomass and consequent influence. Yet these are the animals that are typically ignored in zoos. Thus, without their inclusion in zoo collections, the interpretations that can be given are crippled, and the stories that can be told about management and maintenance of wild habitats are severely compromised.

We all know that the fascination people might have for mice, or pronghorns for that matter, can be an essential hook for attracting attention to the habitat of the animal and to the interrelationships between all components of that habitat. In this way, zoos have enormous potential to be an influential means for educating people about the natural world. Zoos reach across all age groups, all socioeconomic boundaries, and all levels of awareness. If they could also develop the capacity to tell people about ecosystems, not just animals, they could make an enormous and real contribution, through education, toward the conservation of wild places and all wild animals.

Considering that zoos claim to demonstrate the richness and variety of the animal world, they present an alarmingly narrow view. Ninety-five percent of all creatures on earth are smaller than a chicken's egg. The selection criteria for zoo collections, with their emphasis on the bigger, the cuter, and the more spectacular, result in a skewed and narrow view of the animal kingdom (Collins and Thomas 1991).

However, I doubt whether many zoos will consider changing their emphasis. Most will continue to see their role as places that concentrate on exhibiting the typical zoo species that so dominate almost every zoo collection. I fear that this view is leading zoos along a cul-de-sac, and that they will come to a dead end. The countering argument to this scenario is most likely that zoos are performing a vital role in conservation through coordinated breeding programs with several species. I am tempted to say, *only* several species. The AZA's 160-plus member zoos, with annual operating expenses of nearly $800 million a year and a work force of some 16,500 people, are coordinating species survival programs for just one-hundred-plus species of animals (Boyd 1995): two-thirds of these are mammals, and almost all of them can be classified as charismatic megafauna. If, as the World Resources Institute, the World Conservation Union, and the United Nations Environment Programme declare, the problem is conservation of biological diversity, then zoos are not making much of a contribution relative to the combined power of their budgets and the numbers of people they employ, and the millions that visit them (WRI et al. 1992).

My belief is that zoos are not the best places for sustained captive breeding programs, and that they can make their greatest contribution to conservation through education, especially by teaching the importance of saving wild habitats. If one wanted to develop a facility concentrating on the breeding of endangered species, one would not design the place to be a public zoo. Zoos are essentially places for exhibition, and in a zoo operation the resources of space, energy, time, and money for fully fledged breeding programs face too many other competing programs.

It is encouraging that successes in captive breeding have made zoos almost completely self-sufficient in providing animals for their collections, and in a few instances created surpluses, but the task of saving the world's wildlife is too great for even the combined efforts of all the world's zoos. They do not have enough space, nor can they apply sufficient resources. The concentrated breeding of large numbers of endangered species, including all those species that are at risk but are not glamorous enough to qualify for attention from zoos, should best take place in facilities devoted specifically to that task. It is too important, too large, and too complicated a venture to graft onto existing zoo operations. Ideally, the AZA should be energetically lobbying state, federal, and international government agencies to work together in funding and managing endangered species breeding farms and cooperative wildlife reintroduction programs. The Endangered Species Act is ready for a metamorphosis (Mann and Plummer 1995). If its new form includes captive breeding projects, this could fit logically with national and worldwide government-to-government conservation measures in the field. Certainly the loss of wildlife, and especially biological diversity, is a global concern, requiring international legislative coordination.

I plead now for some immediate and major changes in zoos: first, a move away from exhibiting and interpreting only animals; second, better representation of small life forms, to give a more accurate view of the diversity and complexity of the planet's fauna; third, an attempt to show the functional roles that animals play in their ecosystems; fourth, closer attention to the total quality of life of all the animals in the collections; and, fifth, more regional specialization, so that zoo exhibits and education programs can bring directly home the full message of conservation to their local audiences.

In such a mobile population as ours there is little justification for the extensive duplication of species and exhibits in our zoos. A move toward more attention on local habitats would encourage zoo visitors to value more highly the wildlife and wilderness in their own part of the country. My own experiences are limited mainly to the American West, but I know that zoogoers in Oregon, Washington, and British Columbia are told much more about the loss of tropical rain forests than they are about clear-cutting of the forests of the Pacific Northwest, even though the rate of depletion of our temperate forests is equal to or greater than anything happening in those conveniently distant tropics. The millions who visit zoos in California to see elephants and tigers are somberly told that fur coats and ivory trinkets are shameful, but they receive no information on the vast destruction of the chaparral habitats, now almost totally depleted in southern California and replaced with eucalypti, lawns, palm trees, and pansies. Of course, the destiny of rain forests, and of rhinoceroses, is far removed from our own daily lives and easily elicits sympathy, whereas attention on local wildlife issues might cause some uneasiness by requiring paying visitors to question the consequences of their own life-styles.

In summary, then, my own wish for zoos of the next century is that they concentrate on their best potential, which is exhibitry and interpretation. I would like to see a great broadening of the components of the zoo: a focus on all the aspects of the natural world, as opposed to a conscious and conspicuous neglect of the greater part. I would like to see a sharpening of that focus, too, with more attention to some regional elements. Zoos in particular regions could cooperate to avoid duplication. In the Pacific Northwest, for example, Seattle's Woodland Park Zoo could specialize, say, in northern temperate and South American tropical forest exhibits. Zoos in Vancouver, Tacoma, and Portland could then concentrate on other, different biomes. The amount of overlap could be reduced and, between them, all the Pacific Northwest region's wildlife parks could cover a wide range of the world's bioclimatic zones, each in greater depth than at present. If that cooperation could be expanded to include the botanical gardens, science museums, and natural history museums of the region as well, then truly significant progress could be made in conservation education. Similar cooperative planning could be carried out in other regions around the nation, and the sum of the parts could truly be greater than the whole. By working and planning together, and by sharing resources, we could more effectively carry the messages of biological diversity.

Ironically, such specialization would allow greater diversification, broadening the scope and the opportunities for interpretation of ecosystems. Politics and institutional pride, however, would almost certainly conspire to prevent changes of that magnitude. In the meantime there is much that zoos need to do to expand the zoo concept beyond that which we have inherited from the London of 1828.

REFERENCES

Bailey, V. 1929. Life history and habits of grasshopper mice, genus *Onychomys*. *USDA Technical Bulletin* 145:1–19.

Berlin, B. 1973. The relation of folk systematics to biological classification and nomenclature. *Annual Reviews of Ecology and Systematics* 4:259–271.

Boyd, L., ed. 1995. *1994–1995 Zoological Parks and Aquariums in the Americas*. Wheeling, W.Va.: AAZPA.

Brown, J. H., and E. J. Heske. 1990. Control of a desert-grassland transition by a keystone rodent guild. *Science* 250:1705–1707.

Collins, N. M., and J. A. Thomas. 1991. *The Conservation of Insects and Their Habitats*. London: Academic Press.

Ehrlich, P., and A. Ehrlich. 1981. *Extinction*. New York: Random House.

Hancocks, D. 1971. *Animals and Architecture*. New York: Praeger.

Mann, C. C., and M. L. Plummer. 1995. *Noah's Choice: The Future of Endangered Species*. New York: Knopf.

Wilson, E. O. 1992. *The Diversity of Life*. Cambridge: Harvard University Press, Belknap Press.

WRI, IUCN, and UNEP. 1992. *Global Biodiversity Strategy*. Washington, D.C.: World Resources Institute, World Conservation Union, and United Nations Environment Programme.

ARE ZOOS MORALLY DEFENSIBLE?

Tom Regan

Despite important differences, a number of recent tendencies in ethical theory are united in the challenges they pose for well-entrenched human practices involving the utilization of nonhuman animals, including their use in zoos. This essay explores three such tendencies—utilitarianism, the rights view, and environmental holism—and explores their respective answers to the question, Are zoos morally defensible? Both utilitarianism and holism offer ethical theories that in principle could defend zoos, but both, it is argued, are less than adequate ethical outlooks. For reasons set forth below, the third option—the rights view—has implications that run counter to the moral acceptability of zoos, as we know them. The essay concludes not by insisting that zoos as we know them are morally indefensible but, rather, by admitting that we have yet to see an adequate ethical theory that illuminates why they are not.

A great deal of recent work by moral philosophers—much of it in environmental ethics, for example, but much of it also in reference to questions about obligations to future generations and international justice—is directly relevant to the moral assessment of zoos. (Here and throughout I use the word "zoo" to refer to a professionally managed zoological institution accredited by the AZA and having a collection of live animals used for conservation, scientific studies, public education, and public display.) Yet most of this work has been overlooked by advocates of zoological parks. Why this is so is unclear, but certainly the responsibility for this lack of communication needs to be shared. Like all other specialists, moral philosophers have a tendency to converse only among themselves, just as, like others with a shared, crowded agenda, zoo professionals have limited discretionary time, thus little time to explore current tendencies in academic disciplines like moral philosophy. The present book, bringing together, as it does, both ethicists and persons professionally involved with the real-world work of zoos, is especially noteworthy, and as befits the objectives of this book, the present essay attempts to take some modest steps in the direction of better communication between the two professions.

After a brief historical section, three tendencies in contemporary moral philosophy—utilitarianism, animal rights, and holism—are described and some of their implications regarding zoos are explained. Not all these tendencies can be true in every respect (for they contradict each other at crucial places), and perhaps none is true in any. Unquestionably, however, these three tendencies are among the most important options in moral philosophy today, so that how they answer the central question I intend to explore—namely, Are zoos morally defensible?—cannot be irrelevant to an informed moral assessment of zoos.

As will become clear as we proceed, my own moral position is not that of a neutral observer. Of the three tendencies to be considered, I favor one (what I call the "rights view") and disagree rather strongly with the other two. For obvious reasons, my characterizations and assessments of these tendencies are in the nature of rough sketches; for more detailed accounts the reader is referred to my works cited in the references and notes.

CHANGING TIMES

Time was when philosophers had little good to say about animals other than human beings. "Nature's automata," writes Descartes (Regan and Singer 1976, 60). Morally considered, animals are in the same category as "sticks and stones," opines the early twentieth-century Jesuit Joseph Rickaby (179). True, there have been notable exceptions, throughout history, who celebrated the intelligence, beauty, and dignity of animals: Pythagoras, Cicero, Epicurus, Herodotus, Horace, Ovid, Plutarch, Seneca, Virgil—hardly a group of ancient-world animal crazies. By and large, however, a dismissive sentence or two sufficed or, when one's corpus took on grave proportions, a few paragraphs or pages. Thus we find Immanuel Kant, for example, by all accounts one of the most influential philosophers in the history of ideas, devoting almost two full pages to the question of our duties to nonhuman animals, while Saint Thomas Aquinas, easily the most important philosopher-theologian in the Roman Catholic tradition, bequeaths perhaps ten pages to this topic.

Times change. Today even a modest bibliography of the past decade's work by philosophers on the moral status of nonhuman animals (Magel 1989) would easily equal the length of Kant's and Aquinas's treatments combined (Regan and Singer 1976, 122–124, 56–60, 118–122), a quantitative symbol of the changes that have taken place, and continue to take place, in philosophy's attempt to excise the cancerous prejudices lodged in the anthropocentric belly of Western thought.

With relatively few speaking to the contrary (Saint Francis always comes to mind in this context), theists and humanists, rowdy bedfellows in most quarters, have gotten along amicably when discussing questions about the moral center of the terrestrial universe: human interests form the center of this universe. Let the theist look hopefully beyond the harsh edge of bodily death, let the humanist denounce, in Freud's terms, this "infantile view of the world," at least the two could agree that the moral universe revolves around us humans—our desires, our needs, our goals, our preferences, our love for one another. An intense dialectic now characterizes philosophy's assaults on the traditions of humanism and theism, assaults aimed not only at the traditional account of the moral status of nonhuman animals but also at the foundations of our moral dealings with the natural environment, with Nature generally. These assaults should not be viewed as local skirmishes between obscure academicians each bent on occupying a deserted fortress. At issue are the validity of alternative visions of the scheme of things and our place in it. The growing philosophical debate over our treatment of the planet and the other animals with whom we share it is both a symptom and a cause of a culture's attempt to come to critical terms with its past as it attempts to shape its future.

At present moral philosophers are raising a number of major challenges against moral anthropocentrism. I shall consider three. The first comes from utilitarians, the second from proponents of animal rights, and the third from those who advocate a holistic ethic. This essay offers a brief summary of each position with special reference to how it answers our central question—the question, again, Are zoos morally defensible?

UTILITARIANISM

The first fairly recent spark of revolt against moral anthropocentrism comes, as do other recent protests against institutionalized prejudice, from the pens of the nineteenth-century utilitarians Jeremy Bentham and John Stuart Mill. In an oft-quoted passage Bentham enfranchises sentient animals in the utilitarian moral community by declaring, "The question is not, Can they talk?, or Can they reason?, but, Can they suffer?" (Regan and Singer 1976, 130). Mill goes even further, writing that utilitarians "are perfectly willing to stake the whole question on this one issue. Granted that any practice causes more pain to animals than it gives pleasure to man: is that practice moral or immoral? And if, exactly in proportion as human beings raise their heads out of the slough of selfishness, they do not with one voice answer 'immoral' let the morality of the principle of utility be forever condemned" (132). Some of our duties are direct duties to other animals, not indirect duties to humanity. For utilitarians, these animals are themselves involved in the moral game.

Viewed against this historical backdrop, the position of the influential contemporary moral philosopher Peter Singer can be seen to be an extension of the utilitarian critique of moral anthropocentrism (Singer 1990). In Singer's hands utilitarianism requires that we consider the interests of everyone affected by what we do, and also that we weigh equal interests equally. We must not refuse to consider the interests of some people because they are Catholic, or female, or black, for example. Everyone's interests must be considered. And we must not discount the importance of equal interests because of whose interests they are. Everyone's interests must be weighed equitably. Now, to ignore or discount the importance of a woman's interests because she is a woman is an obvious example of the moral prejudice we call sexism, just as to ignore or discount the importance of the interests of African or Native Americans, Hispanics, etc. is an obvious form of racism. It remained for Singer to argue, which he does with great vigor, passion, and skill, that a similar moral prejudice lies at the heart of moral anthropocentrism, a prejudice that Singer, borrowing a term coined by the English author and animal activist Richard Ryder, denominates speciesism (Ryder 1975).

Like Bentham and Mill before him, therefore, Singer denies that humans are obliged to treat other animals equitably in the name of the betterment of humanity and also denies that acting dutifully toward these animals is a warm-up for the real moral game played between humans or, as theists would add, between humans and God. We owe it to those animals who have interests to take their interests into account, just as we also owe it to them to count their interests equitably. In these respects we have direct duties to them, not indirect duties to humanity. To think otherwise is to give sorry testimony to the very prejudice—speciesism—Singer is intent upon silencing.

UTILITARIANISM AND THE MORAL ASSESSMENT OF ZOOS

From a utilitarian perspective, then, the interests of animals must figure in the moral assessment of zoos. These interests include a variety of needs, desires, and preferences, including, for example, the interest wild animals have in freedom of movement, as well as adequate nutrition and an appropriate environment. Even zoos' most severe critics must acknowledge that in many of the most important respects, contemporary zoos have made important advances in meeting at least some of the most important interests of wild animals in captivity.

From a utilitarian perspective, however, there are additional questions that need to be answered before we are justified in answering our central question. For not only must we insist that the interests of captive animals be taken into account and be counted equitably, but we must also do the same for all those people whose interests are affected by having zoos—and this involves a very large number of people indeed, including those who work at zoos, those who visit them, and those (for example, people in the hotel and restaurant business, as well as local and state governments) whose business or tax base benefits from having zoos in their region. To make an informed moral assessment of zoos, given utilitarian theory, in short, we need to consider a great deal more than the interests of those wild animals exhibited in zoos (though we certainly need to consider their interests). Since everyone's interests count, we need to consider everyone's interests, at least insofar as these interests are affected by having zoos—or by not having them.

Now, utilitarians are an optimistic, hearty breed, and what for many (myself included) seems to be an impossible task, to them appears merely difficult. The task is simple enough to state—namely, to determine how the many, the varied, and the competing interests of everyone affected by having zoos (or by not having them) are or will be affected by having (or not having) them. That, as I say, is the easy part. The hard (or impossible) part is actually to carry out this project. Granted, a number of story lines are possible (for example, stories about how much people really learn by going to zoos in comparison with how much they could learn by watching National Geographic specials). But many of these story lines will be in the nature of speculation rather than of fact, others will be empirical sketches rather than detailed studies, and the vital interests of some individuals (for example, the interests people have in having a job, medical benefits, a retirement plan) will tend not to be considered at all or to be greatly undervalued.

Moreover, the utilitarian moral assessment of zoos requires that we know a good deal more before we can make an informed assessment. Not only must we canvass all the interests of all those individuals who are affected, but we must also add up all the interests that are satisfied as well as all the interests that are frustrated, given the various options (for example, keeping zoos as they are, changing them in various ways, or abolishing them altogether). Then, having added all the pluses and minuses—and only then—are we in a position to say which of the options is the best one.

But (to put the point as mildly as possible) how we rationally are to carry out this part of the project (for example, how we rationally determine what an equitable trade-off is between, say, a wild animals' interest in roaming free and a tram operator's interest in a steady job) is far from clear. And yet unless we have comprehensible, comprehensive, and intellectually reliable instructions regarding how we are to do this, we will lack the very knowledge that, given utilitarian theory, we must have before we can make an informed moral assessment of zoos. The suspicion is, at least among utilitarianism's critics, the theory requires knowledge that far exceeds what we humans are capable of acquiring. In the particular case before us, then, it is arguable that utilitarian theory, conscientiously applied, would lead to moral skepticism—would lead, that is, to the conclusion that we just don't know whether or not zoos are morally defensible. At least for many people, myself included, this is a conclusion we would wish to avoid.

In addition to problems of this kind, utilitarianism also seems open to a variety of damaging moral criticisms, among which the following is representative. The theory commits us to withholding our moral assessment of actions or practices until everyone's interests have been taken into account and treated equitably. Thus the theory implies that before we can judge, say, whether the sexual abuse of very young children is morally wrong, we need to consider the interests of everyone involved—the very young child certainly, but also those of the abuser. But this seems morally outrageous. For what one wants to say, it seems to me, is that the sexual abuse of children is wrong independently of the interests of abusers, that their interests should play absolutely no role whatsoever in our judgment that their abuse is morally wrong, so that any theory that implies that their interests should play a role in our judgment must be mistaken. Thus, because utilitarianism does imply this, it must be mistaken.

Suppose this line of criticism is sound. Then it follows that we should not make our moral assessment of anything, whether the sexual abuse of children or the practice of keeping and exhibiting wild animals in zoos, in the way this theory recommends. If the theory is irredeemably flawed—and that it is, is what the example of child abuse is supposed to illustrate—then its answer to any moral question, including in particular our question about the defensibility of zoos, should carry no moral weight, one way or the other (that is, whether the theory would justify zoos or find them indefensible). Despite its historic importance and continued influence, we are, I think, well advised to look elsewhere for an answer to our question.

THE RIGHTS VIEW

An alternative to the utilitarian attack on anthropocentrism is the rights view. Those who accept this view hold that (1) the moral assessment of zoos must be carried out against the backdrop of the rights of animals and that (2) when we make this assessment against this backdrop, zoos, as they presently exist, are not morally defensible. How might one defend what to many people will seem to be such extreme views? This is not a simple question by any means, but something by way of a sketch of this position needs to be presented here (Regan 1983).

The rights view rests on a number of factual beliefs about those animals humans eat, hunt, and trap, as well as those relevantly similar animals humans use in scientific research and exhibit in zoos. Included among these factual beliefs are the following: These animals are not only in the world, but they are also aware of it—and of what happens to them. And what happens to them matters to them. Each has a life that fares experientially better or worse for the one whose life it is. As such, all have lives of their own that are of importance to them apart from their utility to us. Like us, they bring a unified psychological presence to the world. Like us, they are somebodies, not somethings. They are not our tools, not our models, not our resources, not our commodities.

The lives that are theirs include a variety of biological, psychological, and social needs. The satisfaction of these needs is a source of pleasure, their frustration or abuse, a source of pain. The untimely death of the one whose life it is, whether this be painless or otherwise, is the greatest of harms since it is the greatest of losses: the loss of one's life itself. In these fundamental ways these nonhuman animals are the same as human beings. And so it is that according to the rights view, the ethics of our dealings with them and with one another must rest on the same fundamental moral principles.

At its deepest level an enlightened human ethic, according to the rights view, is based on the independent value of the individual: the moral worth of any one human being is not to be measured by how useful that person is in advancing the interests of other human beings. To treat human beings in ways that do not honor their independent value—to treat them as tools or models or commodities, for example—is to violate that most basic of human rights: the right of each of us to be treated with respect.

As viewed by its advocates, the philosophy of animal rights demands only that logic be respected. For any argument that plausibly explains the independent value of human beings, they claim, implies that other animals have this same value, and have it equally. Any argument that plausibly explains the right of humans to be treated with respect, it is further alleged, also implies that these other animals have this same right, and have it equally, too.

Those who accept the philosophy of animal rights, then, believe that women do not exist to serve men, blacks to serve whites, the rich to serve the poor, or the weak to serve the strong. The philosophy of animal rights not only accepts these truths, its advocates maintain, but also insists upon and justifies them. But this philosophy goes further. By insisting upon the independent value and rights of other animals, it attempts to give scientifically informed and morally impartial reasons for denying that these animals exist to serve us. Just as there is no master sex and no master race, so (animal rights advocates maintain) there is no master species.

ANIMAL RIGHTS AND THE MORAL ASSESSMENT OF ZOOS

To view nonhuman animals after the fashion of the philosophy of animal rights makes a truly profound difference to our understanding of what we may do to them. Because other animals have a moral right to respectful treatment, we ought not reduce their moral status to that of being useful means to our ends. That being so, the rights view excludes from consideration many of those factors that are relevant to the utilitarian moral assessment of zoos. As explained earlier, conscientious utilitarians need to ask how having zoos affects the interests people have in being gainfully employed, how the tourist trade and the local and state tax base are impacted, and how much people really learn from visiting zoos. All these questions, however, are irrelevant if those wild animals confined in zoos are not being treated with appropriate respect. If they are not, then, given the rights view, keeping these animals in zoos is wrong, and it is wrong independently of how the interests of others are affected.

Thus, the central question: Are animals in zoos treated with appropriate respect? To answer this question, we begin with an obvious fact—namely, the freedom of these animals is compromised, to varying degrees, by the conditions of their captivity. The rights view recognizes the justification of limiting another's freedom but only in a narrow range of cases. The most obvious relevant case would be one in which it is in the best interests of a particular animal to keep that animal in confinement. In principle, therefore, confining wild animals in zoos can be justified, according to the rights view, but only if it can be shown that it is in their best interests to do so. That being so, it is morally irrelevant to insist that zoos provide important educational and recreational opportunities for humans, or that captive animals serve as useful models in important scientific research, or that regions in which zoos are located benefit economically, or that zoo programs offer the opportunity for protecting rare or endangered species, or that variations on these programs insure genetic stock, or that any other consequence arises from keeping wild animals in captivity that forwards the interests of other individuals, whether humans or nonhumans.

Now, one can imagine circumstances in which such captivity might be defensible. For example, if the life of a wild animal could be saved only by temporarily removing the animal from the threat of human predation, and if, after this threat had abated, the animal was reintroduced into the wild, then this temporary confinement arguably is not disrespectful and thus might be justified. Perhaps there are other circumstances in which a wild animal's liberty could be limited temporarily, for that animal's own good. Obviously, however, there will be comparatively few such cases, and no less obviously, those cases that satisfy the requirements of the rights view are significantly different from the vast majority of cases in which wild animals are today confined in zoos, for these animals are confined and exhibited not because temporary captivity is in their best interests but because their captivity serves some purpose useful to others. As such, the rights view must take a very dim view of zoos, both as we know them now and as they are likely to be in the future. In answer to our central question—Are zoos morally defensible?—the rights view's answer, not surprisingly, is No, they are not.

HOLISM

Although the rights view and utilitarianism differ in important ways, they are the same in others. Like utilitarian attacks on anthropocentrism, the rights view seeks to make its case by working within the major ethical categories of the anthropocentric tradition. For example, utilitarians do not deny the moral relevance of human pleasure and pain, so important to our humanist forebears; rather, they accept it and seek to extend our moral horizons to include the moral relevance of the pleasures and pains of other animals. For its part, the rights view does not deny the moral importance of the individual, a central article of belief in theistic and humanistic thought; rather, it accepts this moral datum and seeks to widen the class of individuals who are thought of in this way to include nonhuman animals.

Because both the positions discussed in the preceding use major ethical categories handed down by our predecessors, some influential thinkers argue that these positions, despite all appearances to the contrary, remain in bondage to anthropocentric prejudices. What is needed, these thinkers believe, is not a broader interpretation of traditional categories (for example, the category "the rights of the individual") but the overthrow of these very categories themselves. Only then will we have a new vision, one that liberates us from the last vestiges of anthropocentrism.

Among those whose thought moves in this direction, none is more influential than Aldo Leopold (1949). Leopold rejects the individualism so dear to the hearts of those who build their moral thinking on the welfare or rights of the individual. What has ultimate value is not the individual but the collective, not the part but the whole, meaning the entire biosphere and its constituent ecosystems. Acts are right, Leopold claims, if they tend to promote the integrity, beauty, diversity, and harmony of the biotic community; they are wrong if they tend contrariwise. As for individuals, be they humans or other animals, they are merely "members of the biotic team," having neither more nor less value in themselves than any other member—having, that is, no value in themselves. What value individuals have, so far as this is meaningful at all, is instrumental only: They are good to the extent that they promote the welfare of the biotic community.

Traditional forms of utilitarianism, not just the rights view, go by the board given Leopold's vision. To extend our moral concern to the pleasures and pains of other animals is not to overcome the prejudices indigenous to anthropocentrism. One who does this is still shackled to those prejudices, supposing that those mental states that matter to humans must be the measure of what matters morally to the world at large. Utilitarians are people who escape from one prejudice (speciesism) only to embrace another (what we might call sentientism). Animal liberation is not nature liberation. In order to forge an ethic that liberates us from our anthropocentric tradition we must develop a holistic understanding of the community of life and our place in it. The land must be viewed as meriting our equal moral concern. Waters, soils, plants, rocks—inanimate, not just animate, existence—must be seen to be morally considerable. All are equal members of the same biotic team.

Holists face daunting challenges when it comes to determining what is right and wrong. That is to be determined by calculating the effects of our actions on the life community. Such calculations will not be easy. Utilitarians, as noted earlier, encounter a serious problem when they are asked to say what the consequences will be if we act in one way rather than another. This problem arises for them despite the fact that they restrict their calculations to sentient life. How much more difficult it must be, then, to calculate the consequences for the entire biosphere!

But perhaps the situation for holists is not as dire as I have suggested. While it is true that we often lack detailed knowledge about how the biosphere is affected by human acts and practices, we sometimes know enough to say that some of the things we are doing are unhealthy for the larger community of life. For example, we do not know exactly how much we are contaminating the water of the earth by using rivers and oceans as garbage dumps for toxic wastes, or exactly how much protection afforded by the ozone layer is being compromised by our profligate use of chlorofluorocarbons. But we do know enough to realize that neither situation bodes well for marine and other life forms as we know them.

Let us assume, then, what I believe is true, that we sometimes are wise enough to understand that the effects of some human practices act like insatiable cancers eating away at the life community. From the perspective of holism, these practices are wrong, and they are wrong because of their detrimental effects on the interrelated systems of biological life.

It is important to realize that holists are aware of the catastrophic consequences toxic dumping and the ever widening hole in the ozone layer are having on individual animals in the wild—on elephants and dolphins, for example. It would be unfair to picture those who subscribe to holism as taking delight in the suffering and death of these individual animals. Holists are not sadists. What is fair and important to note, however, is that the suffering and death of these animals are not morally significant according to these thinkers. Morally, what matters is how the diversity, sustainability, and harmony of the larger community of life are affected, not what happens to individuals.

To make the holists' position clearer, consider the practice of trapping fur-bearing animals for commercial profit. Holists find nothing wrong with this economic venture so long as it does not disrupt the integrity, diversity, and sustainability of the ecosystem. Trappers can cause such disruptions, if they over-trap a particular species. The danger here is that the depletion of a particular species will have a ripple effect on the community as a whole and that the community will lose its diversity, sustainability, and integrity. The overtrapping (and hunting) of wolves and other predatory animals in the northeastern United States often is cited as a case in point (Baker 1985). Once those natural predators were removed, other species of wildlife—deer in particular, it is asserted—are said to have overpopulated, so that today these animals actually imperil the very ecosystem that supports them. All this could have been avoided if, instead of rendering local populations of natural predators extinct by overtrapping and overhunting, the humans had trapped or hunted more judiciously, with an eye to sustainable yield. Although a significant number of individual animals would have been killed, the integrity, harmony, and sustainability of the ecosystem would have been preserved. When and if commercial trappers achieve these results, holists believe they do nothing wrong. From the perspective of holism, the inevitable suffering and untimely death of individual furbearing animals do not matter morally.

HOLISM AND THE MORAL ASSESSMENT OF ZOOS

Holism's position regarding the ethics of zoos in particular is analogous to its position regarding the ethics of our other interactions with wildlife in general. There is nothing wrong with keeping wild animals in permanent confinement if doing so is good for the larger life community. But it is wrong to do this if the effects on the community are detrimental. Moreover, because one of the indices of what is harmful to the life community is a reduction in the diversity of forms of life within the community, holism will recognize a strong prima facie duty to preserve rare or endangered species. To the extent that the best zoos contribute to this effort, holists will applaud their efforts, even if keeping individual animals who belong to threatened species in captivity is not in the best interests of those particular animals. In that and other respects (for example, the moral relevance of the educational and research functions of zoos), the implications of holism are very much at odds with those of the rights view and much closer to those of utilitarianism.

Some people who accept a holistic ethic are skeptical of the real contributions zoos make to species protection. It is appropriate for all of us to press this issue since, despite the claims sometimes made on behalf of zoo programs whose purpose is to reintroduce endangered species into their native habitats, for example, the rate of success might be far less than the public is led to believe. Philosophically, however, there are deeper, more troubling questions that need to be considered. Of the many that come to mind, only one will be discussed here.

Holism—or, to speak more precisely, the unqualified, unequivocal version of holism sketched above[1]—takes a strong moral stance in opposition to whatever upsets the diversity, balance, and sustainability of the community of life. Unquestionably, it is the human presence and the effects of human activities that have by far the most adverse effects on the diversity, balance, and sustainability of the life community. Now, as we have seen, the holist's response to such effects when these are allegedly caused by nonhumans (for example, by an overabundance of deer) is to recommend a limited hunting season, to cull the herd, and thereby restore ecological balance. Why, then, should holists not advocate comparable policies in the face of human depredation of the life community? In other words, why should holists stop short of recommending that the human population be culled using measures no less lethal than those used in the case of controlling the population of deer? Granted, the latter is legal, the former not. But legality is not a reliable guide to morality, and the question before us is a question of morals, not a question of law. And it is the moral question that needs to be pressed.

Given the major tenets of their theory, holists cannot meet the challenge this question poses by insisting that humans are in a different moral category from deer and other wild animals. Like every species, each individual is a member of the biotic team, and no species—anymore than any individual—is of greater importance in the ecological scheme of things than any other. It is therefore a palpable double standard to permit killing deer, who (let us assume) cause some environmental damage, and to prohibit killing humans, who cause much, much more.[2] In other words, either holists mean what they say, or they do not. If they do not, then there is no reason to take them seriously. If they do, then they cannot avoid embracing the draconian implications to which their position commits them.

Let us assume that holists mean what they say and that they should be taken seriously. Our question, then, is whether to agree with them. One can only hope that few will do so. One would also hope that a moral position authorizing policies that have all the markings of species genocide will find few partisans. Granted, the environmental crisis is a crisis of monumental proportions, and granted, human beings are the major cause; nevertheless, a morally acceptable approach to this crisis needs to rest on some basis other than the biocentric egalitarianism that helps define holism.

The relevance of the preceding to our central question is analogous to the earlier discussion of utilitarianism. As was true in that earlier case, it is no good attempting to defend zoos in particular by appealing to a moral outlook that is morally unacceptable in general. Thus, because holism is not a morally acceptable outlook, it is not an acceptable basis for assessing the moral justification of zoos.

Those who believe that zoos, as they presently exist, are morally defensible, therefore, will have to find a moral outlook that parts company both with holism and with utilitarianism. The rights view, of course, is a third major option. But that view, for reasons advanced in the preceding, is highly critical of zoos, on grounds that they violate the right of wild animals to be treated with respect. This essay concludes, therefore, on the following cautionary note—if or as one hopes to marshal a moral defense of zoos, one will have to articulate, defend, and competently apply some theory other than the three surveyed on this occasion.

NOTES

Parts of this essay previously appeared in Tom Regan, *The Thee Generation: Reflections on the Coming Revolution* (Philadelphia, Pa.: Temple University Press, 1991).

1. Some of the major problems faced by J. Baird Callicott's revised holism (1993)—in particular, questions about trade-offs between such disparate sorts of values as the good of one's family and the good of an ecosystem—are analogous to those faced by utilitarianism. Space prevents a detailed discussion of these views on this occasion.

2. Holists might reply that unlike deer, humans have free choice and can be educated so that they choose with an ecological conscience; thus, we should wait until massive educational efforts have been made, then see whether people change their behavior appropriately, before instituting lethal solutions. Again, programs that reduce the rate of human population growth might be preferred over those that recommend reducing the population that already exists. A fuller (and fairer) discussion of holism would be obliged to consider just how well founded these options are.

REFERENCES

Baker, R. 1985. *The American Hunting Myth.* New York: Vantage Press.

Callicott, J. B. 1993. In search of an environmental ethic. In *Matters of Life and Death,* ed. T. Regan, 3d ed., 322–382. New York: McGraw-Hill.

Leopold, A. 1949. *A Sand County Almanac.* New York: Ballentine Books.

Magel, C. 1989. *Keyguide to Information Sources in Animal Rights.* London: Mansell.

Regan, T. 1983. *The Case for Animal Rights.* Berkeley: University of California Press.

Regan, T., and P. Singer, eds. 1976. *Animal Rights and Human Obligations.* Englewood Cliffs, N.J.: Prentice-Hall.

Ryder, R. 1975. *Victims of Science.* London: Davis-Poynter.

Singer, P. 1990. *Animal Liberation.* 2d ed. New York: Random House.

ZOOS REVISITED

Dale Jamieson

T he possibility of perpetual reinvention is deeply embedded in the American psyche. Waiters can become movie stars, gangsters can be transformed into respectable businessmen, and corrupt White House officials can return as fundamentalist preachers. One California governor, who signed the most liberal abortion law in the nation, became a fiercely born-again antiabortionist; another former California governor, one of the leading political fund-raisers of his generation, ran for president on a platform that denounced political fund-raising as the root of all evil. Despite its attractions, reinvention is not always successful. In F. Scott Fitzgerald's *Great Gatsby,* the title character emerges from a shady past to assume the life of a Long Island gentleman. Yet despite the trappings of wealth and power, his new identity is fragile. In the end he succumbs to his past.

In their drive to reinvent themselves American zoos are very American. Early zoos were explicitly meant to demonstrate and celebrate the domination of nature by man. They included all sorts of exotics, both human and nonhuman.[1] As the control of zoos moved from rich and powerful individuals to communities and governments, they increasingly were seen as sources of urban amusement. But in these enlightened times many zoo professionals no longer see amusement and entertainment as roles that are worthy of zoos. Indeed, in this spirit, the New York Zoological Society has abolished its zoos; however wildlife conservation parks have risen, phoenixlike, to replace them. In their current reinvention zoos are being pitched as the last best hope for endangered wildlife. For advocates of zoos, as for Jay Gatsby, the past is evil but fortunately always behind us. The present is good, and the future promises to be even better—assuming the money holds out.

Critics of zoos rightly see this attitude as self-serving and disingenuous. Most zoos are still in the business of entertainment rather than species preservation. Despite protestations to the contrary, most zoos are still more or less random collections of animals kept under largely bad conditions. Although the best zoos have been concerned to position themselves as environmental heroes, they have done little to promote this ethic in the zoo industry as a whole. There are many bad exhibits and many bad zoos, but not much is being done to shut them down. Even the best zoos have problems with preventable mortality and morbidity due to accidents or abuse and are too often in league, wittingly or unwittingly, with people whose idea of a good animal is one that turns a quick profit. The rhetoric of science, favored by the best people in the best zoos, has not yet penetrated the reality of most zoos and indeed carries with it new possibilities for abuse. Even now, with the bad old days presumably behind us, there is not much ground for complacency.

Still, it is clear that zoos are changing. They are becoming more naturalistic in environment, focusing more on species preservation and scientific research and less on entertainment. Zoos in the future, at least the better ones, will increasingly become more like parks.

Parks and preserves are changing as well. They are becoming more like zoos. In 1987 Kenya's Lake Nakuru National Park was completely fenced (Conway 1990). It is only a matter of time until large East African mammals are managed in much the same way as domestic animals, as already has been suggested by the World Conservation Union (see Conway, this volume; Hutchins et al., this volume). This tendency toward management is also at work in the national parks in this country.

What will become of wild nature in this proliferation of miniparks or mega-zoos (Conway 1990)? Wild nature may be done for. Human population growth remains out of control. The effects of human consumption and production are modifying fundamental planetary systems in what may be irreversible ways. We are probably already committed to a climate change that will have profound effects on both nature and human society. Extremely remote areas in the arctic and antarctic regions are suffering the effects of human-induced ozone depletion. Today no part of the planet is unaffected by human action. Nature may not yet be tamed, but she is no longer wild (McKibben 1989). The evolution of every animal species, to some degree, is now affected by human action (Borza and Jamieson 1990; Jamieson 1990, 1992).

One of the most dramatic effects of human action is the epidemic of extinctions currently sweeping the earth. Increasingly zoos have attempted to position themselves as the guardians of wild nature, as the boy with his thumb in the dike trying to hold back the floodwaters. I do not believe that zoos can successfully play this role. Establishing genetic warehouses is not the same as preserving wild animals. Highly managed theme parks are not wild nature.

Although in the bad old days zoos may have made their contributions to extinction, they are not responsible for the current wave. Nor are they directly to blame for our pathetic response to it. What is to blame is the peculiar moral schizophrenia of a culture that drives a species to the edge of extinction and then romanticizes the remnants. Until a species is on the brink of extinction it seems to have little claim on our moral sensibility.

Consider the northern spotted owl. Most people probably agree with the Denver newspaper, the *Rocky Mountain News,* which editorialized (16 March 1992) that loggers need jobs as much as the spotted owls need trees. This is what passes for a moderate position, carefully balancing the unsustainable life-style of a few thousand humans against the very existence of another form of life. Once the owl is extinct or a few stragglers have been moved indoors, people will sing a different song. No steps will be too extreme to save this endangered species.

In the bad old days I published a paper with the subtle, highly nuanced title "Against Zoos" (Jamieson 1985). For my effort I was virtually accused of child abuse by a local television station. Its correspondents interviewed children visiting the Denver Zoo, eliciting their reactions to some pointy-headed philosopher who wanted to take their fun away. The responses were predictable. A column in the *Chicago Tribune* (28 April 1991) said that my ideas were so absurd that "only an intellectual could believe them." No less a journal than *Time Magazine* (24 June 1991) called me a "zoophobe" and suggested that I am indifferent to the fate of endangered species.

What I tried to do in that much-maligned paper was to set forth as rationally as possible the case against zoos. I examined the arguments that have been given on their behalf: that they provide amusement, education, opportunities for scientific research, and help in preserving species. I saw some merit in each argument, but in the end I concluded that these benefits were outweighed by the moral presumption against keeping animals in captivity. I also claimed that despite the best intentions of zoo personnel, the profound message of zoos is that it is permissible for humans to dominate animals, for the entire experience of a zoo is framed by the fact of captivity.

Serious people have taken issue with my claims and arguments (see, e.g., Chiszar et al. 1990, Hutchins et al., this volume). Because some of my critics place more weight on the role of zoos in preserving endangered species than I do, I want to discuss that issue in some detail. However, I first want to reconsider whether there is a presumption against keeping animals in captivity, since this claim is foundational to my argument against zoos.[2]

IS THERE A PRESUMPTION AGAINST KEEPING ANIMALS IN CAPTIVITY?

In my 1985 paper I argued that there is a presumption against keeping animals in captivity. My argument was rather intuitive. Keeping animals in captivity usually involves restricting their liberty in ways that denies them many goods including gathering their own food, developing their own social orders, and generally behaving in ways that are natural to them. In the case of many animals captivity also involves removing them from their native habitats and conditions. If animals have any moral standing at all, then it is plausible to suppose that depriving them of liberty is presumptively wrong, since an interest in liberty is central to most morally significant creatures.

My claim that there is such a presumption has recently been challenged (Leahy 1991). If Leahy is correct in thinking that there is no such presumption, then there is no general reason for being opposed to zoos. The acceptability of keeping animals in captivity would turn entirely on a case-by-case examination of the conditions under which various animals are kept. Before considering Leahy's arguments against this presumption, let us first consider the view to which he is committed.

The idea that there is a presumption against keeping animals in captivity implies that it is not a matter of moral indifference whether animals are kept captive. But it carries no implication about how strong the presumption is. People who agree that there is a presumption against keeping animals in captivity can disagree about the strength of the presumption or about whether it is permissible to keep an animal in captivity in a particular case. What Leahy is committed to is the view that everything else being equal, it is a matter of moral indifference as to whether animals are kept in captivity; we might as well flip a coin. I believe that this view is implausible.

Although it is difficult to perform this thought experiment, imagine that we could guarantee the same or better quality of life for an animal in a zoo that the animal would enjoy in the wild. Suppose further that there are no additional benefits to humans or animals that would be gained by keeping the animal in captivity. The only difference between these two cases that might be relevant is that in one case the animal is confined to a zoo and in the other case the animal is free to pursue his or her own life. Would we say that the fact of confinement is a morally relevant consideration? I believe that most people would say that it is, and that it would be morally preferable for the animal to be free rather than captive. In my opinion this shows that most of us believe that there is a moral presumption against keeping animals in captivity. That we believe that there is such a presumption is indicated in various ways. For example, sometimes it is said that keeping an animal in captivity is a privilege that involves assuming special obligations for the animal's welfare. This expresses the sense, I believe, that in confining an animal we are in some way wronging him or her, and thus owe him or her some compensation.

With this result in mind, let us consider Leahy's arguments. He appears to offer two. The first (following Hediger 1964) involves the claim that animals are not truly free in the wild. They are constrained by ecological and social pressures and are "struck down by natural predators and diseases which, quite reasonably, can be said to limit their freedom" (Leahy 1991, 242). Since animals are not truly free in the wild, keeping them in captivity does not deprive them of liberty. The second argument is a conceptual one. According to Leahy, animals do not have language and are not self-conscious; therefore they cannot make choices or raise objections. Since they cannot make choices or raise objections, they cannot be said to live their own lives. Since they cannot live their own lives, they can never really be free. Since animals can never really be free, confining them in zoos does not deprive them of their freedom.

The first argument is intended to show that as a matter of fact animals are not free in their natural habitats while the second argument is intended to show that animals can never be free under any circumstances. There is no presumption against keeping them in captivity because in neither case does captivity deprive them of something that they have in the wild.

We should see first that these arguments do not really question the view that there is a presumption against depriving animals of liberty. What these arguments are supposed to show is that animals do not or cannot have liberty, thus they are not deprived of it by captivity. If it could be shown that animals do have liberty in the wild but not in captivity, then Leahy might agree that there is a presumption against keeping animals in captivity on grounds that it deprives them of liberty. At least he has said nothing that counts against this view.

The core of the issue, then, is the plausibility of the commonsense view that animals lose their liberty when they are removed from the wild and kept in zoos. I affirm the commonsense view; Leahy denies it. Who is right?

Consider Leahy's second argument first. Two steps in the argument that invite objection are these: the claim that animals are not self-conscious, and the claim that self-consciousness or language is required for making choices.

The topic of self-consciousness is a difficult one. Philosophers and psychologists often use this concept in different ways. One approach, characteristic of Descartes and much of the philosophical tradition, associates self-consciousness with the ability to use language or other complex symbol systems. But even if it were agreed that the use of complex symbol systems is required for self-consciousness, it would appear that various primates and cetaceans satisfy this criterion and thus would be excluded from the scope of Leahy's conclusion (Herman and Morrel-Samuels 1990, Savage-Rumbaugh and Brakke 1990). For those animals who use complex symbol systems, Leahy would have no argument for supposing that they are not free in the wild. Thus with respect to those animals at least, my claim that there is a presumption in favor of liberty would appear to survive unscathed. A second approach, characteristic of work in cognitive ethology, regards attributions of self-consciousness as underwritten by such factors as behavior, evolutionary continuity, and structural similarity. Researchers have argued that a wide range of behaviors in a variety of animals involve self-consciousness, including social play, deception, and vigilance (Mitchell and Thompson 1986, Byrne and Whiten 1988, Griffin 1992, Jamieson and Bekoff 1993). Whichever approach is adopted, the claim that only humans are self-conscious appears doubtful.

The second dubious step in this argument involves the claim that self-consciousness is required for making choices. The philosopher's paradigm of choice may involve listing alternatives on a yellow pad with the pros and cons of each fully described in the margins, but this is only one way of making choices. Many of our choices are made without explicitly representing alternatives and totting up pluses and minuses—for example, when we choose coffee rather than tea, hit the brake rather than the gas, or immediately agree to give a lecture in Iowa in response to a telephone call. In these kinds of cases it is hard to see exactly how self-consciousness is supposed to be involved. Moreover, important work on animal behavior has purported to address such topics as mate choice (Bateson 1983), habitat choice (Rosenzweig 1990), and the choice of nest sites (Bekoff et al. 1989). For the most part this work has been done without presupposing that animals are self-conscious. It may be that these researchers misuse the term "choice" or are simply wrong in supposing that animals make choices in these situations. However, I believe that it is more plausible to suppose that it is Leahy's claim that is false and that self-consciousness is not required for choice. Since at least two steps in Leahy's second argument appear dubious, it is plausible to suppose that the argument fails.

Leahy's first argument attempts to show not that animals cannot be free under any conditions but that as a matter of fact they are not free in the wild. The idea is that if they are not free in the wild, then they lose nothing when they are confined in zoos. The evidence for the claim that animals are not free in the wild is that they are constrained by ecological and social pressures and are struck down by natural predators and disease.

If pointing to ecological and social pressures were sufficient for showing that an animal is not free, it would prove too much, for all organisms, including humans, are constrained by ecological and social pressures. The most that this claim could establish is that social and ecological pressures restrict animals to such an extent that they are more free in captivity than they are in the wild.

Are animals more free in zoos than in the wild? On the face of it, this claim is wildly implausible. It is like saying that humans are more free in prison than on the street because they are not subject to the same pressures as people on the street. The argument seems to overlook the fact that social pressures exist in zoos as well as in the wild, and in many cases such pressures are more intense in zoos because individuals are inhibited from responding to them in the ways in which they would in the wild. But more important, even if it could be shown that caged animals, whether human or nonhuman, live longer than those who are uncaged, this would not provide evidence for the claim about freedom. Nor could the claim be established by showing that caged animals are happier than uncaged animals. Liberty is not the same as longevity or happiness, nor does it always manifest in these ways. Moreover, there is very little evidence for supposing that captive animals live longer or are happier in zoos than they are in the wild. It seems plain that most animals have less freedom in zoos than in the wild. Indeed, the very point of systems of confinement is to deprive them of freedom.

For reasons that I have given it seems to me that Leahy's arguments fail. The commonsense position, that everything else being equal it is better for animals to be free, is vindicated. However, there is another line of argument that might be thought to be more challenging than those pursued thus far. It might be granted that there is a presumption of liberty with respect to animals who are born in the wild, but denied that there is any such presumption with respect to those who are born in captivity. It might be argued that captive-bred animals have never known freedom, so they are denied nothing by captivity.

In my view there is a presumption in favor of liberty with respect to all animals, whether bred in captivity or in the wild. Imagine humans who have never known liberty. Would it be plausible to deny that there is a presumption of liberty for them on the grounds that they do not miss what they have never known? An affirmative answer would be absurd. Indeed, we might think that the tragedy of their captivity is all the greater because they have never known liberty. Transferring these intuitions to nonhuman animals, we can see that there is a presumption in favor of liberty even with respect to animals born in captivity. Indeed, the presumption may even be stronger in their case. Still, some people would argue against this presumption, pointing out that many animals bred in captivity would not survive liberation, despite attempts at preparation. Their lives in nature would be nasty, brutish, and short. Even if this is true it fails to show that there is no presumption in favor of liberty for these animals. At most it shows that in these cases the presumption in favor of liberty is outweighed by concerns about the welfare of these animals. The presumption for liberty exists, but it may be wrong to release these animals into the wild.

What I have argued in this section is that a basic claim of my 1985 paper, that there is a presumption in favor of liberty for animals, still stands. The burden of proof rests on those who would confine animals in zoos. The most compelling reason for confining animals in zoos, in some people's eyes, is the need to preserve endangered species. It is to this justification that I now turn.[3]

CAN ZOOS PRESERVE ENDANGERED SPECIES?

There are a number of arguments against zoos as meaningful sites for preserving endangered species. First, such preservation is needed, it is rightly pointed out, because we are losing species at an enormous rate. But although estimates differ and not all the facts are known, it is obvious that not more than a tiny fraction of these species can be preserved in zoos. Ehrlich and Ehrlich estimate that American zoos could preserve about one hundred mammals under the best conditions (1981, 211). Second, only a small number of the species preserved in zoos could ever be reintroduced into their natural habitats. Indeed most attempts at reintroduction have failed (Beck, this volume). For many species, zoos are likely to be the last stop on the way to extinction. Finally, over many generations the genetic structure and behavior of captive populations change. Captivity substitutes selection pressures imposed by humans, either intentionally or inadvertently, for those of an animal's natural habitat. Indeed, under some definitions of domestication, confining animals in zoos and breeding them in captivity transforms them into domesticated animals (see, e.g., Rodd 1990, 113; Clutton-Brock 1992; but see also Norton, this volume). Whether we count zoo animals as domesticated or not, it is clear that in fifty, one hundred, or a thousand years we may not have the same animal that was placed in captivity, much less the animal that would have existed had it evolved in nature. Taken together these arguments show that the role that captive breeding and reintroduction can play in the preservation of endangered species is at best marginal. Thus the benefit of preservation is not significant enough to overcome the presumption against depriving an animal of its liberty.

Against arguments such as these (Jamieson 1985, Varner and Monroe 1991), it is sometimes objected that they are entirely hypothetical. Where are the data? it is sometimes asked, and then we hear anecdotes about species that have been saved by captive breeding programs. Such arguments are made against Varner and Monroe by Hutchins and Wemmer, who go on to assert that "there are many problems facing captive breeding and reintroduction programs, but they are not insurmountable" (1991, 10). But how do they know that? Where are the data that show that such problems are not insurmountable? Is this a scientific statement or the expression of a quasireligious faith in the idea that humans have the ability to technofix everything, even the threatened extinction of other species?[4]

The point is that demands for data can be made by either party to the dispute. The fact is that there are anecdotes on both sides, qualitative material that different people evaluate in different ways, but very little that looks like hard data. The skeptic about captive breeding programs will say that the defender of zoos has the burden to show that such programs really can be successful. If there is a presumption against keeping animals in captivity, then it is wrong to do so unless a case can be made that the benefits outweigh this presumption. From the perspective of a skeptic, an inconclusive argument on this point is one that the skeptic wins.

Defenders of zoos say that the burden is on the other side, for captive breeding keeps options open. True enough. We ought to keep options open, not only for ourselves but for future people as well. But at what cost? Unless the presumption that animals should not be kept in captivity can be overcome by the moral case for keeping options open, this observation does not carry much weight. It certainly does not establish a burden of proof.

There is another dimension to this dispute. The critics of zoos point out that breeding and reintroduction programs can be extremely invasive, involving not just denials of liberty but sometimes pain and suffering for individuals. Defenders of zoos sometimes say that this suffering is for the good of the species. This is the maneuver that in my 1985 paper I called sacrificing the interests of the lowercase gorilla for those of the uppercase Gorilla.

There is a lot of confusion about the concept of species and its proper role in our biological and moral thinking (Ereshefsky 1992, Hargrove 1992). Yet law, policy, and common morality take the concept very seriously. An animal that is part of an endangered species may have millions spent to protect her, but if she is a member of an endangered subspecies or a hybrid she may be exterminated as a pest. Some of these issues are explored by May (1990), O'Brien and Mayr (1991), Vane-Wright et al. (1991), Geist (1992), and Rojas (1992).

One confusion in our biological thinking concerns the relation between variability and species diversity. Species diversity is one kind of variability but not the only kind. Within most species there is an enormous amount of variability—think of dogs or coyotes, for example (Bekoff and Wells 1986). The evolutionary story requires variability, but it is not clear that it requires a very strong conception of species. Dawkins writes that " 'the species' [is] an arbitrary stretch of continuously flowing river, with no particular reason to draw lines delimiting its beginning and end" (1986, 264). Darwin himself was quite conventionalist about the concept of species, writing that "I look at the term species, as one arbitrarily given for the sake of convenience to a set of individuals closely resembling one another" (Darwin 1958, 67). The demotion of the concept of species from the exalted role that it played in Aristotelian biology was one consequence of the Darwinian revolution. Like other consequences of the Darwinian revolution, we are still struggling to grasp its full significance (Dewey 1910, Rachels 1990).

Variability is important to us as well as to the evolutionary process. We value variability, but just as we often focus on the charismatic megafauna and overlook other creatures that are as important to nature, so we often fix on species variability as the only kind of diversity that matters. We compound the problem when we think that it is species to which we have obligations rather than the creatures themselves. This is an instance of the general fallacy of attributing to species the properties of individual creatures. Individual creatures have hearts and lungs; species do not. Individual creatures often have welfares, but species never do. The notion of a species is an abstraction; the idea of its welfare is a human construction. While there is something that it is like to be an animal there is nothing that it is like to be a species.

I am a Darwinian about the concept of species, but I am not callous about the survival of nature. I am as concerned about saving wild nature as any defender of zoo breeding programs. But I believe that the only hope for doing this is to put large tracts of the earth's surface off-limits to human beings and to alter radically our present life-styles.[5] I agree with Ehrenfeld that "the true prospects for conservation ultimately depend not on the conservation manipulations of scientists but on the overarching consideration of how many people there will be in the world in the next century, the way they live, and the ways in which they come to regard and use nature" (1991, 39).

I believe further that attempts at preserving wild nature through zoo breeding programs are a cruel hoax. If zoo breeding programs are successful they will not preserve species but rather transform animals into exhibits in a living museum. "This is what used to exist in the wild," we can say to our children while pointing at some rare creature alienated from her environment, "before the K-Mart and the biotechnology factory went in." Zoo professionals like to say that they are the Noahs of the modern world and that zoos are their arks. But Noah found a place to land his animals where they could thrive and multiply. If zoos are like arks, then rare animals are like passengers on a voyage of the damned, never to find a port that will let them dock or a land in which they can live their lives in peace and freedom. If we are serious about preserving wild nature we must preserve the land, and not pretend that we can bring nature indoors.

In my darker moments I believe, not just that zoos are in the business of perpetuating fraud with their rhetoric about preserving animals, but that, knowingly or not, they are deeply implicated in causing the problem that they purport to be addressing. Zoo professionals are often eager to remove animals from the wild to more controlled environments where they can be studied. But as more and more animals are taken out of the wild, the case for preserving wild nature erodes. Why save a habitat if there is nothing to inhabit it? Advocates of zoos like to point out that they are not just in the business of removing animals from the wild, but increasingly they are also involved in trying to preserve animals in nature as well. Although zoos boast of their programs in the developing world, very few can withstand scrutiny.[6] The truth is that very few zoos make meaningful attempts to preserve animals in nature, and most zoos spend more on publicity and public relations than they do on programs involving animals. This is especially appalling because in many cases programs to preserve animals in situ are relatively cheap. For example, the Bonobo Protection Fund estimates that the bonobo population in Zaire could be effectively protected for an initial investment of $185,000 and $60,000 per year thereafter. This is a small amount to spend for the protection of the rarest of ape species.

In my opinion we should have the honesty to recognize that zoos are for us rather than for the animals. Perhaps they do something to alleviate our sense of guilt for what we are doing to the planet, but they do little to help the animals we are driving to extinction. Our feeble attempts at preservation are a matter of our own interests, values, and preoccupations rather than acts of generosity toward those animals whom we destroy and then try to save. Insofar as zoos distract us from the truth about ourselves and what we are doing to nature, they are part of the problem rather than part of the solution.

SUMMARY AND CONCLUSIONS

Much of what I have said may sound like an aggravating stew of idealism and curmudgeonliness. Hutchins and Wemmer speak for many people when they say that philosophers seem "more concerned with logical arguments than with practical solutions to real problems" (1991, 5). Although I wish nonphilosophers were more concerned with logical arguments, I sympathize with their sentiments. It is a fact that despite the arguments that I and others have given, zoos are not going to go away. It is easier to try to change large institutions that are adept at fund-raising than it is to abolish them. At any rate we are responsible for the lives of a great many animals, and more are being bred all the time. Given that zoos exist, there is a great difference between good ones and bad ones. I would like to close by expressing some of my hopes and fears about how zoos may develop in the future.

As I have already said, the best zoos in the future will be increasingly indistinguishable from small parks. The conditions under which animals will be kept for breeding purposes and scientific study will be naturalistic. While the idea of a *naturalistic* environment should not be confused with a *natural* environment, it is clear that human-designed naturalistic environments rule out some of the worst of the abuses to which captive animals traditionally have been subject. For example, naturalistic environments would not permit animals to be constantly observed by hundreds of small boys who feed them Cracker Jacks and hurl various objects at them. This obviously would be an improvement over many exhibits that exist today.

In my opinion there will be increasing tension between what zoos do to gain public support (entertain) and what they must do in order to justify themselves (preserve species). This tension will emerge within zoos as those who are interested in animals and science will increasingly come into conflict with those whose charge is budgets and public relations. This conflict already prevents zoos from being as good as they can be, and it will become more pronounced in the future. This is a fear.

One hope that I have for the future is that we will recognize that if we keep animals in captivity, then what we owe them is everything. Whatever else we may believe about the morality of zoos, I hope we can come to a consensus that these animals are in our custody through no wish or fault of their own. They are refugees from a holocaust that humans have unleashed against nature. There should be no question of culling these animals or trading off their interests against those of humans. If we are to keep animals in captivity, then we must conform to the highest standards of treatment and respect. My hope is that zoo professionals will accept this principle and that an enlightened and aggressive public will keep them to it, for the animals themselves have no voice in human affairs, and as nature recedes their voices become ever more silent.

ACKNOWLEDGMENTS

Discussions with many people have affected my views about zoos. I thank the participants in the conference Animal Welfare and Conservation: Ethical Paradoxes in Modern Zoos and Aquariums, at which a version of this essay was presented, especially Bryan Norton, Michael Hutchins, and Terry Maple for inviting me to participate, and Don Lindburg, who was an intellectual and moral inspiration. Over the years I have learned a great deal about various topics touched on in this paper from Marc Bekoff, Anna Goebbel, Sue Townsend, and John Wortman. Despite my debts to all of these people, I alone am responsible for the views that I have expressed.

NOTES

1. For a moving account of an African Pygmy who was confined to the New York City Zoo, see Bradford and Bloom 1992.

2. Although I prefer to avoid the language of rights, my work on zoos has been greatly influenced by Rachels (1976). For a good discussion of the concept of freedom, see Taylor 1986, 105–111. It should also be noted that for the purposes of this essay I use the terms "liberty" and "freedom" interchangeably.

3. In a recent book Bostock agrees that there is a presumption against keeping animals in captivity, but claims that "we can go a long way towards providing good conditions in zoos" (1993, 50). For the presumption to be overcome however, it must be shown that the benefits of confining animals in zoos are greater than the burdens. This is not established by speculative claims about the possibility of creating good conditions for animals in zoos.

4. Proponents of zoos seem especially given to making unsubstantiated, sweeping claims. Wolfe attacks my 1985 paper for failing to consider "that one function of zoos may be to help children make symbolic sense of the world around them." He then goes on to conclude, "Children learn to use their powers of fantasy and imagination—to love animals—by going to the zoo. Strip them of this rich source of their interpretive life and, as adults, they will likely be more unfeeling, not less" (Wolfe 1991, 116–117). This is all very nice, high-minded rhetoric, and may even be true. But what I claimed in my 1985 paper is that there are very few data to support the educational claims that are made on behalf

of zoos. Whether zoos indeed have the uplifting effects on children claimed by Wolfe is an empirical question. I ask again: where are the data?

5. For this reason I endorse the general concepts put forward by the Wildlands Project, P.O. Box 5365, Tucson, Arizona 85703.

6. *Newsweek* (12 April 1993) documents the ineffectiveness and corruption of various programs to save endangered species, several of them involving major zoos. For a case study, see Schaller 1993.

REFERENCES

Bateson, P. 1983. *Mate Choice*. New York: Cambridge University Press.

Bekoff, M., A. Scott, and D. Conner. 1989. Ecological analyses of nesting success in evening grosbeaks. *Ecologia* 81:67–74.

Bekoff, M., and M. Wells. 1986. Social ecology and behavior of coyotes. *Advances in the Study of Behavior* 16:251–338.

Borza, K., and D. Jamieson. 1990. *Global Change and Biodiversity Loss: Some Impediments to Response*. Boulder: Center for Space and Geosciences Policy, University of Colorado.

Bostock, S. 1993. *Zoos and Animal Rights*. London: Routledge.

Bradford, P., and H. Bloom. 1992. *Ota Benga: The Pygmy in the Zoo*. New York: St. Martin's Press.

Byrne, R., and A. Whiten, eds. 1988. *Machiavellian Intelligence: Social Expertise and the Evolution of Intellect in Monkeys, Apes, and Humans*. Oxford: Oxford University Press.

Chiszar, D., J. B. Murphy, and W. Iliff. 1990. For zoos. *Psychological Record* 40:3–13.

Clutton-Brock, J. 1992. How the beasts were tamed. *New Scientist*, 15 February: 41–43.

Conway, W. 1990. Miniparks and megazoos: From protecting ecosystems to saving species. Thomas Hall Lecture, presented at Washington University, St. Louis.

Darwin, C. 1958. *The Origin of Species*. New York: Penguin.

Dawkins, R. 1986. *The Blind Watchmaker*. New York: Norton.

Dewey, J. 1910. *The Influence of Darwinism on Philosophy and Other Essays in Contemporary Thought*. New York: Henry Holt & Co.

Ehrenfeld, D. 1991. The management of diversity. In *Ecology, Economics, Ethics: The Broken Circle*, ed. F. Borman and S. Kellert, 26–39. New Haven, Conn.: Yale University Press.

Ehrlich, P., and A. Ehrlich. 1981. *Extinction: The Causes and Consequences of the Disappearance of Species*. New York: Random House.

Ereshefsky, M., ed. 1992. *The Units of Evolution: Essays on the Nature of Species*. Cambridge: MIT Press.

Geist, V. 1992. Endangered species and the law. *Nature* 357:274–276.

Griffin, D. 1992. *Animal Minds*. Chicago: University of Chicago Press.

Hargrove, E., ed. 1992. *The Animal Rights, Environmental Ethics Debate: The Environmental Perspective*. Albany: State University of New York Press.

Hediger, H. 1964. *Wild Animals in Captivity*. New York: Dover.

Herman, L., and P. Morrel-Samuels. 1990. Knowledge acquisition and asymmetry between language comprehension and production: Dolphins and apes as general models for animals. In *Interpretation and Explanation in the Study of Animal Behavior. Vol. 1:*

Interpretation, Intentionality, and Communication, ed. M. Bekoff and D. Jamieson, 283–312. Boulder, Colo.: Westview Press.

Hutchins, M., and C. Wemmer. 1991. Response: In defense of captive breeding. *Endangered Species Update* 8:5–6.

Jamieson, D. 1985. Against zoos. In *In Defense of Animals,* ed. P. Singer, 108–117. New York: Harper & Row.

———. 1990. Managing the future: Public policy, scientific uncertainty, and global warming. In *Upstream/Downstream: Essays in Environmental Ethics,* ed. D. Scherer, 67–89. Philadelphia, Pa.: Temple University Press.

———. 1992. Ethics, public policy, and global warming. *Science, Technology, and Human Values* 17:139–153.

Jamieson, D., and M. Bekoff. 1993. On aims and methods in cognitive ethology. In *PSA 1992,* Vol. 2, ed. D. Hull, M. Forbes, and K. Okruhlik, 110–124. Lansing, Mich.: Philosophy of Science Association.

Leahy, M. 1991. *Against Liberation: Putting Animals in Perspective.* New York: Routledge.

May, R. 1990. Taxonomy as destiny. *Nature* 347:129–130.

McKibben, B. 1989. *The End of Nature.* New York: Random House.

Mitchell, R., and N. Thompson, eds. 1986. *Deception: Perspectives on Human and Nonhuman Deceit.* Albany: State University of New York Press.

O'Brien, S. J., and E. Mayr. 1991. Bureaucratic mischief: Recognizing endangered species and subspecies. *Science* 251:1187–1188.

Rachels, J. 1976. Do animals have a right to liberty? In *Animal Rights and Human Obligations,* ed. T. Regan and P. Singer, 205–223. Englewood Cliffs, N.J.: Prentice-Hall.

———. 1990. *Created from Animals: The Moral Implications of Darwinism.* New York: Oxford University Press.

Rodd, R. 1990. *Biology, Ethics, and Animals.* Oxford: Oxford University Press.

Rojas, M. 1992. The species problem and conservation: What are we protecting? *Conservation Biology* 6:170–178.

Rosenzweig, M. 1990. Do animals choose habitats? In *Interpretation and Explanation in the Study of Animal Behavior. Vol. 1: Interpretation, Intentionality, and Communication,* ed. M. Bekoff and D. Jamieson, 157–179. Boulder, Colo.: Westview Press.

Savage-Rumbaugh, S., and K. Brakke. 1990. Animal language: Methodological and interpretive issues. In *Interpretation and Explanation in the Study of Animal Behavior. Vol. 1: Interpretation, Intentionality, and Communication,* ed. M. Bekoff and D. Jamieson, 313–343. Boulder, Colo.: Westview Press.

Schaller, G. 1993. *The Last Panda.* Chicago: University of Chicago Press.

Taylor, P. 1986. *Respect for Nature: A Theory of Environmental Ethics.* Princeton, N.J.: Princeton University Press.

Vane-Wright, R., C. Humphries, and P. Williams. 1991. What to protect? Systematics and the agony of choice. *Biological Conservation* 55:235–254.

Varner, G., and M. Monroe. 1991. Ethical perspectives on captive breeding: Is it for the birds? *Endangered Species Update* 8:27–29.

Wolfe, A. 1991. Up from humanism. *American Prospect,* 112–125.

PART TWO

THE TARGETS OF PROTECTION

GENES OR INDIVIDUALS OR POPULATIONS OR SPECIES OR ECOSYSTEMS

I t is widely accepted that we should protect biological systems and resources both because of their inherent interest and for their value to future human generations. What, exactly, are we obligated to protect? Nature is experienced by us as a multilayered and almost infinitely complex system; each layer—at least each biotic layer—has its functions in the larger whole, and its own methods of communicating with its environment. The immediately attractive answer—that we should save all of the layers from genes to habitats—is unfortunately not a realistic possibility. To accomplish this would be to freeze natural systems, whereas their very essence is to change and evolve. The problem is to separate acceptable changes from unacceptable ones and to develop a plan of environmental management that is mindful of all the levels, but which recognizes that change—including anthropogenic change—is an inevitable part of the process. The separation of acceptable from unacceptable changes will therefore require some means to decide which elements will receive priority protection in situations of conflict among levels. This conflict can be of two types: In some cases it is impossible to protect one level without introducing very invasive changes on another level, as when we must sacrifice some animal in order to protect its species or when one species undergoes a population explosion and threatens the structure of a valued ecosystem. In other cases, it might be possible in theory to protect all levels of a system simultaneously, but societal resources may not be available to give equal priority to all elements and processes of natural systems. While recognizing that the latter problems of insufficient resources and the need to set priorities may in some cases be most important, this book concentrates on one special type of the first conflict—if protection of species and the wild processes they embody requires invasive procedures affecting individual animals, which set of obligations and responsibilities should override the others? Or to put the point simply, what constraints are placed on efforts to save species and habitats by the unquestionably important obligations we have to treat animals humanely?

In this section we have asked a variety of thinkers to examine the question of conservation targets as objectively as they can, but to do so from the diverse perspectives they represent. Dale Jamieson, an ethicist who has studied animal behavior and awareness, builds on his argument in Part One by showing specifically how the presumption in favor of keeping wild animals wild places a heavy burden of proof on those who would protect composite entities such as species or ecosystems at the expense of the lives and freedom of individual specimens. Writing from the viewpoint of a geneticist, Robert Vrijenhoek outlines the case that it is ecological processes that must be saved. This emphasis is supported by the behavioral ecologist Valerius Geist and by the philosopher Bryan Norton. Geist cites the rapid deterioration of habitats and urges quick action to protect species even if this requires significant meddling in the daily lives of animals, including removing species temporarily from the wild. Norton supplements his advocacy of programs to protect processes by suggesting that moral quandaries that are apparently based on unavoidable conflicts can often be clarified by emphasizing differences in the contexts in which humans interact with animals as the main determinants of human responsibility.

Again, consensus was not gained on the broad, theoretical issues addressed in Part Two. Those participants whose personal concerns lie mainly in conservation emphasize the importance of processes as the sustenance for all elements of biological systems, and they recommend an emphasis on protecting these ecological and evolutionary processes as the main targets of an adequate program for protecting biological resources. Those whose ethical principles have been developed in a context of concern for individual animals, on the other hand, tend to emphasize obligations to protect both the welfare and dignity of individual animals and to leave them free in their natural habitat. The range of opinions here is again quite large; nevertheless, we think it is fair to say that there is agreement that efforts to protect species and ecosystems are justified and that most disagreement concerns what methods—especially what invasive procedures affecting individual animals—are appropriate to attain that goal.

WILDLIFE CONSERVATION AND

INDIVIDUAL ANIMAL WELFARE

Dale Jamieson

I t should be obvious that there are conflicts between people who manage wildlife and people who advocate the interests of individual animals. At least to a point, there is nothing wrong with such conflicts. The dream of a world without conflict is the dream of a world in which one set of values dominates to the exclusion of others. Conflicts arise when people become aware of the fact that legitimate values can conflict with each other. A generation ago wildlife managers were unquestioned authorities; questions of individual animal welfare were largely ignored. Conflict is the price of taking animals seriously. Just as there are moral conflicts between humans, so there are moral conflicts between animals, and between humans and animals. Hard choices have to be made in all of these cases. Animal protectionist philosophy, rather than being limiting, can help us to make these hard choices. It can help reduce and resolve conflict. However, in my opinion, no philosophy can make all these conflicts disappear. I believe that conflicts of value are intrinsic to wildlife management.

This chapter has three sections. I first try to explain why conflicts arise, and then provide some suggestions on how conflicts may be reduced. Finally I claim that some conflicts will remain, and state some general principles that may figure in their resolution.

WHY DO WE HAVE CONFLICTS?

One reason we have conflicts is because of the history of our relations with nonhuman animals. Many species are endangered directly or indirectly because we have annihilated their populations. Now when we are interested in preserving them, our actions are often ill-advised. We create parks and refuges that do not respect ecological boundaries. We do not plan for the effects of natural disasters or the possibility of climate change. Conflicts arise in part because of those decisions that we have made in the past. Conflicts also arise because of our present practices. As we continue to overpopulate and overconsume we increasingly come into conflict with remaining animal populations. If we were serious about reducing conflict between wildlife conservation and individual and animal welfare we would change human behavior. We would reduce our numbers and curtail our consumption.

As conflicts become more intense and we become more ideologically committed to species preservation, the welfare of individual animals can suffer. If species are what are to be preserved and they are identified with gene pools, then living, breathing animals can come to be seen as mere means to species preservation, dispensable once they have reproduced. I believe that this attitude is exemplified in some proposals to euthanize surplus zoo animals. This attitude that it is species that matter also contributes to conflicts between wildlife management and individual animal welfare.

HOW CAN CONFLICT BE REDUCED?

We can reduce conflict by rethinking the goals of wildlife management. I think that wildlife management should have two goals: preservation of diversity, and preservation of wildness. In order to reach these goals we must appreciate the holistic, ecological, and demographic factors that affect the success of our efforts.

Notice that these goals make no reference to species preservation. Species diversity is one kind of diversity, but so is diversity of populations and individuals. We should be concerned with preserving species, but we should also be concerned with preserving populations and individuals in order to preserve diversity. We should also be concerned to preserve wildness. In my view wildness cannot be preserved in artificial environments dominated by humans. For this reason wildlife management should focus on preserving nature, because this is the only way to preserve wildness.

Another factor that may help to reduce conflict is our increasing realization that to some extent the welfare of the environment, of humans, and of individual animals is bound up together. Popular works such as those by Robbins (1987) and Rifkin (1992) have demonstrated the extent to which industrial animal agriculture destroys the environment and endangers human health, as well as imposing great suffering on animals. Moving away from our addiction to beef is in the interests of almost all living creatures. Were we to change the goals of wildlife management and move away from industrial animal agriculture, our attitudes toward individual animals would inevitably change. We might begin to see ourselves more in partnership with them rather than as their dominators. Conflicts would be reduced, but they would not disappear.

HOW CAN WE MANAGE THE CONFLICTS THAT REMAIN?

Apparently irreducible conflicts between wildlife management and individual animal welfare would remain even if we were to take the aforementioned steps. Conflicts recently discussed include those between excessive goat populations and endangered plant species in the Channel Islands off the California coast, and the need to feed live prey to predators in captive environments in order to prepare them for reintroduction (e.g., Hutchins and Wemmer 1987; Rolston 1988; Beck, this volume). The following three principles may help to guide our thinking about these conflicts.

Clarify the Issues

We are often told that it is a necessity when animals are killed or captured. There are very few real necessities in nature. Most necessities are necessary in order to achieve various goals. When we consider harming an animal we should explicitly consider the goal that is supposed to be furthered and not hide behind the idea that it is necessary to do this. Goals come into conflict and not all goals can be achieved, certainly not by the most efficient means. If our goal is to improve our understanding of cancer, there is little doubt that experimenting on humans is the most effective means. We refrain from saying that it is a necessary means because we recognize that there are moral constraints on what we may do to humans even in the pursuit of worthy goals. Similarly, we should recognize that there are moral constraints on what we can do to animals even in the pursuit of worthy goals. We will not all agree on what these constraints are, but in order for them to be productively discussed we must clarify the issues that are involved, and not invoke the idea of necessity as if there were nothing more to say.

Become Moral Darwinists

James Rachels has recently argued that our culture has not yet learned the moral lessons of Darwinism (1990). Almost everyone agrees that in almost all cases natural selection operates on individuals. Yet although we view animals as individuals for the purposes of scientific explanation, we tend to view them only as members of a species for purposes of moral decision-making. But it is as misleading to speak of the welfare of a nonhuman species as it is to speak of the welfare of the human species. It is individual organisms that have welfares. As the philosopher Jeremy Bentham wrote nearly two centuries ago, the welfare of a community is simply the sum of the welfares of individuals (1948, 3). It is lions that feel pain, not the lion, just as it is Americans who need better education, not the American. This may seem like a trivial semantic point, but there is a deep discontinuity in the way that we think about humans and nonhumans. With humans our concern is focused on the individuals. We would never talk about culling human populations or forcibly manipulating their breeding patterns. But with nonhumans we focus on the species and consider individuals only as a means to the welfare of the species. In the light of the Darwinian revolution, this moral privileging of humans should be no more acceptable than the scientific privileging that we no longer accept. We should simply acknowledge that killing or confining an animal to preserve a species is not a conflict between the interest of the individual animal and the interest of the species; it is a conflict between the interest of the animal and the human desire to preserve the species. In the case of both humans and nonhumans there may be cases in which sacrificing the interests of some in order to promote some goal that we care about is morally acceptable, but there is little evidence that animals other than humans care about species preservation.

Become Less Arrogant

It may be that many species will not survive without human intervention. But this does not mean that they will survive with human intervention. Indeed, our track record as planetary managers is deplorable: we generally make things worse when we try to make them better.

Vaclav Havel has said that the fall of communism has lessons not just for people in the East but for people everywhere (1991). Communism, according to Havel, was the ultimate human attempt to manage everything. It failed because it could manage nothing. Our attempts to manage nature are, if anything, more arrogant than communist attempts to manage human societies. We need some humility, and to recognize that our attempts to make things better may well make things worse.

To embrace humility is not to adopt "therapeutic nihilism" (Hargrove 1989). We have no reason to believe that this is the best of all possible worlds or that Mother Nature is always right. Rather, humility involves being skeptical about our motives and about our ability to intervene successfully in large, complex systems.

Even if these principles are followed, conflicts between wildlife management and individual animal welfare will remain. This should not be surprising. We have irreducible moral conflicts between humans: for example, conflicts between providing the best possible medical care for those who can afford it and access to basic medical care for everyone, conflicts between those who favor affirmative action as a response to past injustices and those who oppose it, and conflicts between those who believe that abortion is murder and those who believe that it is grounds for murder. The reason these conflicts are so difficult is because in most of these cases there are legitimate values on each side. The fact that these conflicts will remain, however, does not mean that we cannot make better or worse decisions in these cases. We have traditions in moral philosophy that can help to shed light on these conflicts; these are the same traditions that in the last twenty years have been focused on our treatment of nonhuman animals (for a review see Jamieson 1993). Recognizing that conflicts between wildlife management and individual animal welfare will remain is not in itself a solution to these conflicts. But if we see that these conflicts are not different in kind from the moral conflicts that we face in other areas of life, and that we have resources for reflecting on them, we may have the beginning of moral progress. One thing we can be sure of is this: these conflicts cannot be resolved by technocratic appeals to economics, management, or science. We are all going to have to become philosophers.

REFERENCES

Bentham, J. 1948. *The Principles of Morals and Legislation.* New York: Hafner Press.

Hargrove, E. 1989. *Foundations of Environmental Ethics.* Englewood Cliffs, N.J.: Prentice-Hall.

Havel, V. 1991. *Open Letters: Selected Writings 1965–1990.* Ed. P. Wilson. New York: Knopf.

Hutchins, M., and C. Wemmer. 1987. Wildlife conservation and animal rights: Are they compatible? In *Advances in Animal Welfare Science 1986–1987,* ed. M. W. Fox and L. D. Mickley, 111–137. Washington, D.C.: Humane Society of the United States.

Jamieson, D. 1993. Ethics and animals: A brief review. *Journal of Environmental and Agricultural Ethics* 6 (Special Supplement 1): 15–20.

Rachels, J. 1990. *Created from Animals: The Moral Implications of Darwinism.* New York: Oxford University Press.

Rifkin, J. 1992. *Beyond Beef: The Rise and Fall of the Cattle Culture.* New York: Dutton.

Robbins, J. 1987. *Diet for a New America.* Walpole, N.H.: Stillpoint.

Rolston, H. 1988. *Environmental Ethics: Duties to and Values in the Natural World.* Philadelphia, Pa.: Temple University Press.

NATURAL PROCESSES, INDIVIDUALS,

AND UNITS OF CONSERVATION

Robert Vrijenhoek

C onservation efforts should focus on protecting processes that connect all biological levels of organization: genes, individuals, species, and eco- systems. The present focus on endangered species affects a small number of charismatic organisms, while failing to protect fundamental physical, chemical, ecological, and evolutionary processes. Herein, I examine hierarchical relationships among these levels of biological organization and the processes that connect them. I consider individuals and species conduits for critical physical, ecological, and evolutionary processes that form these connections.

Conservationists are criticized by animal welfare advocates, who question the ethics and morality of programs that would violate the rights of individual animals to achieve some greater good for a species or ecosystem (Regan 1983, Hutchins and Wemmer 1987). Thus, they question programs involving the extirpation of introduced exotic species and the captive breeding of animals in zoos and aquar- iums. As a scientist, I will leave the issue of the rights of individuals (human and animal) to those better trained in philosophy. I focus instead on the role of individuals as connectors in the natural hierarchy of life.

What do we want to preserve when we talk about saving biodiversity? Genes? Individuals? Species? Ecosystems? Or processes? The public focuses its emotional and economic attention on saving a few charismatic species (e.g., whales, pandas, rhinos, condors) while generally failing to understand the tattered threads that form the connections in the web of life. For whom do we want to preserve this diversity? For our present enjoyment? For our great-grandchildren? Or for its own sake? As a biologist, I see humanity as an integral part of this web— supported by it and thus responsible for the whole of it. All the units (genes, individuals, species, ecosystems) are important for biodiversity because of the processes that connect them. I cannot separate ecosystems from the physical, chemical, and biological processes that regulate them, nor can I separate individuals and species from the genetic processes that define them. Unfortunately, anthropogenic disturbances, rampant overpopulation, and greed have torn this complex web. To concentrate on a few charismatic species while failing to preserve the processes that connect them seems futile. To argue that individuals have rights that transcend those of the species overstates the importance of individuals in the hierarchy of life.

Clearly, our primary goal must be to preserve habitats and the ecosystem processes that govern them. Simultaneously, to avoid demographic extinction of threatened and endangered species, we must secure remnant populations even if we have to raise them in zoos and aquariums. Yet, in our efforts to save species, we should carefully design survival plans that do no further genetic harm. Captive breeders must constantly be aware that decisions made for economic or demographic reasons today may have long-term impacts on the survival and adaptive potential of their charges.

Nevertheless, efforts to preserve species ex situ are empty without corresponding attempts to reclaim and restore damaged habitats. Assuming that humanity will one day curb its appetite for exploiting and destroying natural habitats, captive breeding programs should be aimed at the eventual release of endangered species throughout their former ranges. Managers of zoos and aquariums can play a pivotal role, not only in maintaining the genetic health of their charges but also in ensuring opportunities for the processes of adaptation and evolution. I return to these issues below.

To understand the value of genes, individuals, species, ecosystems, and processes to the conservation effort, we must ask the following questions: (1) Are we dealing with static or dynamic units? (2) What constitutes natural change versus anthropogenic disturbance? (3) Is it the opportunity for natural change and turnover that we should attempt to preserve? (4) Are we smart enough to know what to correct and what we should leave alone? For a broad-ranging discussion of the last question, the reader should refer to an enlightening book entitled *The Arrogance of Humanism,* by Ehrenfeld (1978).

CONSERVING PROCESSES

Processes are by definition dynamic. However, natural processes need not produce visible change on a scale that is discernible within our lifetimes. Similarly, we will not observe change if there is a balance between conflicting forces. For example, the normal birth and death rates of a population will—all other things being equal—lead to a population that is stable in terms of the numbers of individuals (N) and their age-class distribution. Of course, for seriously endangered populations, the death rate exceeds the birth rate, and equilibrium occurs when N equals 0 (i.e., extinction).

PHYSICAL PROCESSES

The geological, chemical, and climatic processes that affect the spatial and temporal dynamics of our planet are largely outside our scope for management. The best we can hope for is to prevent further damage resulting from massive deforestation, wars, and global pollution. For example, an appropriate solution to the problem of ozone depletion in our atmosphere is to stop producing and releasing more CFCs. I am extremely skeptical of technological solutions to the problem given our poor understanding of atmospheric dynamics.

The strong interplay between geological and chemical processes and life is elegantly illustrated by the specialized organisms associated with deep-sea hydrothermal vents. Discovered only seventeen years ago, these rich biological communities flourish in the absence of sunlight and photosynthesis (Grassle 1985). Life is driven by chemosynthesis. Hydrogen disulfide (H_2S) dissolved in superheated water is used by sulfur-oxidizing bacteria to generate the energy necessary for the manufacture of sugar from carbon dioxide and water. Sulfur-oxidizing bacteria also occur as symbionts in the tissues of clams, mussels, and giant tubeworms. These communities are highly ephemeral. They have short lifetimes as volcanism opens and closes vents in the newly formed ocean bottom. Since I have visited these sites in *Alvin,* a small submarine capable of diving as deep as four kilometers to the bottom, my reverence for the sanctity of life has only been strengthened. My respect for the extremes under which life can flourish has grown. Yet, I fear for the survival of these alien and ephemeral communities. Rich in minerals, the hydrothermal vents are ripe for commercial exploitation (Malahoff 1985). I hope the vents will remain beyond the reach of drilling and dredging technologies, because if we destroy the vents, we surely will be incapable of fixing them.

Anthropogenic disturbances of natural processes can have rapid and dramatic effects on an ecosystem. The desert streams of the American Southwest provide examples of rampant habitat destruction (Minckley and Deacon 1968). Stream diversion, dam construction, and groundwater pumping have drastically affected the flow regimes and abundance of water, and have left fish populations isolated in remnant springs and stream segments. Introduction of numerous exotic species provided the coup de grace for many native fish species living in streams associated with the Colorado River basin of Arizona and New Mexico. Similarly, the construction of dams for the production of power has led to severe reduction of native salmon populations in the American Northwest (Williams et al. 1992). The deleterious effects of these changes in stream hydrology can be reversed or ameliorated in many situations. Remarkably, there are current plans to dismantle two dams that block Pacific salmon migrations in the Elwha River in the state of Washington. If the dams are removed, we will have a wonderful opportunity to learn more about the biological and economic benefits of restoration of rivers and their fauna (Williams et al. 1992).

One must distinguish such anthropogenic disturbances from natural processes that lead to extinction. Highly endangered fish populations also exist in isolated springs, associated with valleys left by large Pleistocene lakes like the Death Valley system (Pister 1981). Natural drying during the past ten thousand years has reduced these lakes to a few surviving springs and marshes. If the drying continues, the native fish species will eventually become extinct. Many species and subspecies have already disappeared. Perhaps these fish are doomed. Perhaps we shouldn't try to save them. Yet, we should not canonize weakly based "scientific" predictions. To do so would only serve to justify the building of another retirement community or golf course on these delicate and beautiful habitats. We cannot predict climatic trends a few weeks into the future, much less decades or centuries. Instead, we need to prevent further anthropogenic insults that would accelerate the natural demise of these sites (Meffe and Vrijenhoek 1988). The patient might be dying, but we need not deliver the death blow.

ECOLOGICAL PROCESSES

Processes that shape ecosystems, communities, and populations are dynamic; some lead to stable equilibria and others function under nonequilibrium conditions. For example, ecological succession is a nonequilibrium process, characterized by progressive change in species composition and energy flow. Each assemblage of species in a succession modifies the environment in such a way as to make it unfavorable for themselves, paving the way for a new assemblage that can exploit the changed conditions.

Disease and predator-prey interactions also affect community structure and population dynamics. Controlling these processes in zoos and aquariums might help to rebuild a captive population, but there is a cost. Resistance to disease and avoidance of predators are consequences of immune surveillance systems and complex behaviors that are shaped by natural selection. Over the long term, protection of captive populations against such threats may produce strains that are debilitated when challenged by natural predators and disease.

DEMOGRAPHIC PROCESSES

Finally, captive breeders and preserve managers can control demographic processes (birth rates, death rates, age-class structure, etc.) to maximize the recovery of an endangered population. Clearly, demographic factors are of major importance in saving endangered species (Lande 1988). Species survival plans should consider the density-dependent factors, social behaviors, and processes that affect and facilitate breeding in the target species. To avoid demographic extinction, managers of endangered species must strive to maintain a population size that is large enough to hedge against stochastic (random and unpredictable) processes like disease epidemics, bad weather, or just a poor year for reproduction and survival.

BIOTIC EVOLUTION

Evolution as a process involves genetic adjustments to changes in the internal and external environments experienced by a population. Figure 1 illustrates its adaptive and diversifying aspects. Evolution within an organismal lineage (anagenesis) involves the accumulation of both adaptive genetic changes as a consequence of natural selection and random changes that might be adaptively neutral. However, the concept of an ancestor-descendant lineage is an abstraction, a cartoon of evolution. What really evolves is a cloud of genes. New mutations arise in the cloud, many are lost, and some replace ancestral genes. Sometimes, a cloud splits (cladogenesis) and gives rise to independent lineages, or even new species (speciation). Together, cladogenesis and anagenesis are responsible for the diversification that we associate with the evolutionary process. In contrast, discrete clouds sometimes hybridize and fuse (reticulate) into one. Reticulate evolution can be viewed as anticladogenesis. Apparently, many of the temperate grasses in the New World had hybrid origins (Grant 1981).

Traditionally, extinction has been portrayed as a failure of adaptation. As students, we were taught that dinosaurs went extinct because they couldn't keep up with a rapidly deteriorating environment. Nowadays, revisionists argue that

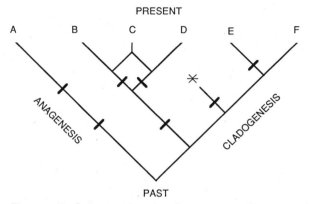

Figure 1. Evolutionary processes. Anagenesis is adaptive evolution within a lineage. Cladogenesis consists of splitting events that give rise to new lineages. Six modern taxa are represented (A through F); one lineage (*) is extinct. Lineage C is the product of reticulate evolution (i.e., hybridization between lineages B and D). The hatch marks represent unique traits that arose within lineages and thereby mark each lineage.

dinosaurs were exquisitely well adapted; they went extinct during a cataclysmic mass extinction that had little to do with their adaptive potential (Gould 1984). Regardless of the causes, extinction of species creates new evolutionary opportunities. Nature abhors a vacuum, and the organismal gaps left by extinction permit niche expansion and diversification by survivors. The rapid diversification of mammals during the Cenozoic might have resulted from opportunities that appeared only after the demise of the dinosaurs.

Nevertheless, the present trauma of mass extinction is a consequence of anthropogenic disturbance, human overpopulation, and greed; not a cometary collision or post-Pleistocene drying. Although we are not morally or ethically responsible for natural extinctions, we must consider opportunities to limit humanity's role in the current cataclysm.

CONSERVING ECOSYSTEMS AND HABITATS

Preservation and restoration of habitats would contribute to the survival of constituent species. Unfortunately, we don't have an endangered habitat law in the United States; we have the Endangered Species Act of 1973 (ESA). In some cases, species may be hard to define, but I suspect they will be legally more defensible than the boundaries of many ecosystems. Legislation like the Wetlands Act protects some aquatic ecosystems, yet governments can redefine wetlands to suit their needs. We need to remind ourselves that only a few years ago marshes were considered useful only as potential sites for landfills or international airports.

Protection of ecosystems upon which endangered species depend is a stated purpose of the ESA (as amended 16 USC 1531 et seq.). Designating the northern spotted owl *(Strix caurina occidentalis)* as endangered provides legal protection to vast areas of the Pacific Northwest. Or so we thought. On 14 May 1992, the Bush administration issued a plan to allow restricted logging on 688 hectares (1,700 acres) within the spotted owl's habitat—to save jobs!

In addition to the ESA, we need legislation that protects ecosystems and habitats for their own value—the dynamic processes and species that define them. Certain habitats like the Florida Everglades or a desert spring are characterized by unique communities of plants and animals. Even if the species themselves are not endangered, the uniqueness of the community might deserve protection. Yet, some communities cannot be preserved as static entities. They represent only a passing phase in a dynamic succession. In these cases, ecological processes must be protected. Sometimes, physical disturbances like fire and hurricanes or biological disturbances like grazing can hold a community in a permanent nonequilibrium state. Maintenance of the New Jersey Pine Barrens requires forest fires to release seeds from the ordinarily closed cones of the pitch pines. Some burning should be allowed to preserve this unique community. The immediate consequences might seem cruel for animals living in the barrens, but the process is natural.

Many communities have become polluted with exotic species. Exotic birds like the European starling *(Sturnus vulgaris)* have become serious crop pests in North America, and other exotic species have introduced avian diseases that decimated native birds in Hawaii (Temple 1992). The mosquitofish *(Gambusia affinis)* has been released globally to control mosquitoes. Unfortunately, this fish played a major role in the demise of native fishes because it is also a voracious predator on fish fry (Arthington and Lloyd 1989, Courtenay and Meffe 1989). Efforts to restore natural communities will require programs aimed at extirpation of many exotic pests. Clearly, conservation efforts will run up against concerns for animal welfare.

CONSERVING SPECIES

The ESA has focused our attention on preserving species. Unfortunately, legal interpretations of the ESA treat species as if they were static entities. The term "species" is similar to "cancer"—both represent syndromes that can have a thousand different causes. Evolutionary biologists recognize the dynamic nature of species. A species is an ancestor-descendant lineage that shares in a common gene pool. Speciation (the process of splitting into two new species) requires the origin of barriers sufficient to impede genetic exchange that would destroy the cohesiveness of the descendant lineages (Templeton 1989). Some evolutionists have argued that gene flow is the antithesis of speciation, that even a small amount of gene flow will result in homogeneity between populations (Mayr 1963). However, complete reproductive isolation is not necessary to preserve the integrity of species. Considerable gene exchange can occur between units that exhibit the morphological and ecological cohesiveness typically associated with the taxonomic application of species status (Carson 1975, Templeton 1989).

I view species as dynamic processes—clouds of genes that can occasionally be connected to other such clouds in time and space. The cohesiveness of individual clouds and the areas of low density between clouds define species (see Templeton 1989 for an elaboration of the cohesiveness concept). For example, gray wolves *(Canis lupus)* and coyotes *(Canis latrans)* hybridize in nature (Wayne and Jenks 1991). Despite the opportunity for gene flow, they exhibit different social be- haviors, and they probably have been morphologically identifiable for two mil- lion years (Nowak 1978). We must avoid the typological thinking that portrays species as static entities or Platonic ideals (i.e., one point in the center of the cloud). Instead, species are genetically variable entities that are connected gene- alogically with other species that still exist and those that have disappeared. Thus, the existence of hybrids does not argue against species status for the hybridizing entities. On the contrary, we should be more confident in the designation of species status if selective processes are sufficiently strong and genetic processes are sufficiently cohesive to counteract the homogenizing effects of gene flow. Species are manifestations of the evolutionary process. Sometimes they split apart, some- times they reticulate, and sometimes they just disappear.

Treating a species or a subspecies as a static entity creates problems with applications of the ESA. For example, the endangered status of the Florida panther *(Felis concolor coryi)* has been questioned because it apparently contains alien mitochondrial DNA that is typical of a South American subspecies (O'Brien and Mayr 1991). Great care must be taken, however, before one infers that hybridization is the source of such alien DNA. First, it is necessary to exclude the hypothesis that the alien genes are ancestral characters shared by both the Florida and South American subspecies. Regardless of the source of these genes, the presumed hybrid panthers appear to exhibit increased vigor as opposed to pure Florida panthers, which seem to be suffering from inbreeding (O'Brien et al. 1990).

Should the Florida panthers be removed from protection under the ESA because they are no longer genetically pure? The notion of purity is the problem here, not the panthers. Hybridization and gene flow are natural parts of the evolutionary process. These processes don't necessarily destroy species or sub-species. Instead, they might rescue populations like the Florida panther from the serious consequences of inbreeding depression. Hybridization and reticulate speciation contribute significantly to the origin and maintenance of biological diversity. Static, antievolutionary ideas based on the purity of species should not be used to discriminate against endangered taxa (see Dowling et al. 1992 for a discussion of this problem).

Most species are not distributed continuously throughout their range. For management purposes, the ESA also recognizes geographical populations as threatened or endangered (Waples 1992). Often, discrete geographical populations or groups of populations are designated as subspecies. Should we strive to preserve species as a whole, or should we focus on the constituent subspecies and populations? Can we afford to maintain the five remaining subspecies of tigers *(Panthera tigris)* given the limited space in zoos and preserves (Seal et al. 1987, Wemmer et al. 1987)? These are not trivial questions if the goal is to preserve the evolutionary process, for our cloud of genes can be composed of many smaller clouds. Management decisions on such important issues cannot be made for emotional reasons or out of economic convenience. They require a solid base of information on the distribution of genetic diversity within and differences among the populations that constitute a species.

CONSERVING GENETIC DIVERSITY

Individual whole organisms can adapt behaviorally and physiologically to accommodate new environmental challenges. We adapt to a hotter climate by sweating more, by reducing heat-producing activities, and by seeking shade. However, these adjustments are temporary. Long-term adaptation to heat stress would occur through natural selection of genetic variants that are more tolerant of heat. Genes are the dynamic units of adaptation and evolution.

As discrete entities, genes are replicated relatively faithfully between generations. Mutations that alter the function of a typical gene locus (the position a gene occupies on a chromosome) seem to occur about once in a hundred thousand to once in a million replications (Mukai et al. 1972, Simmons and Crow 1977). Despite the faithful replication of individual genes, the multilocus genotypes that define individuals are dynamic assemblages of genes. The chromosomes are reshuffled like a deck of cards during gamete formation, and every egg and sperm gets dealt a new hand. For example, humans have twenty-three pairs of chromosomes. Mendelian segregation and assortment of these chromosomes during meiosis can produce 2^{23} or about 8.4 million different kinds of gametes. Yet, that is not all. Figure 2 illustrates how crossing over exchanges genes between sister chromosomes, increasing the diversity many fold. If only a single crossover occurs per chromosome pair, the number of unique combinations increases to 70 trillion (7×10^{13}). Consequently, the number of unique offspring that a single pair of humans can produce is the square of this number, 5×10^{27}. With about 5 billion people on this planet, it is obvious (and fortunate) that none of us will achieve our genetic potential. But it does explain why each of us is genetically unique, with the exception of monozygotic twins.

Coupled with new mutations, this incredible diversity is the raw material for natural selection. However, genetic diversity is not distributed homogeneously throughout a species. At the simplest level, total diversity in a species (H_T) can be partitioned into that which exists within populations (H_S) and the differences between populations (D_{ST}). Loss of diversity within populations (H_S) can have immediate deleterious consequences. Inbreeding depression is commonly manifested in zoo populations (Ralls and Ballou 1983). The rate of loss of H_S depends on the number of breeding adults in each generation (N_e). The smaller N_e, the more rapidly H_S declines. Demographic and social factors affecting N_e and the loss of genetic variation in zoo populations have been discussed extensively elsewhere (Frankel and Soulé 1981, Allendorf 1986, Fuerst and Maruyama 1986). Captive breeders and preserve managers should attempt to keep the genetically effective size of their populations at levels that would avoid the loss of genetic diversity over the short term. However, a risky alternative exists for captive

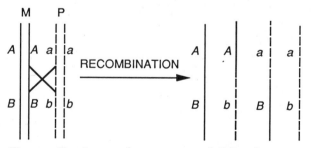

Figure 2. Crossing over between maternal (M) and paternal (P) chromosome sets. The two original chromosome sets (M and P) produce four genetic combinations as a consequence of one crossover event (AB, Ab, aB, and ab).

strains that are very small and are already manifesting symptoms of inbreeding depression (Templeton and Read 1983). One might purposefully inbreed further within lines and attempt to purge the deleterious genes through purifying selection. Nonetheless, such a program runs the risk of producing domesticated strains that are unlikely to perform well if released in the wild.

The differences among populations (D_{ST}) also are important to a species. Different populations of a species might carry variants of specific genes (alleles) that increase adaptation to local conditions. However, not all the differences between populations have adaptive significance. Some differences might have become fixed in different populations by chance alone. For example, we don't know if the striping patterns on the bodies and tails of tiger subspecies have any adaptive significance, but we should not ignore the possibility that these patterns might be correlated with physiological or behavioral differences that are of adaptive significance. If we do not have sufficient space in zoos to preserve genetic variation in all five subspecies of tigers, we might consider hybridizing the subspecies to create an archetypal tiger. This strategy would potentially maintain the total genetic diversity (H_T) by converting D_{ST} to H_S. Mixing might benefit the tigers if it reverses inbreeding depression within populations, as it apparently has for Florida panthers (O'Brien et al. 1990). Unfortunately, in the absence of strong diversifying selection, such mixing will obliterate intergenic correlations that existed within each of the subspecies. Thus, mixing may lead to outbreeding depression, a lowering of fitness in the hybrids and their descendants (Templeton 1986). D_{ST} will be lost, and intergenic correlations will never sort out again in precisely the same way, even if the mixed tigers could be released in their former ranges. The differences we see between populations are unique consequences of an interplay between natural selection and random processes like mutation and genetic drift. No easy solutions exist for maintaining the total genetic diversity in endangered species like tigers. Yet, attempts to resolve potential conflicts between the maintenance of H_S and D_{ST} should be basic goals of Species Survival Plans (Meffe and Vrijenhoek 1988).

THE NATURE OF INDIVIDUALITY

Animal rights activists argue that the individual rights of sentient animals should not be violated for the greater good of species or ecosystems; to do so is "environmental fascism" (Regan 1983). They recognize no special rights for species or ecosystems. The debate between conservationists and rights activists is not about the desirability of saving endangered species; it is about why the effort is important (moral reasons versus biological reasons) and for whom it is important (individuals versus species or ecosystems). As a biologist, I admit that I cannot find the dividing line between what is and what is not a sentient animal, but animal rightists seem to have no problem in this regard. For the sake of debate, I pose the argument that individuals deserve no special biological consideration. They are only conduits in the hierarchy of life, actors in an evolutionary process that takes information from genes and sculpts species and ecosystems.

The individuality of organisms is not always easy to define. Figure 3 illustrates a fish. It appears to be an individual. Boundaries between fish and water are clear and distinct. So what is the problem? This Mexican topminnow is a member of a clone. It is genetically identical to tens of thousands of fish in nature that are exactly the same (Moore 1977, Quattro et al. 1992b). My colleagues and I have captured the same clones in the same places, year after year, for the past twenty-five years, and I suspect that some clones have existed in these streams for thousands of years (Quattro et al. 1992a). So what constitutes an individual—the fish in Figure 3, or all the fish of the same genotype for the past several thousand years? This issue is even less clear for many colonial plants and animals (e.g., strawberries and some sea anemones) that reproduce by budding. Defining individuals in a genetic and evolutionary sense is not always easy (Buss 1983).

Figure 3. *Poeciliopsis 2 monacha-lucida,* an all-female species of fish that reproduces by cloning.

For sexually reproducing species, defining individuals is not a problem. The individual is a dynamic ecological unit—it is born, it eats, it grows, it reproduces, it ages, it dies. However, its genotype is unique. It will never appear again. From an evolutionary point of view, individuals are static entities. An individual's relevance is in its Darwinian fitness, its capacity to transmit its genes to the next generation. Unlike clones, however, sexual reproducers pass on only copies of their genes, not exact replicates of the whole multilocus genotype. Ultimately, each gene is valued for its contribution to fitness as averaged against the variable background of all other genes in the gene pool.

When considered across generations, individuals are little more than temporary vesicles that express and replicate genes. The processes of essentially faithful DNA replication and natural selection are all that remain permanent. Some have argued that a chicken is just an egg's way of making more eggs, and similarly we might argue that a human is just a gene's way of making more genes (Dawkins 1976). Although this inglorious view of individuals is not very comforting, we must avoid the imposition of anthropocentric views and moral dogma on critical management decisions necessary for the preservation and restoration of endangered species.

UNITS OF NATURAL SELECTION

Isn't it well established, however, that individuals are the fundamental units of natural selection? Although individuals are important as phenotypic manifestations of genes, they are only part of a hierarchy of biological levels at which selection can act. Lewontin provided a lucid discussion of the units of selection: genes, organelles, cells, gametes, individuals, kin, social groups, populations, and species (1970). I briefly paraphrase the relevant portions of his arguments. Natural selection requires only three conditions: (1) phenotypic variation (differences exist among individuals), (2) differential fitness (different phenotypes have different birth and death rates—i.e., turnover rates), and (3) heritability of fitness (a correlation in fitness exists between parent and offspring). As long as the three criteria hold, natural selection can act at that level. At the most elementary level, selection must have acted on the primitive soup in which the first self-replicating molecules arose (Dawkins 1976). Some molecules must have been more efficient replicators (higher birth rates), and other molecules may have been more stable (greater survival). Eventually, those molecules combining efficient replication with greater survival would accumulate at the expense of others. From this thermodynamic competition emerged the nucleic acid system that defines all living things on this planet, from viruses to humans. Interested readers should refer to Eigen et al. (1982) for an exciting review of selection experiments on nucleic acids in abiotic systems. Chemical evolution works in test tubes, and it has remarkable properties of self-organization.

Certain organelles within cells (e.g., mitochondria, chloroplasts, and kinetosomes) contain nucleic acid genomes that are subject to selection (Wallace 1982, Birky 1991). Selection also occurs among cells within a multicellular organism. Programmed cell death is an important part of the developmental process in multicellular organisms. The rapid spread of cancerous cells can be considered to be a manifestation of intercellular selection (Lewontin 1970). Selection also occurs among gametes produced by multicellular organisms. Gametic selection is best known in plants, where it occurs among pollen grains that bear different genotypes (Mulcahy 1975). Adult multicellular individuals are products of competition between numerous differentiating cell lines, of cell death, and of intracellular selection among organelles and genes.

Whether selection acts effectively on groups of unrelated individuals has been hotly debated among evolutionary biologists (Williams 1966, Dawkins 1976). In the broad sense, intergroup selection requires only differential turnover of groups and heritability of the group's collective phenotype (Lewontin 1965). Defining heritability of a group's phenotype has been difficult, but not impossible (Wilson 1980). In a narrower sense, group selection has been restricted to situations in which a trait is beneficial for the group but detrimental to individuals (e.g., alarm calls and altruistic behaviors). Evolution of altruistic traits would appear to be unlikely, because the turnover rate of individuals (birth and death) is much higher than of groups. Thus, even weak selection favoring selfish individuals should prevail over group selection. However, altruistic traits that benefit a kinship group can evolve, because beneficiaries and benefactor are also likely to carry many of the same genes, including the altruism genes (Haldane 1932). Other scenarios for narrow-sense group selection have been explored elsewhere (Williams 1966).

Selection also acts among partially isolated populations (demes) of the same species. If you interpret the population's phenotype as its growth rate and dispersal abilities, some demes will increase in size and spread while others shrink. Sewell Wright invoked interdemic selection as the basis for his "shifting balance" model of evolution (1977). In his view, the genetic variance among demes (D_{ST}) is a critical part of the total genetic variance necessary for adaptive evolution. The lucky demes that reach the highest peaks on the Darwinian fitness landscape will increase in size and spread their optimal genotypic combinations through the entire population via interdemic selection. In Wright's view, genetic variance within demes, differences among demes, dispersal, and chance are primary factors affecting the rate of adaptive evolution.

Species selection has received considerable attention from paleontologists (Stanley 1979) who see patterns of increase and decline in groups of related species (clades). If one clade has a higher rate of speciation or lower rate of extinction than another, then assuming some competition for resources, the former will increase and the latter will decline. Natural selection has also been invoked as an organizing process at the level of communities and ecosystems (Dunbar 1971). These higher levels of organization might meet criteria 1 and 2 as listed above, but identifying a form of heritability for community structure or ecosystems processes is more difficult to imagine.

INDIVIDUALS AND CONSERVATION

Although individuals are only grist for the evolutionary mill, each individual of an endangered species might be critical to a species preservation plan. The imminent probability of demographic extinction makes each individual count as a potential reproductive unit. Furthermore, each remaining individual of a sexual species is unique, representing a significant portion of the total genetic diversity. Clearly, great care must be exercised in captive breeding programs so that this remnant pool of variation is not lost (Foose et al. 1986, Fuerst and Maruyama 1986, Lacy 1987).

In many cases, we need to capture and quickly breed a sufficient number of individuals to avoid extinction. The California condor provides an excellent example. This bird was rapidly declining because of its inability to produce viable eggs in the wild. The last fourteen wild individuals were captured and brought in to the San Diego Zoo. The immediate goals were, first, to stabilize these individuals (none died at the hands of zookeepers) and, second, to initiate a breeding program that would build the population as quickly as possible. The captive adults and their offspring now number fifty-two. In January 1992, two offspring were released in southern California (Ryder and Thompson 1992).

Now that the population has been stabilized, any further breeding should be designed to avoid matings between relatives. Thus, geneticists examined DNA fingerprints of the fifty-two condors. DNA fingerprints of relatives are more similar than those of unrelated individuals, so statistical estimates of relatedness can be used to design a breeding program that avoids matings between relatives. Does that mean there might be some genetically surplus individuals who have already reproduced and should not be allowed to reproduce again? Surplus individuals are a likely outcome of any breeding program that seeks to avoid inbreeding. Clearly, such a program would appear to be at odds with animal rights activists who see captive breeding as a violation of individual rights. Yet, not only is avoidance of inbreeding good for the group, it is also for the evolutionary good of the individual. Inbred progeny typically suffer from slow growth, decreased fertility, poor survival, and increased developmental problems (Lerner 1954, Falconer 1981). Since an individual's Darwinian fitness is a measure of its success in transmitting its genes to subsequent generations, the outbreeding individual might produce more viable and fertile progeny and hence transmit more of its genes. How then do we balance the presumed rights of an individual to pick its mate and breed versus the health of its progeny? Once we have taken the animals into captivity, we must also accept responsibility from birth to death for the health of our charges and for the genetic and evolutionary health of their future offspring. This responsibility requires us to design breeding programs that to the best of our knowledge will do no further harm.

SUMMARY AND CONCLUSIONS

I see no moral or ethical problems with the captive breeding programs of zoos and aquariums, as long as we base them on sound genetic designs and evolutionary principles. Similarly, I have no problem with government policies that help to limit the reproduction of humans. What I find morally and ethically reprehensible is the inability of most political and religious leaders to recognize the real problem facing this globe—too many people!

The goal of conservation biology should be to preserve and protect natural processes. We need to stop thinking parochially about static units as if they were the building blocks of nature. Instead, we need to consider processes that shape the diversity of life on this planet, including our own lives. Physical, chemical, and geological processes create the opportunities for life. Ecosystem processes and species interactions shape habitats and communities. Most important, evolution connects these processes in the complex web of life.

ACKNOWLEDGMENTS

I thank my students and research associates (Clark Craddock, Michael Black, and Stephen Karl) who offered helpful criticisms of the manuscript. I am indebted to my colleagues at Rutgers (Jim Miller, Don Caccamise, and John Kuser) and elsewhere (Robin Waples and Phil Pister) who helped direct me to relevant literature. This is contribution 2-67175-5-94 of the New Jersey Agricultural Experiment Station, and contribution 94-18 of the Institute of Marine and Coastal Sciences.

REFERENCES

Allendorf, F. W. 1986. Genetic drift and the loss of alleles versus heterozygosity. *Zoo Biology* 5:181–190.

Arthington, A. H., and L. N. Lloyd. 1989. Introduced poeciliids in Australia and New Zealand. In *Ecology and Evolution of Livebearing Fishes (Poeciliidae)*, ed. G. K. Meffe and F. F. Snelson, Jr., 333–338. Englewood Cliffs, N.J.: Prentice-Hall.

Birky, C. W., Jr. 1991. Evolution and population genetics of organelle genes: Mechanisms and models. In *Evolution at the Molecular Level*, ed. R. K. Selander, A. G. Clark, and T. S. Whittam. Sunderland, Mass.: Sinauer Associates.

Buss, L. W. 1983. Evolution, development, and the units of selection. *Proceedings of the National Academy of Science* 80:1387–1391.

Carson, H. 1975. The genetics of speciation at the diploid level. *American Naturalist* 109:73–92.

Courtenay, W. R., Jr., and G. K. Meffe. 1989. Small fishes in strange places: A review of introduced poeciliids. In *Ecology and Evolution of Livebearing Fishes (Poeciliidae)*, ed. G. K. Meffe and F. F. Snelson, Jr., 319–331. Englewood Cliffs, N.J.: Prentice-Hall.

Dawkins, R. 1976. *The Selfish Gene*. Oxford: Oxford University Press.

Dowling, T. E., B. D. DeMarais, W. L. Minckley, M. E. Douglas, and P. C. Marsh. 1992. Use of genetic characters in conservation biology. *Conservation Biology* 6:7–8.

Dunbar, M. J. 1971. The evolution of stability in marine environments: Natural selection at the level of the ecosystem. In *Group Selection*, ed. G. S. Williams. Chicago: Aldine-Atherton.

Ehrenfeld, D. 1978. *The Arrogance of Humanism*. New York: Oxford University Press.

Eigen, M., W. Gardiner, P. Schuster, and R. Winkler-Oswatitsch. 1982. The origin of genetic information. In *Evolution Now: A Century after Darwin*, ed. J. M. Smith. San Francisco, Calif.: W. H. Freeman & Co.

Falconer, D. S. 1981. *Introduction to Quantitative Genetics*. London: Longman.

Foose, T. J., R. Lande, N. R. Flesness, G. Rabb, and B. Read. 1986. Propagation plans. *Zoo Biology* 5:139–146.

Frankel, O. H., and M. E. Soulé. 1981. *Conservation and Evolution*. Cambridge: Cambridge University Press.

Fuerst, P. A., and T. Maruyama. 1986. Considerations on the conservation of alleles and of genic heterozygosity in small managed populations. *Zoo Biology* 5:171–179.

Gould, S. J. 1984. The cosmic dance of Siva. *Natural History* 93:14–19.

Grant, V. 1981. *Plant Speciation*, 2d ed. New York: Columbia University Press.

Grassle, J. F. 1985. Hydrothermal vent animals: Distribution and biology. *Science* 229:713–717.

Haldane, J. B. S. 1932. *The Causes of Evolution.* New York: Longman.

Hutchins, M., and C. Wemmer. 1987. Wildlife conservation and animal rights: Are they compatible? In *Advances in Animal Welfare Science 1986–1987,* ed. M. W. Fox and L. D. Mickley, 111–137. Washington, D.C.: Humane Society of the United States.

Lacy, R. 1987. Loss of genetic diversity from managed populations: Interacting effects of drift, mutation, immigration, selection, and population subdivision. *Conservation Biology* 1:143–158.

Lande, R. 1988. Genetics and demography in biological conservation. *Science* 241:1455–1460.

Lerner, I. M. 1954. *Genetic Homeostasis.* London: Oliver & Boyd.

Lewontin, R. C. 1965. Selection in and of populations. In *Ideas in Modern Biology,* ed. J. A. Moore. New York: Natural History Press.

————. 1970. The units of natural selection. *Annual Review of Ecological Systems* 1:118.

Malahoff, A. 1985. Hydrothermal vents and polymetallic sulfides of the Galapagos and Gorda/Juan de Fuca Ridge systems and of submarine volcanoes. *Biological Society of Washington Bulletin* 6:19–41.

Mayr, E. 1963. *Animal Species and Evolution.* Cambridge, Mass.: Belknap Press.

Meffe, G. K., and R. C. Vrijenhoek. 1988. Conservation genetics in the management of desert fishes. *Conservation Biology* 2:157–169.

Minckley, W. L., and J. E. Deacon. 1968. Southwestern fishes and the enigma of "endangered species." *Science* 159:1424–1432.

Moore, W. S. 1977. A histocompatibility analysis of inheritance in the unisexual fish *Poeciliopsis 2 monacha-lucida. Copeia* 1977:213–223.

Mukai, T., S. T. Chigusa, L. E. Mettler, and J. F. Crow. 1972. Mutation rate and dominance of genes affecting viability in *Drosophila melanogaster. Genetics* 72:335–355.

Mulcahy, D. L., ed. 1975. *Symposium on Gametic Competition in Plants and Animals.* New York: Elsevier.

Nowak, R. M. 1978. Evolution and taxonomy of coyotes and related canids. In *Coyotes, Biology, Behavior, and Management,* ed. M. Berkoff, 3–16. New York: Academic Press.

O'Brien, S. J., and E. Mayr. 1991. Bureaucratic mischief: Recognizing endangered species and subspecies. *Science* 251:1187–1188.

O'Brien, S. J., M. E. Roelke, N. Yuhki, K. W. Richards, W. E. Johnson, W. L. Franklin, A. E. Anderson, O. L. Bass, Jr., R. C. Belden, and J. S. Martenson. 1990. Genetic introgression within the Florida panther *Felis concolor coryi. National Geographic Research* 6:485–494.

Pister, E. P. 1981. The conservation of desert fishes. In *Fishes of the North American Deserts,* ed. R. J. Naiman and D. L. Soltz, 411–445. New York: John Wiley & Sons.

Quattro, J. M., J. C. Avise, and R. C. Vrijenhoek. 1992a. An ancient clonal lineage in the fish genus *Poeciliopsis* (Atheriniformes: Poeciliidae). *Proceedings of the National Academy of Science* 89:348–352.

————. 1992b. Mode of origin and sources of genotypic diversity in triploid fish clones (*Poeciliopsis:* Poeciliidae). *Genetics* 130:621–628.

Ralls, K., and J. Ballou. 1983. Extinction: Lessons from zoos. In *Genetics and Conservation: A Reference for Managing Wild Animal and Plant Populations,* ed. C. M. Schonewald-Cox, S. M. Chambers, B. McBryde, and W. L. Thomas, 164–184. Menlo Park, Calif.: Benjamin/Cummings.

Regan, T. 1983. *The Case for Animal Rights*. Berkeley and Los Angeles: University of California Press.

Ryder, O. A., and E. Thompson. 1992. Molecular and mathematical approaches in genealogical analysis and strain identification. Presented to a symposium at Rutgers University, 10–11 January 1992.

Seal, U. S., P. Jackson, and R. L. Tatum. 1987. A global tiger conservation plan. In *Tigers of the World: The Biology, Biopolitics, Management, and Conservation of an Endangered Species,* ed. R. L. Tilson and U. S. Seal, 487–498. Park Ridge, N.J.: Noyes Publications.

Simmons, M. J., and J. F. Crow. 1977. Mutations affecting fitness in *Drosophila* populations. *Annual Review of Genetics* 11:49–78.

Stanley, S. M. 1979. *Macroevolution, Pattern and Process*. San Francisco, Calif.: Freeman.

Temple, S. A. 1992. Exotic birds: A growing problem with no easy solution. *Auk* 109:395–397.

Templeton, A. 1986. Coadaptation and outbreeding depression. In *Conservation Biology: The Science of Scarcity and Diversity,* ed. M. E. Soulé, 105–116. Sunderland, Mass.: Sinauer Associates.

———. 1989. The meaning of species and speciation: A genetic perspective. In *Speciation and Adaptation,* ed. D. Otte and J. Endler, 3–27. Sunderland, Mass.: Sinauer Associates.

Templeton, A. R., and B. Read. 1983. The elimination of inbreeding depression in a captive herd of Speke's gazelle. In *Genetics and Conservation: A Reference for Managing Wild Animal and Plant Populations,* ed. C. M. Schonewald-Cox, S. M. Chambers, B. McBryde, and W. L. Thomas, 241–261. Menlo Park, Calif.: Benjamin/Cummings.

Wallace, D. G. 1982. Structure and evolution of organelle DNAs. In *Endocytobiology,* ed. H. E. A. Schenk and W. Schweeler, 87–100. New York: De Gruyter.

Waples, R. S. 1992. Pacific salmon, *Oncorhynchus* spp., and the definition of "species" under the Endangered Species Act. *Marine Fisheries Review* 53:11–22.

Wayne, R. K., and S. M. Jenks. 1991. Mitochondrial DNA analysis implying extensive hybridization of the endangered red wolf *Canis rufus. Nature* 351:565–567.

Wemmer, C., J. L. D. Smith, and H. R. Mishra. 1987. Tigers in the wild: The biopolitical challenges. In *Tigers of the World: The Biology, Biopolitics, Management, and Conservation of an Endangered Species,* ed. R. L. Tilson and U. S. Seal, 395–405. Park Ridge, N.J.: Noyes Publications.

Williams, G. S. 1966. *Adaptation and Natural Selection*. Princeton, N.J.: Princeton University Press.

Williams, J. E., J. A. Lichatowich, and W. Nehlsen. 1992. Declining salmon and steelhead populations: New endangered species concerns for the West. *Endangered Species Update* 9:1–8.

Wilson, D. S. 1980. *The Natural Selection of Populations and Communities*. Menlo Park, Calif.: Benjamin/Cummings.

Wright, S. 1977. *Evolution and the Genetics of Populations. Vol. 3: Experimental Results and Evolutionary Deductions*. Chicago: University of Chicago Press.

Rescuing Species and Ecosystems

Valerius Geist

G lobal events may bypass contentious debates, and the question of whether zoos should be legitimate oases of nature conservation or whether they should be dismissed in favor of saving native ecosystems in situ (Varner and Monroe 1991) is a case in point. Two future factors make debate about the appropriateness of these options irrelevant.

One factor is the possible global depletion of the ozone layer in the next eight to fifteen years, with a possible, but not necessarily probable, recovery in about one hundred years. High-energy ultraviolet radiation (UV-B) is expected to do significant damage to living biota while the ozone layer is thin. Although this is controversial and treated with due scholarly caution by Tevini's collection of scientific essays (1993), any plan for conservation needs to consider seriously the possibility of serious ecological damage in high latitudes and altitudes due to UV-B burnout.

The other factor is the push of human population growth to infinity around 2026, as illustrated by the doomsday equation of Foerster et al. (1960). This is a conservative equation, as human population growth today is about four years, or some 500 million people, ahead of schedule (Umpleby 1987). It suggests that with the human world population growing exponentially and heading toward infinity between 2020 and 2026, we shall face considerable difficulties trying to conserve nature about twenty-five to thirty-five years hence. Throughout history, ancient and recent, social upheavals resulting in armed conflict have destroyed biota, be it by the starving populations who kill whatever is available, or by armed men who take what they need from the land and destroy the rest, or by a populace who are prohibited from hunting but engage in it with a vengeance when the opportunity arises (Stahl 1979). The land may suffer acts of war, such as the massive burning or chemical defoliation of Vietnam's forests, or the destruction of the enemy's food, a strategy the U.S. Army pursued in the depletion of the American bison to defeat western Indian tribes (Ambrose 1975), or the killing of wildlife for sport or for profit by vandals in uniform, as exemplified in the recent history of the ivory trade (Parker and Amin 1983). It suggests that current conservation efforts notwithstanding, we need to look for realistic fail-safe solutions to nature conservation, and abandoning artificial propagation in the hope that in situ preservation will work is a case in point.

Current approaches to human population problems and conservation, emphasizing national planning efforts with international cooperation and using science as a tool, are illustrated in *Limits to Growth* (Meadows et al. 1972), in the subsequent *World Conservation Strategy* (IUCN 1980), in the Brundtland Report (1987), in the second version of *World Conservation Strategy* or *Caring for the Earth* (IUCN 1991), in the discourse on Gaia by Myers (1984), and in Lester Brown's annual *State of the World* reports (e.g., Brown et al. 1988). Textbooks in ecology (e.g., Stiling 1992) and in environmental sciences (e.g., Cunningham and Saigo 1990, Nebel and Wright 1993) address current global problems in conservation, while conservationists and economists wrestle with sustainability and the difficulty of implementing it (Elder 1991). Whether a rational international effort in conservation can be managed in view of emotional tribalism and irrational self-destruction by warring factions, as currently seen in the many small-scale wars fought globally, is to be doubted.

In addition to those international conservation efforts, one should also consider a more radical approach, namely, a Noah's ark approach to conservation. The problems of a thinning ozone shield have made the matter more urgent. It is necessary to put in place a regional Noah's ark approach, in particular one that aims to salvage a fraction of the earth's high-latitude and high-altitude biota.

The outlines of the ozone depletion problem are emerging (see Tevini 1993). While there is an expectation of worldwide thinning, some areas will be affected more than others. Most affected will be areas directly under the polar ozone holes, and again in the high mountains. Reports have already been published of domestic sheep, wild hares, and spawning salmon going blind in the extreme south of South America, of people developing cataracts, of plants showing growth deformities, etc. (Ozonfrass 1992). A synopsis of current work suggests severe damage by high-energy ultraviolet radiation in high-mountain ecosystems beginning ten to twenty years from now. Even if the ozone-destroying chlorofluorocarbons (CFCs) are themselves destroyed in the stratosphere a decade after being last released at ground levels, which an optimist would put three to five years into the future, it is likely that about one hundred years from today the ozone shield will be reconstituted to an effective thickness.

One can safely proceed on the assumption that ozone thinning will continue for another fifteen or more years. A technological miracle may yet, somehow, restore the ozone shield despite CFC release. If such an unforeseen solution comes along, little has been lost by experimenting with a Noah's ark solution. We shall have gained experience and a small breathing space to prepare for the ultimate of Noah's arks, one designed to carry endangered species and ecosystems over the human population wave. If a technical solution to the ozone depletion problem does not materialize, however, then we shall have acted responsibly and ethically toward our planet's biota, as ozone depletion is, after all, a problem of our own making.

Since the biota from endangered areas will need physical protection from high-energy UV radiation, there is little option but to go for a solution that includes physical shielding of a seed stock of species and ecosystems from UV-B radiation within limited, confined spaces. Like it or not, this means using buildings of some kind; that is, large roofed structures to house terraria, ponds, zoos, aviaries, aquariums, herbaria, etc. Therefore, solutions for conservation such as those that have been worked out in zoos, aquariums, etc. may be our only option. The option to conserve in situ will not be available in areas burned out by UV-B (and other regions such as those affected by severe pollution and irreversible habitat loss). Also, with global warming expected to disrupt climatic systems, in situ conservation in national parks, ecological reserves, wildlife refuges, etc. will eventually become impossible because of the extinction or geographic shift of habitats. Artificial means of propagating biodiversity will be required until the biota can be safely reintroduced to uninhabited lands. To save biodiversity, we shall require in the immediate future more, not fewer, high-tech solutions and thus the imposition of artificial environments on beings, sentient or otherwise.

While cognizant of global efforts to contain environmental damage, create sustainable and equitable economies, curb human population growth, and so forth, and while wishing these efforts success, one needs to plan, nevertheless, for the possibility that these efforts will fail. A Noah's ark approach to conservation is then merely a safeguard against failure by other endeavors, in particular since a human population explosion is probably nearing. It is not unlikely that this human population surge will come and go between fifty and eighty years from now. Problems generated by global human population growth are likely to increase social and economic dislocations severely ten to thirty years hence. They include factors such as the spread of armed local warfare with destruction of the economic, institutional, and political infrastructures, including programs dedicated to national conservation. One also expects famine, disease epidemics, and efforts to deal with its many victims while maintaining a viable economy; global warming causing unpredictable climatic shifts followed by crop failures and more famine; increased global pollution caused by the industrial catch-up of Third World and former communist nations; ozone-related problems for agriculture and public health; oceanic pollution due to corroding containers in oceanic dumps and leaching from terrestrial landfill sites; and increasing depletion of common oceanic resources. Moreover, massive social and economic dislocations will diminish attention to environmental issues as problems of short-term survival dominate agendas. The challenge will be, therefore, to design a Noah's ark that floats on safely, even when its human crew has abandoned it.

While we may not be able to do much about population growth, ethnic warfare, disease epidemics, megapollution, and mass emigrations—that is, we might not be able to do much about the expected Malthusian decline—we may be able to influence the subsequent beginning. After each end there is, surely, a beginning. The future of earth's biota, and with it humanity's future, will depend then on how much of the biota has been saved past the human population wave, just as there might be a return of a significant fraction of the biota to UV-B-singed landscapes after the ozone shield is restored.

Captive propagation, with all its blemishes, is central to a Noah's ark solution. We expect those with concerns for the captive biota's well-being to keep the zoo community under scrutiny, insuring humane treatment of all biota, not only of sentient creatures.

THE KUMPAN CONCEPT

Missing in the debate about how to treat wild creatures in captivity is the notion of the Kumpan in the life of captive wildlife, to paraphrase a Lorenzian term (Lorenz 1935). Ethologists have experienced how humans can make a real difference in the life of captives, and even of free-living, wild animals that have been tamed in situ. My mountain sheep, closely acquainted with me in day-to-day contacts, gave every sign of exuberance upon seeing me return to them. At times they tried to follow me as I left the mountains, or to hold me back physically, preventing me from leaving them, and they searched for my whereabouts with determination and cleverness when I hid from them (see Geist 1971, 1975, 1993).

We can, as keepers of the captive biota, give the pleasure of intelligent, appropriate companionship to sentient beings, captive or otherwise, raising life to higher levels of appreciation for both parties. We can develop such bonds knowingly, with purpose, and use such in management or reintroductions. Ethological studies teach us that life in captivity must not be a prison sentence for wildlife.

As problematic as zoos and comparable institutions of captive propagation may appear to some philosophically, captive propagation is the only hope for a significant part of the flora and fauna to be safeguarded till a new recolonization of earth's damaged surface by the biota may begin. The alternative to captive propagation is to willfully condemn species and ecosystems to certain and, one may point out, very cruel death. That is ethically not an acceptable option, not from the perspective of John Passmore's anthropocentrism (1974), or of Tom Regan's sentientism (1983), or of Aldo Leopold's ecological holism (1949), or of Albert Schweitzer's reverence for life (1923). Moreover, some species already exist only in captive condition, and a return to the native state is not in sight because, currently, their geographic home no longer contains secure, adequate habitat. As long as they live, there is still hope that someday they will again live free under unrestrained conditions. As long as they reproduce and successfully raise young we may be reasonably certain that their life is not empty of deep instinctive satisfactions.

A PLEA FOR PLANTS

One may be permitted a brief excursion into another philosophical matter, one involving captive plants. Neither Schweitzer's concept of reverence for life nor the discovery in the life sciences of the fundamental unity of all life allows the division of life into beings of lower and higher value, as is proclaimed by the division into sentient and nonsentient beings. Such a distinction has no basis in science. It is an arbitrary characterization, which excludes a plethora of sensory and communication modes as practiced in species of plants, animals, and protists in favor of those close to humans phylogenetically. Differences in biology are differences due to adaptation, which cannot be judged as better or worse. Evolution imposes no natural values; we do, for reasons of our own.

Take the matter of suffering: A field biologist, experienced in evaluating the suitability of habitats for plants by their growth forms, cannot label near-death manifestations in plants as anything else but suffering. Plants do respond to damage inflicted and to unfavorable conditions of growth by morphological deformations and by alterations in patterns of reproduction. Plants, like animals, have antipredator responses, that is, means of minimizing damage to their bodies. However, these are means appropriate to sessile life forms, such as the local production of toxins, resins, thorns, and protective bark. An understanding of how these function is accessible to us only through detailed studies and through our intellect, not through sights and sounds that evoke emotive responses, as do the cries and writhing of a wounded large mammal. Plants are different forms of life, but as deserving of our attention and sympathy as any form of life closer to us in phylogeny.

HOW TO APPROACH A SOLUTION

Essential to any solution safeguarding biota from UV-B radiation due to ozone depletion will be the technical expertise of zoos, aquariums, and herbaria, which have successfully engaged in propagation for conservation. The infrastructure for international action may be already in place, thanks to the efforts of the World Conservation Union (IUCN). This group appears to be in a fortunate position to initiate action on the ozone problem. Its expertise might be supplemented by expertise in the design and management of large-scale, integrated structures as used in greenhouse agriculture. Experts who understand the often tricky requirements in the ecology and demography of species slated for reintroduction to the free state are needed (Lyles and May 1987, Lande 1988), as well as specialists in ecological landscape rehabilitation. IUCN needs to encourage a fairly rapid development of conservation sites in nations most likely to be affected by UV-B burnout.

IUCN could initiate the process with a quick call for an international conference on the ozone problem in relation to conservation. An assessment in hard, technical terms is essential to any intelligent mitigation. That should be followed by a rapid inventory of essentially safe genetic resources now in captivity, followed by a plan to augment such resources, generating some redundancy against the possibility of extinction of breeding stock. That assessment could lead to a plan to order priorities in immediate conservation, so that the most conservation could be done with the least amount of resources. Criteria could be (1) identifying taxa for triage to save as quickly as possible that which, realistically, can be saved; (2) identifying linked ecosystem components that need to be saved so that some semblance of the former ecosystems can reasonably be reconstituted; and (3) involving the IUCN Species Survival Commission taxonomic specialist groups as quickly as possible as active partners and asking them to identify priorities.

After those preliminary steps are taken, technical think tanks must be established to develop potential solutions and commence pilot projects. Such projects need urgently to be initiated. Pilot projects are vital mechanisms for rapid learning and for training the necessary work force. Also, ways of protecting existing captive breeding stock from increased UV-B need to be found. All captive breeding sites will be affected to some extent by UV-B radiation. The hottest time of the day may need to be taken into account in setting daily routines (e.g., a prohibition on being outside between 9 A.M. and 3 P.M.). Or the oryx might need eye shades! Overall, coordination of these steps requires an inventory of the expertise necessary to get the job done, as well as a clearinghouse for experts.

A first intellectual effort on how to conserve biota threatened by UV-B burnout is essential. However, we also need to focus political power and legislation quickly on solving the problem. As Hutchins and Wemmer point out (1991), not only habitats are endangered, but species may also be extirpated despite available habitat. The security of plant and animal species is much undermined by luxury markets trading in biota. To put faith in CITES, however, is to ignore the problems that are caused when the most powerful occidental institution, the marketplace, offers rewards for dead wildlife or wild plants. We are aware that the marketplace, turned around to reward entrepreneurship based on living wildlife, is a boon to conservation (Geist 1988). In short, we need not be fearful of economic consequences in terminating the exploitation of wildlife and wild plants as a commodity.

Cash flow must be created to finance a Noah's ark venture. This will require close cooperation with the business community. Just as the business community decided after World War II to preclude another war in Europe by becoming instrumental in establishing a European common market and, now, a common country within which former enemies carry the same passport, so we would benefit from the organizational talent, fiscal savvy, and political clout embedded in the business community. Rapidly saving whole ecosystems from UV-B burnout is a task so large and difficult that it will be accomplished only if the world's business community is with us.

As for publicity and the generation of public support, it is best to resist the temptation to run to the news media. The consequences of ozone depletion, when they are recognized, are highly disturbing. Prudence demands getting prepared to act on reasonable expectations. There is, however, still a small chance that some mitigating factor might crop up. If so, crying wolf too early discredits scientists as alarmists and may preclude effective action on a Noah's ark solution to the impending peaking of human populations about thirty years from now.

SUMMARY AND CONCLUSIONS

Global events, such as the thinning of the ozone shield with potential burnout of the biota in high mountains and below ozone holes by UV-B radiation, or the looming human population explosion, make artificial confinement and propagation of biota for future reintroductions unavoidable. There is no choice between artificial and in situ propagation, not when landscapes are severely degraded and natural habitats disappear. A Noah's ark approach to conservation would be good practice for the much more demanding tasks twenty to thirty-five years from now, protecting the seed stocks of earth's biota in stormy social times during the human population eruption. We should plan now for new beginnings of regional biotas where there are reasonable expectations of devastation. There are no reasonable alternatives to the artificial propagation of biotas from regions endangered by UV-B radiation, desertification, pollution, global warming, or devastation by local warfare and the concomitant collapse of economic and administrative infrastructures. As much of the biota as reasonable needs to be saved, so that each new, regional beginning may be the richer for it. Captive life for sentient beings need not be drudgery, not with humans as intelligent participants or companions (Kumpan) in their lives. Humans can strike up very positive relations with free-living creatures, relationships that appear to be thoroughly enjoyed by all parties concerned. There is a need for broad involvement by influential, farsighted decision makers. The basic international machinery to conduct the necessary pilot projects is already in place in the worldwide networks of cooperating zoological parks, as well as in the organizations of the IUCN. To begin with, a conference should be held on the effects of ozone thinning on ecology and conservation.

REFERENCES

Ambrose, S. E. 1975. *Crazy Horse and Custer*. New York: Meridian, New American Library.

Brown, L. R., et al. 1988. *State of the World: A Worldwatch Institute Report on Progress toward a Sustainable Society*. New York: W. W. Norton.

Brundtland, G. H. (chair). 1987. *Our Common Future: The World Commission on Environment and Development*. Oxford: Oxford University Press.

Cunningham, W. P., and B. W. Saigo. 1990. *Environmental Science: A Global Concern*. Dubuque, Iowa: William C. Brown.

Elder, P. S. 1991. Sustainability. *McGill Law Journal* 36(3): 822–852.

Foerster, H. von, P. M. Mora, and L. W. Amiot. 1960. Doomsday: Friday, 13 November, A.D. 2026. *Science* 132:1291–1295.

Geist, V. 1971. *Mountain Sheep*. Chicago: University of Chicago Press.

———. 1975. *Mountain Sheep and Man in the Northern Wilds*. Ithaca, N.Y.: Cornell University Press.

———. 1988. How markets in wildlife meat and parts, and the sale of hunting privileges, jeopardize wildlife conservation. *Conservation Biology* 2(1): 15–26.

———. 1993. *Mountain Sheep Country*. Minocqua, Wisc.: North Word Press.

Hutchins, M., and C. Wemmer. 1991. Response: In defense of captive breeding. *Endangered Species Update* 8:5–6.

IUCN. 1980. *World Conservation Strategy: Living Resource Conservation for Sustainable Development*. Gland, Switzerland: World Conservation Union.

———. 1991. *Caring for the Earth*. Gland, Switzerland: World Conservation Union.

Lande, R. 1988. Genetics and demography in biological conservation. *Science* 241:1455–1460.

Leopold, A. 1948. *A Sand County Almanac*. London: Oxford University Press.

Lorenz, K. 1935. Der Kumpan in der Umwelt des Vogels. *Ebenda* 83:137–213, 289–413.

Lyles, A. M., and R. M. May. 1987. Problems in leaving the ark. *Nature* 326:245–246.

Meadows, D. H., D. L. Meadows, J. Randers, and W. W. Behrens III. 1972. *Limits to Growth: A Report for the Club of Rome's Project on the Predicament of Mankind*. New York: Universe Books.

Myers, N., ed. 1984. *Gaia: An Atlas of Planetary Management*. New York: Anchor Books, Doubleday.

Nebel, B. J., and R. T. Wright. 1993. *Environmental Science*. 4th ed. Engelwood Cliffs, N.J.: Prentice-Hall.

Ozonfrass—letzter Akt? 1992. *Spiegel*, 10 February, 202–212.

Parker, I., and M. Amin. 1983. *Ivory Crisis*. London: Chatto & Windus.

Passmore, J. 1974. *Man's Responsibility for Nature*. New York: Charles Scribner's Sons.

Regan, T. 1983. *The Case for Animal Rights*. Berkeley: University of California Press.

Schweitzer, A. 1923. *Civilization and Ethics*. London: A. & C. Black.

Stahl, D. 1979. *Wild, Lebendige Umwelt*. Freiburg and Munich: K. Alber.

Stiling, P. 1992. *Introductory Ecology*. Englewood Cliffs, N.J.: Prentice-Hall.

Tevini, M., ed. 1993. *UV-B Radiation and Ozone Depletion*. Boca Raton, Fla.: Lewis Publishers.

Umpleby, S. A. 1987. World population: Still ahead of schedule. *Science* 235:1555–1556.

Varner, G. E., and M. C. Monroe. 1991. Ethical perspectives on captive breeding: Is it for the birds? *Endangered Species Update* 8(1): 27–29.

CARING FOR NATURE

A Broader Look at Animal Stewardship

Bryan Norton

The naturalist writer Annie Dillard poignantly poses a central moral issue of our time in a chapter on the fecundity of nature (Dillard 1974). She notes that the rock barnacles on a short stretch of rocky seashore pour forth a million million larvae, of which only a few will survive to maturity. Nature, she notes, apparently cares not a whit about individuals. But all modern Western systems of ethics value individuals above all. "Either this world, my mother, is a monster, or I myself am a freak," Dillard laments. Dillard's point might be stated more calmly as a concern about how broadly to apply traditional concepts of justice when our actions have sometimes terrible impacts on members of other species.

As the crisis of biodiversity unfolds over coming decades, conservationists will advocate more and more invasive methods to save in captivity species that are threatened in the wild. As habitat alteration and fragmentation continue, more and more interventions in the lives of wild animals will be required if their wild populations are to be protected and kept in ecological balance. Actions taken to protect species and ecological communities may not be in the interest of individual members (Hutchins and Wemmer 1987). The recommendations of captive breeders and conservation biologists therefore conflict in many cases with the recommendations of animal welfare advocates (Ehrenfeld 1991).

The excruciating policy questions addressed in this book stem from Dillard's dilemma: committed conservationists tend to concentrate on the protection of populations, species, ecosystems, and processes. When they do so, individuals become the means to carry genetic information from generation to generation, and the commitments of conservation seem to dictate that we follow nature in deemphasizing individuals. But biology has also emphasized the common heritage of human and nonhuman species. As Thoreau noted, "The hare in its extremity cries like a child" (Thoreau 1946), and modern science has in general agreed with modern sensibilities that countless human actions severely affect the interests of individual animals. Hence this book. Our goal is to achieve as much consensus as possible regarding ways to balance the actions required by a concern to save wild species against the equally unquestioned good of reducing animal suffering, discomfort, and invasions of autonomy.

There exists no consensus, among experts or the public, regarding one of the fundamental questions in biological conservation: what should be the central focus (target) of protection efforts? I am assuming that there does exist a broad consensus supporting the goal of protecting biodiversity, but that the meaning of that consensus, especially its meaning for preservation priorities, has not been worked out. Should our goal be to protect genes? Individuals? Populations? Species? Communities? Ecosystems? All of the above? None of the above? To answer these questions intelligently we need some sense of what it is we value in nature. That we cannot answer them in a way that commands wide policy consensus has led to confusion in management practice—confusions that point out just how perplexing are our obligations to protect the natural world. We here seek answers that will help us to avoid these confusions in the future and, we hope, to reduce tension between the well-intentioned persons in the environmental, humane, and zoo communities.

I would like to offer a broader perspective, one that considers not only our obligations as they arise in particular contexts, such as within zoos, in captive breeding programs, on farms, or in environmental restorations—but also one that looks at all of these perspectives simultaneously, seeking general patterns that structure our thinking in all of these areas. My method will be pluralistic in the sense that I do not assume that all values can be measured in a common moral metric such as rights, individual welfare, or ratios of pleasure and pain. In this respect my method differs from that of most attempts to describe an ethic for the treatment of nonhuman animals, in that they usually emphasize the similarities between humans and members of nonhuman species. In these approaches, some characteristic, such as rationality, sentience, or teleological action, is designated as the basis of "intrinsic/inherent value" and "moral considerability" (Goodpaster 1978, Regan 1983). All creatures with this designated characteristic have rights or intrinsic value and their interests must be included in any moral calculations.

In some cases, this approach is coupled with the somewhat dogmatic assertion that whatever has inherent value has it equally, which leaves open the question of how far from humans one should locate the sharp cutoff (Regan 1983; see

Callicott 1989 for a criticism of Regan's view). Other theorists conclude that moral-making characteristics emerge gradually from evolutionary complexity; ethical obligations to members of other species therefore ripple outward in fainter and fainter copies of human ethics, as the species in question get less humanlike. Whatever the differences in outcomes, these approaches share a conceptional strategy: to reduce all of our obligations regarding nature to as few types as possible in search of a common metric for valuing nature and its elements.

Underlying these reductionistic approaches is a drive toward moral monism, the view that all moral value can and should be traced to a single, foundational principle. It is apparently thought that expressing all of our values in a common metric will make policy decisions less confusing and difficult. While I have learned much from those who demand a monistic criterion, I have become extremely skeptical that these ethical approaches will guide us toward an intelligent protection policy.

I think we need an alternative approach to thinking about our obligations to nature, an approach that emphasizes differences rather than similarities—especially differences in context. Moral pluralism is the view that we value many things in different ways, and that these differing values are sometimes in conflict. Further, these values may be incommensurate, so that they cannot be weighed in a common metric, because they exist at different scales of analysis—individuals, species, ecosystems, etc. The best we can do in these situations is to honestly seek a fair balance point at which we have done our best to minimize harms for which we are responsible (Donnelley 1992). Rather than trying to make these values commensurate, the contextualist emphasizes different obligations according to situation, or context (Norton 1991). The focus then shifts to developing second-order rules determining which moral imperatives should be emphasized in given situations. A direct assault on the problem of protection priorities may now be possible, even though we are not armed with a single moral yardstick for setting our social priorities.

I believe that we have an obligation to minimize the suffering of individual animals in some situations and that we have obligations to emphasize species protection in other situations. The problem is to explain coherently and effectively how to tell the difference between these situations. Let me start by exploring what we value in wild animals and by explaining what I think we should be protecting in wild, natural systems. Then, I will contrast our obligations to wild animals with our obligations to domesticated animals, showing that very different rules apply in these different contexts of interspecific interactions. The key to understanding moral treatment of zoo animals and moral obligations regarding captive breeding programs is to understand how to weave an ethic that respects the integrity of the individual and also recognizes the role of an individual in the broader processes that involve its species. I will then provoke a consideration of the possibility of animal altruism—a concept that challenges our traditional conceptualizations and hastens us toward a more integrated ethic for the treatment of nonhuman animals.

WILD ANIMALS

I believe that wild animals, considered as individuals, are valuable in an important sense, but that in most situations they are not morally considerable to humans. Individual animals are valuable to humans—they are often valued aesthetically and sentimentally, for example—but they do not exist in a relevant moral sphere for us. Let me explain this assertion, because it is sure to be controversial.

While wild animals *may* be considered morally, this depends on the situation. In the most straightforward situation—when individual wild animals live largely undisturbed by human activities in their natural habitat—humans accept no responsibility for animals as individuals.[1] This very general claim of nonresponsibility is not justified by any absolute claim about the intelligence or sensitivity of those individuals. It is rather a manifestation of a decision to respect the animal individuals *as wild*. By deciding to respect their wildness, we have agreed not to interfere in their daily lives, or deaths. We value them, but we value their wildness more; to respect their wildness is, in effect, to refrain from placing a *moral* value on their welfare or their suffering (see Callicott 1989). It is to treat them as a separate community, one with which we limit our interactions in order to encourage its autonomy from our own society. We also value wild animals as part of natural processes. I believe that our interactions with animals in the wild take on a moral dimension only at the population and species level, not at the individual level.[2]

Strachan Donnelley has introduced a useful distinction between what he calls ontological value and moral value (Donnelley 1992). Because we know that all animals strive to perpetuate their lives and their genetic line, there exists a good that results when these strivings are rewarded and a loss when they are thwarted and an organism dies. We can perhaps determine grades of ontological goodness corresponding to the mental complexity, self-awareness, etc. of the individual organism. Although it is difficult to ascertain the exact level of mental awareness experienced by animal organisms, we recognize it because we have experienced the same struggle between life and death ourselves (Donnelley 1992).

It is clear that animals have varying levels and acuities of experience, including varied levels of self-awareness. But it is equally clear that the experience we share carries with it both a striving to perpetuate that experience and the inevitability of diminution in some situations, and certain death in the long run. Ontological goodness, and grades of it, correspond to the strength and vitality of the experience of individual organisms. But this ontological goodness does not in itself entail moral goodness or moral obligations to interfere to protect it.[3] Humans cannot and ought not to accept responsibility to avoid deaths of individual animals in the wild. Building on Donnelley's distinction between ontological and moral value, I suggest it is not this *content* of animal experience but the *context* in which we encounter it that determines the strength and type of our obligations to animals and other natural objects. The content of the experience of some animals is surely rich enough to make them candidates for moral considerability. Nevertheless, I feel justified in killing feral goats on an island to protect the indigenous plant communities there, even though I have no doubt that the individual goats have a greater ontological value than the plants. Ontological value is morally relevant in some situations—as when animals are in captivity—but it is morally irrelevant in the wild, because the maintenance of the animal's wildness (appropriate behaviors in a wild context) prohibits our manipulating that experience. The forbearance we exercise here is very similar, psychologically and perhaps morally, to the attitude of wise parents who, after the time of maturity, let their children live their own lives.

While it may be necessary to manage wild populations to protect them against acute impacts of human disturbance, the goal of that management should be temporary and intended to remove the need for further interventions. Once an animal is brought into the human context, its ontological value, which corresponds to richness of experience of individual specimens, becomes extremely important in determining our moral responsibilities. I will return to permissible intrusions into the lives of captive and tamed animals below; first, I must say more about obligations that derive from our legitimate concerns for future generations.

OBLIGATIONS TO SUSTAIN NATURAL PROCESSES

The obligation to sustain biological diversity stretches far into the future, far beyond the point at which we can identify individuals (Norton 1984, 1991) and requires that we shift our sights from individuals and the various elements of biodiversity toward concern for the *processes* of nature (Norton and Ulanowicz 1992, Vrijenhoek, this volume).[4] In this larger resolution, individuals can only be seen as parts, functional elements in larger processes (Koestler 1967, Allen and Starr 1982). I will now explore our obligations to wild species in this longer, intergenerational context.

Thermodynamic models provide a better starting point for process-oriented models than do element-oriented models. If we assume that natural processes unfold in hierarchical order, and follow general systems theory in assuming that smaller systems change at a more rapid rate than do the larger systems they partly constitute, we can introduce some conceptual order into the confusing problems of scale in environmental policy. We experience nature from within a complex dynamic in the sense that there exist many dynamic, divergent forces on many scales. Since we are affected by, and affect, many dynamics, any ethic that is to integrate all of our obligations must provide a second-order guide to balancing our varied obligations within these varied dynamics.

When our attention shifts from concern for individuals—whether the legitimate regret we feel when we unavoidably hit an animal on the highway or the sentimental valuing as expressed in the Bambi syndrome—to concern for biodiversity, we shift from an individual to an intergenerational temporal scale and to a landscape-sized spatial scale. In this lengthened and broadened scale, our concern shifts to interactions among populations and species, not individuals. As individuals become rarer, we value individuals more and may accept responsibility to protect individuals as a means to protect species and the processes they perpetuate. The worldwide threat to biological diversity results not so much from the behavior of any individual human but from huge trends in human population and land use. I agree with those who believe that humans have important responsibilities to reverse the trends that are destroying biological diversity,[5] but this is a responsibility that has its natural habitat in discussions of extinction, concern for long-term biological conservation, and in concerns about intergenerational inequities in the allocation of resources. On this scale, individuals and local populations are regarded as instrumental in perpetuating creative and competitive processes that drive ecology and evolution. This concern for the creative activity of natural processes is not well expressed as entailing obligations to any particular *elements* of the system—it is a process-oriented moral concern and it directs our attention to the structures and functional interactions that constitute dynamic ecological systems.

Some of us call the creative drive of ecological systems "autopoiesis" (from the Greek: "self-making")—it is the mysterious driving force that creates, through dissipation of energy in open systems, a kind of growth or development, as order is created out of chaos (Maturana and Varela 1980, Ulanowicz 1986, Rees 1990, Costanza et al. 1992, Lewin 1992, Norton and Ulanowicz 1992). As pluralists, we do not assume that this value can be measured on a metric common with human rights or concerns about animal welfare.

Looked at in this larger scale, humans have made mistakes in the past, by intentionally and unintentionally introducing exotic species and by causing extinctions. Among these, I believe it was a mistake to eliminate predators on wilderness ranges—the policy of predator eradication destroyed a crucial (keystone) process in those ecosystems, and it has saddled subsequent generations of wildlife managers with the onerous task of destroying individuals of prey species who overpopulate and degrade their ranges. A failure to understand a crucial process led our forefathers to destroy that process. The moral responsibility exists not so much on the individual level as at the interpopulation level. I am suggesting, then, that we make moral decisions in different spheres, and that differing considerations should dominate at different scales. Managing to protect biological diversity—and this will more often mean managing humans, not wildlife—occurs on an intergenerational scale on which populations and species interact, not on an individual level (Norton and Hannon in preparation). Our moral decision to value wild animals *as wild* isolates us from moral obligations to wild animals as individuals.

It is now possible to split two moral questions that exist in two separate spheres. The question of whether to cull a herd, or to begin a captive breeding program, is a population-level question: are ecological processes as now configured adequate to perpetuate the health and integrity of the essential elements and structures of the system? The question of which method is permissible, or effective, for reducing overpopulations of ungulates, or what treatment is due captive breeding stock of endangered species, can be addressed in a separate context, which might be called interspecific ethics—the ethics of our relations with individuals of other species. The integrated pluralist believes we will judge our actions affecting animal organisms differently depending on the context of those actions (Midgley 1983, Callicott 1990). Once we have made a decision that we must interfere on the population level in ways that deeply affect individual animal lives, of course, we must accept responsibility for humane implementation of our environmental goals.

This contextual separation of two dynamics and two somewhat independent ethics applying at the populational and individual levels does not, of course, resolve all issues. By separating the question into populational levels, we separate the judgment of whether herd reduction is necessary or whether captive breeding is necessary from the question of what *means* are justified in such interventions. Especially, authors of this book do not agree regarding the means that are permissible at the individual level to achieve objectives that emerge on the population and ecosystem level of the scalar hierarchy. Here, I believe that the long-term value of keeping wild animals wild, especially in national parks and national forests but on private lands as well, dominates our obligation to minimize suffering of individuals, but I expect this position to be controversial. I can understand an alternative position, for example, that would accept the separation of questions but argue that we may never cause unnecessary pain in herd regulation—development of safe and effective birth control devices for wild animals would then become a conservation priority.[6]

Before proceeding further, let me mention several controversial issues that seem to form a coherent class. (1) Should we remove overabundant ungulates from predator-free ranges, and under what situations (to reduce destruction to ecosystems or to reduce competition with domestic grazers, for example)? (2) Suppose an epidemic of disease were attacking the last wild populations of a primate such as the chimpanzee, and that we desperately needed information that could be obtained only from extremely invasive and painful tests on living chimpanzees. Would we sacrifice several chimpanzee lives to save their species? (3) When, as seems almost inevitable, we get down to the last few mountain gorillas, should we live-capture the last few as breeding stock for a captive population? (4) Should we sacrifice a few Andean condors to gain essential information in order to save the California condor? What is common to all of these cases is that they pose policy choices that set individual members of a species against the well-being of some larger composite such as a species, a community, or an ecosystem. That is, a policy choice that may seem attractive if our goal is to perpetuate species or retain intact systems requires the suffering or death of some individual animals in an effort to save wild species. These issues all, in their own way, highlight the problems of determining what the target of protection efforts should be. And they all turn on the question, under what conditions will we sacrifice individual animals to perpetuate a larger good, either social or ecological? We should be especially interested in these cases because I believe that recommendations on these cases may represent an intellectual watershed in our deliberations. Those who, like Tom Regan, emphasize the similarities between our obligations to animals and our obligations to humans will prohibit any animal altruism toward composites, while those who emphasize that we have obligations, first of all, to protect species and ecological systems will tend to favor animal altruism or sacrifice in at least some cases. We can now characterize, both exactly and quite generally, what is at the heart of disagreements between environmentalists and animal welfare advocates, and especially animal liberationists. Individualists, who predominate in the liberationist tradition (but are much less dominant in the animal welfare tradition), unalterably oppose any nonvoluntary sacrifice of animal individuals for broader goods. Holists, by contrast, are united by their insistence that, at least in many cases, interests of individuals are over-ridden by the interests of composites; some form of holism is therefore attractive to conservationists.

ANIMALS IN HUMAN COMMUNITIES

Most of us can agree, I think, that our obligations to domestic animals—farm animals, pets, and other animals that are integrated into our lives on a day-to-day basis—are far more extensive than our obligations to wild animals. The context in which we interact with these animals implies a contract to look after them. By living together with them, we have brought them into our community, and we are obliged to feed them, to care for them if they are injured, and so forth. We also accept responsibility for controlling their populations. My point is that it is the context of our interactions—the responsibilities we have taken—that deter-mines our moral obligations more than the characteristics of the individuals involved; some of these responsibilities rest on us because of unfortunate choices and actions of our predecessors and ancestors. The morally relevant fact is not usually the *content* of experience of an individual creature but the *context* of our interactions with it. If I encounter the neighbor's cat about to get a songbird in my back yard, I would intervene if possible. If I were hiking in the wilderness and I were fortunate to see a wolf pack run down and kill a deer, intervention would be profoundly inappropriate. The crucial moral fact that decides cases like this has nothing to do with the relative mental or moral capacities of songbirds and deer and everything to do with the context of the experiences.

The more we meddle in the daily lives of wild animals, the more we accept moral responsibility for the consequences of our actions in their individual lives. The context in which we interact with domesticated animals implies a contract to look after them. No such contract exists with wild animals; for this reason, we have no moral obligations to individual members of wild species who remain in their natural habitat.

WILD ANIMALS IN ZOOS

I have argued that it is mainly the context, and not the content, of our interac-tions with animals that determines our moral obligations to them, and I have argued that our obligations to wild animals generally emerge at the population level, where our policy decisions affect large trends in ecological systems and the processes that sustain them. These obligations are quite distinct and are based on values very different from those underlying our obligations to domestic animals. The obligation to sustain biodiversity intergenerationally is a process–oriented ethic; this process orientation is unavoidable because of the complex, and usually unknown, interactions among species in ecosystems. I do not treat it as commensurate with the interspecific ethic, which is based on the complexity of individual experience, or with ethical systems that state, preemptively, that all objects with independent moral value have that value equally (Regan 1983). A pluralistic, but contextual, ethic has promise to organize our thinking as we struggle to make clear the means by which we decide difficult cases.

But can these distinctions help us to deal with our present quandaries? Zoo animals are neither fully domesticated nor are they fully wild; they exist in a mixed context (Midgley 1983). Accordingly, humans have some obligations to zoo animals that originate in both contexts. Our taking them from the wild obligates us to protect them in ways they cannot protect themselves, and this clearly involves obligations to individuals. But we also hope and intend that they contribute, in one way or another, to the perpetuation of wild processes, of which their species is a part. We therefore respect their wildness and even hope that they or their descendants can be returned to a wild context free of our manipulation.

Many of the quandaries we face in this book ultimately depend on which of these two aspects of our interactions with zoo animals we emphasize. I certainly do not have all of the answers, but I want to focus on one crucial flash point where the two moral systems—intergenerational sustainability of natural processes and concern for the autonomy and well-being of individual animals— intersect. Consider the broad class of cases, exemplified above, in which environmental managers believe it is necessary to harm severely one or more individual animals in the interests of protecting their species, some other species, or some ecological system. Can we ever justify what we might call "animal altruism," the sacrifice of significant interests—such as the interest in freedom or life—of individual animals in the interests of a long-term goal such as balancing populations, preserving genetic diversity, or protecting the integrity of an ecosystem? In many of the daily decisions of environmental managers, especially those most concerned with biological resources, this is an unavoidable question (Kleiman 1989). If we agree with Dale Jamieson (who argues that the removal of animals from the wild and placing them in zoos harms them) that there is a moral presumption for leaving wild animals wild, it follows that we must provide a countervailing moral principle adequate to override this harm (Jamieson, this volume; Regan, this volume).

We have therefore arrived at a central issue—the issue of animal altruism—that provides an intellectual map of the confusing territory of ethics between the species. Many well-intentioned people seek a middle-ground position between moral individualism and literal extensionism of human individualistic morality to animals on the one hand and a holistic ethic that values individuals only as repositories of genes and essential parts of system processes on the other hand. But defenders of this middle ground are impaled on the horns of a dilemma: how can they accept the obvious value that exists in the experience of complex animals and at the same time make difficult decisions that will make some animals casualties in the struggle to save biological diversity?

ANIMAL ALTRUISM

It may be useful to contrast these cases in which conservationists believe it necessary to sacrifice animals with cases of human altruism and self-sacrifice. I think most of us believe, with regard to voluntary donors of organs, voluntary participation in the defense of one's country, and countless other selfless acts that we approve and applaud, that acceptance of risk of harm, even predictable loss of life, can be justified in the interest of the life of a loved one or of a higher good such as national defense, provided the action is undertaken and the risk accepted *voluntarily*. The sacrifice of individuals for a higher good is in this case permissible and in many cases laudatory because the requirement of voluntarism guarantees the moral integrity of the individual, rational moral agent and transforms an act that is simply foolish from the perspective of individual survival into an act of heroism when it is viewed in the larger perspective of family or community well-being. What is lacking in the idea of animal altruism, if it is structured to mimic human altruism, is this crucial act of voluntary self-sacrifice that gives nobility to free and selfless community-oriented service.

In individualist approaches to ethics, the key idea is unquestionably consent and voluntary choice, and so we must recognize the systematic ambiguity created by discussing animal altruism or sacrifice. If the moral status of animals strictly parallels the moral status of humans, animal altruism is conceptually and morally impossible. Moral altruism among animals is no more than a euphemism for involuntary animal sacrifice. The sacrifice of individuals to a higher good—such as protecting a crucial process (killing ungulates on predator-free ranges), invasive and risky procedures on individual animals to gain knowledge that will help to save their species, and removal of exotics and feral populations to protect fragile ecological communities—is ruled out as a priori impossible.

I doubt these issues can be resolved so simply. We should therefore address as directly as we possibly can the following question: are there any conditions under which we should sacrifice individual animals in the interests of protecting a composite of which they are a part? If so, what, exactly are those conditions?

Field studies of social behavior of wild animals of course reveal *behaviors* very analogous to altruism, but our reluctance to say unqualifiedly that such behavior is voluntary (a requirement that is currently at issue) causes us to pause before calling it truly altruistic. But when the dominant male in a gorilla band stays behind and sacrifices himself in order to protect the band from poachers, the analogy seems strong and we have no problem seeing the heroism involved; nobody would quarrel, I think, with calling this animal altruism. Further, analogous behavior can be recognized in far less complex creatures than gorillas— army ants sacrifice themselves by the thousands by marching into a stream to create a bridge of dead bodies for their advancing comrades behind them.

But consider another case: Suppose human keepers of mountain gorillas realize that a healthy adult must be sacrificed to gain knowledge that would further a captive breeding program and perhaps save the species from extinction. The gorilla individuals cannot deliberate and cannot volunteer (even in the anthropomorphized sense of the gorilla hero), because they do not deal in abstractions and they probably cannot consider the importance of a long-range goal such as saving their species. If we think of animal altruism as strictly analogous to human altruism in that it requires unequivocally voluntary compliance with decisions to face risk, pain, or certain death, we can only conclude that no gorilla can be altruistic toward an abstraction such as a species and hence that sacrificing an individual to save its species is without question wrong in every case. If this argument is convincing, then it would seem to follow that other cases of sacrifice of animals (for human health experiments or to benefit another, endangered species) will also be ruled out.

Extensionist approaches to interspecific ethics, which emphasize similarities across species and base all moral decisions on a single value metric, are therefore likely to conclude that sacrifice of individuals for species survival is always wrong because the individuals cannot fulfill the key requirement of voluntary acceptance of risk.

This seems to me paradoxical. An ethic that begins by emphasizing similarities among human and nonhuman individuals ends in greater willingness to sacrifice human individuals than nonhuman individuals because humans have a genetically conditioned and morally developed ability to deliberate and choose heroism. Lacking these genetic and moral prerequisites, animals are disqualified from serving their kind unless they face a palpable and concrete threat to their immediate family and associates. It seems to me that Tom Regan, for example, would be committed by his monistic rights theory to this very general principle, which would have pervasive consequences for decisions by environmental managers (Regan and Francione 1992). If so, I cannot resist asking two more, admittedly rhetorical, questions: Does the conclusion that an individual nonhuman animal can never act heroically in the service of its species not involve a kind of anthropomorphism? Does this conclusion not rest on an implicit requirement that nonhuman animals must in some sense fulfill human criteria in order to extend their own sense of heroism, so evident in the case of the gorilla's self-sacrifice for his group, to include sacrifice for the abstract extension of his commitment to his group, a commitment to perpetuate his genes for many generations?

I am fully aware that there is a certain circularity in the positions we take here. It is perhaps because I side with holistic environmental managers that my intuitions point the way they do, but it seems obviously wrong to define altruism in a way that requires any altruist to act voluntarily and then conclude immediately that animals, because they are incapable of voluntarism in service of an abstraction, can never act altruistically. I hope that individualists are not ruling out animal altruism or sacrifice based only on this linguistic sleight of hand. Since I believe that cases such as this are crucial to our discussions, let me explain in general terms how I would decide them.

The goal is to forge a concept of animal altruism that both recognizes the inherent differences between human and nonhuman consciousness and at the same time reflects appropriate levels of respect for individual organisms. I believe we can, but I doubt we can if we conceptually straitjacket ourselves by valuing only individuals and their experiences. I therefore advocate not a middle-of-the-road approach but a both-sides-of-the-road approach based on moral pluralism.

To accomplish this goal, let us return to Donnelley's distinction between ontological and moral value (Donnelley 1992). Earlier I used this distinction to explain how we do not have obligations to interfere to reduce naturally occurring pain (during acts of predation in the wild, for example), even though we recognize high degrees of ontological value in the prey's struggle to survive. This ontological value makes animals candidates for moral status, depending on context. In the wild, we only express our awe at the power and cruelty of nature, and we avoid intervening whenever possible. But once the individual animal is brought by us into the human community, new constraints are imposed upon human treatment of individuals—we have taken responsibility for the animal's well-being. The humans in question have therefore committed themselves to act as responsible stewards for the now helpless animal, and they must devise a new ethic appropriate to this different situation.

This line of reasoning explains the gravity of our decision to remove wild animals from the wild; it also explains the seriousness of the responsibilities we come to owe individuals once we become their moral guardians. Sometimes this responsibility is thrust upon us, as when a baby animal is orphaned as an unforeseen outcome of human activities; at other times, we grasp this responsibility, recognizing that a given species, which we value, is threatened and that extraordinary interventions are necessary to prevent permanent rents in the fabric of natural systems. In either case we must act responsibly, recognizing to the extent possible the impacts our actions will have on individual lives, and recognizing also the gravity of the situation.

As a moral pluralist, I perceive the ethical problem as one of facing at least three broad moral responsibilities (in addition to the obvious obligation to treat our family and associates justly). One obligation is to our fellow humans of this generation who suffer from famine, war, and poverty. But there is a second obligation, to future generations. In this I count prominently the obligation to sustain nature's bounty and to provide the future with a livable ecological context (niche) that will leave them options for a pleasant life. A third obligation is to respect the other creatures that exist with us on this planet, which is both fragile and resilient. I recognize that intelligent and fair-minded people differ regarding which of these responsibilities is dominant; I therefore advocate what I call a convergence strategy, which places priority on pursuing policies that will simultaneously support our goals in all three areas of obligation. According to the convergence hypothesis, we assume that in the long run, what is good for our species will also be good for other species, taken as species (Norton 1991). We therefore seek policies that protect ecological systems from disruption due to our increasingly ubiquitous activities. Similarly, we place high priority on attempts to raise the standard of living of the poorest humans, recognizing that improvements in standard of living normally lead to reduction of human birth rates and an eventual positive impact on protecting biodiversity.

If human population growth continues and virtually all wildness is wiped out, we will have harmed both future generations of humans and the processes of nature. What we have in common with other animals is the ceaseless striving to exist and reproduce, the source of ontological value. It is essential that we respect that ontological value; it imposes obligations on us when we accept responsibility for the animals we bring into our community. But the community we bring them into is not the human community; it is a *mixed* community (Midgley 1983). And this community cannot be identified with those animals existing now—it is rather a hierarchically organized system that changes on many different scales and according to multiple dynamics. Respect for wild animals as wild requires that we see them as participants in both a personal struggle to survive and in a struggle to perpetuate their kind. The individual drive to survive is of course not just accidentally connected to the individual's contribution to the gene pool of the species. Just as human beings are both social beings who feel sentiment toward others and self-preserving organisms in their own right, we cannot do justice to animal ethics without recognizing that the struggle of all animals to survive has meaning only if it is a part of their correlative struggle to perpetuate their species in the long run.

Now consider again the situation of a captive animal that will be used as a part of a captive breeding program. This animal is a wild animal, but we have accepted responsibility for its care and brought it into the context of the conservation community, and in that context we have obligations to respect its considerable ontological value, which requires that we respect the sentience of the animal by limiting its pain, especially in the extremity of death. But full respect for a truly wild animal, an animal that exists with integrity in its traditional niche and still inhabits the least disturbed areas of its traditional range, requires that we recognize also the intergenerational aspect of the great striving toward life. The struggle of animals to exist in the wild is both a struggle to survive individually and a struggle to perpetuate their species. A reasonable concept of animal altruism must account for the natural instinct of animals toward individual survival and, in addition, their natural instinct to perpetuate their species.

It is an awkward truth that humans must decide which members of the community will be sacrificed in the furtherance of their species and of the ecological community that constitutes their niche. The problem of exploding human population and destruction of habitats for other species is in a profound sense a human problem. It is anthropogenic in its causes, and we must accept responsibility for our past and present actions that inevitably shape the future. But the solution is surely not to domesticate all animals. If we accept responsibility for a wild animal and then reduce that animal to a creature incapable of noble acts in service of its community, this too is an act of disrespect, because in the wild the animal acted both to survive and to perpetuate its species. We can now see the error in individualistic and extensionist positions on interspecific ethics. By denying that animal altruism in service of higher ideals can ever be justified because the crucial condition of voluntarism cannot be fulfilled, the individualists respect only the drive toward individual preservation and ignore the equally powerful drive of animals to perpetuate their species. Individual members of every social species (which includes at least every sexually reproducing species) live an existential paradox—the drives to survive and to reproduce one's kind, usually mutually reinforcing, can conflict, forcing a choice between individual preservation and contribution to species perpetuation.

This second aspect of the striving of all living things can also require self-sacrifice, as when a parent sacrifices or risks its life to protect its young. I therefore conclude that we can treat animal altruism as a conceptual extension of the striving to live and perpetuate one's species. These animals may therefore be enlisted in a noble cause; they cannot do so voluntarily, but their willingness to sacrifice themselves is implicit (it cannot be explicit because they are incapable of language) in their struggle to perpetuate their species. Following this line of reasoning, individual animals can justifiably be sacrificed, provided (1) the sacrifice contributes toward a goal that is implicit in the life struggle of the animal, (2) the animal is treated appropriately within the context of the struggle to save the natural world community of which humans and all other species are constitutive members, and (3) necessary means are taken to reduce pain and suffering to the extent possible and commensurate with the ontological status of the animal under human care.

Again, context is important, just as it is in human ethics. Suppose my country is at war and I enlist in the defensive forces; I thereby enter a new context. Suppose that the fortunes of war place my regiment at a crucial point in the battlefield, and the general in charge reluctantly concludes that my regiment must be sacrificed to plug a hole in the lines and allow orderly retreat. My voluntary enlistment declared my willingness to be a soldier. It is at this excruciating moment that that voluntary act is consummated. I cannot with honor decline to fight. I have voluntarily entered a context in which I have given up the right to decide whether to retreat or fight. Although his behavior may be more genetically mediated, the dominant male gorilla defends his harem with a similar resignation.

If we think of animals in captive breeding programs as implicitly accepting the goal of perpetuating their species, then they may unfortunately be called upon to make the ultimate sacrifice. Making this decision, however, requires that we, who are capable of seeing the gravity of the situation facing them and other species, respect their striving in all its complexity. We accept the awesome responsibility to manage the lives of other species even to the extent that we accept the responsibility to sacrifice some of those animals as a part of protecting their species and their niche. Our responsible treatment of them then becomes analogous to the decision of the military commander as we reluctantly place individual animals in the breach to hold off the forces that would interrupt its survival as a species in the wild. This is an act not to be taken lightly; the grave responsibility we have accepted by our experiments with change in natural systems should be sobering, and the decisions we make must be sensitive to the complexities of the situation—we must weigh all of the relevant information and we must seek to meet all of the sometimes conflicting responsibilities to the fullest degree possible. The decision maker must accept that the animal in question has great ontological value, value that expresses itself both individually and collectively as the means by which individuals contribute to the perpetuation of their species. As an individual member of a living species, each animal exists on two planes, as a striver to save itself and on the larger scale in which it tries to perpetuate its genes; in this larger scale, the individual life gains existential meaning in a larger context. And it is not simple cruelty to sacrifice an animal for the good of that animal's species.

The question of whether a member of one species can be sacrificed for the survival of another species seems to me more problematic, but I suggest for discussion the following position: a member of one species can be sacrificed for the well-being of another if that sacrifice can be justified as a necessary act to protect natural communities and the habitats that will make possible the perpetuation of all species.

The remaining difficulty lies in the asymmetry between the human role and that of other species. A large part of this asymmetry results from an undeniable fact. Humans, in our own struggle to control nature, are the culprits who cause most of the dilemmas. Then we, in our wisdom, decide to sacrifice individuals to save species and processes that our own excesses have drawn into risk. It is the march of human domination that threatens the habitat of other species, and our own, as well. Can we, as the only species (as far as we know) capable of scientific projections and perception of long-term dangers, inflict upon individual members of other species the ultimate sacrifice in order to avert consequences of our own mistakes? I do not have, I admit, a fully satisfactory answer to this question; I can, however, offer two considerations that put the question in a context in which it is perhaps less perplexing. First, note that viewed in the long run, nature's processes are more basic than individuals. It is processes that create and sustain individuals. If we take seriously the obligation to protect and sustain biological diversity, we must act to protect the wild processes and the species that embody them. In accepting the responsibility to sustain biodiversity, we ipso facto accept the responsibility to choose, when it is unavoidable, to save those creative and sustaining processes even at the expense of individuals. Second, while it is undeniable that the current crisis in biodiversity is human in origin, our species has, at least until recently, acted without intentional malice in expanding our range. Until we developed, only in the last century or so, an awareness of our ability to destroy natural processes on a huge scale, our striving to perpetuate and expand our own population was morally on a par with the actions of members of other species who strive for analogous goals for their own species. Once one recognizes that sustaining biodiversity will require caring for processes within a dynamic system, and that all species change the larger dynamic as they strive to survive and reproduce their kind, it seems to follow that currently existing humans cannot be held morally responsible for the actions of our forebears. The moral question is, What will we do now that we are beginning to understand the incredible complexity of the factual and moral situation we face? I have argued that we will begin to act in morally acceptable ways only if we develop a complex and pluralistic understanding of our multiple moral obligations in our complex situation.

SUMMARY AND CONCLUSIONS

I have developed a general approach for balancing our moral responsibilities to animals of other species. This approach is pluralistic in the sense that it recognizes more than one type of value and more than one source of obligations, but it is an integrated pluralism, specifically a contextual pluralism, which manages these multiple obligations by emphasizing them differently in different situations. Wild animals have ontological value, but our decision to leave them wild precludes our accepting moral responsibility for their well-being in the wild. Once they are made a part of the mixed communities of zoos and captive breeding programs, we accept expanded responsibility for their welfare, and our obligations will be governed by the comparative level of self-awareness and complexity that constitutes their ontological good.

I am not sure how far this line of reasoning will carry us, but it does hold promise for developing a more encompassing position—one that places high value on individual wild animals while recognizing that wild animals strive to perpetuate their species, allowing, at least in principle, the possibility of animal altruism in service of the long-term good of wild species and their habitats. This line of reasoning may justify us in enlisting animals in causes that they cannot understand, because the goal of perpetuating their species and other species is implicit in their struggle to perpetuate their genetic line as an element in the fabric of nature, and of the experiment of life more generally.

NOTES

1. Because I see the problem as one of balance, in pluralistic terms, I do not think this rule should be applied narrowly, independent of other considerations. For example, Alston Chase discusses a case in which the Park Service refused to help a bison that had fallen through ice into the Yellowstone River (1986). Because the bison's life has positive value, I do not take the general preference for wildness to *require* that we let the bison die in these cases. I would reason that while I have no moral obligation to save the bison, a general preference for saving things of ontological value might in this case override the minor cost to the authenticity of the struggle for existence.

2. This is the point of Aldo Leopold's dramatic argument that we must "think like a mountain," which means "thinking on the temporal and spatial scale of the mountain" (Leopold 1949; Norton 1988, 1990). I have argued that it is total diversity over large geographic areas that should be the target of protection efforts (Norton 1987). Emphasis on within-habitat diversity would mandate too much concern for populations (which are naturally ephemeral on some scale of time), whereas concern for cross-habitat diversity would not support enough concern for the genetic diversity that comes from distinctive species and subspecies that evolve in specialized habitats. While I still believe that protecting total diversity over landscape-sized geographical areas is the best general measure of how we are doing over years and decades, I now doubt that any approach to biodiversity policy that emphasizes elements rather than processes will serve to characterize biodiversity in the long run (see Norton and Ulanowicz 1992).

3. Hargrove has insightfully emphasized the importance of this distinction in the naturalist tradition, which has shown little concern for individual specimens of plentiful species but has condemned wanton destruction of individuals—destruction for no good cause (1989). Since what is considered wanton killing of individuals may well depend partly on their rarity, Hargrove links current concern for species and the processes they represent with a long and respected tradition in science, literature, and art. See Hutchins and Wemmer 1987 for a survey of recent thinking on the justifiability of herd reductions.

4. Ulanowicz and I have provided a precise and potentially quantifiable argument that system characteristics at different scales depend on different dynamics (Norton and Ulanowicz 1992). This independence of dynamics is crucial in understanding environmental policy and management; it implies that policies affecting economics will not all have equivalent impacts on ecological systems that provide their context. It is therefore possible in principle to have development that supports biodiversity.

5. See Jamieson 1985 for a discussion of the moral gravity of the role we have assumed.

6. See Hutchins and Wemmer 1987 for a discussion of the strengths and weaknesses of the case for contraception as opposed to killing.

REFERENCES

Allen, T. F. H., and T. B. Starr. 1982. *Hierarchy: Perspectives for Ecological Complexity.* Chicago: University of Chicago Press.

Callicott, J. B. 1989. *In Defense of the Land Ethic.* Albany: State University of New York Press.

———. 1990. The case against moral pluralism. *Environmental Ethics* 12:99–124.

Chase, A. 1986. *Playing God in Yellowstone: The Destruction of America's First National Park.* Boston: Atlantic Monthly Press.

Costanza, R., B. G. Norton, and B. D. Haskell, eds. 1992. *Ecosystem Health: New Goals for Environmental Management.* Covelo, Calif.: Island Press.

Dillard, A. 1974. *Pilgrim at Tinker Creek.* New York: Harper & Row.

Donnelley, S. 1992. *Bioethical Troubles: Animal Individuals and Human Organisms.* Jerusalem: Hans Jonas Symposium.

Ehrenfeld, D. 1991. Conservation and the rights of animals. *Conservation Biology* 5:1–3.

Goodpaster, K. 1978. On being morally considerable. *Journal of Philosophy* 75:306–325.

Hargrove, E. 1989. *Foundations of Environmental Ethics.* Englewood Cliffs, N.J.: Prentice-Hall.

Hutchins, M., and C. Wemmer. 1987. Wildlife conservation and animal rights: Are they compatible? In *Advances in Animal Welfare Science 1986–1987,* ed. M. W. Fox and L. D. Mickley, 111–137. Washington, D.C.: Humane Society of the United States.

Jamieson, D. 1985. Against zoos. In *In Defense of Animals,* ed. P. Singer, 108–117. New York: Blackwell.

Kleiman, D. G. 1989. Reintroduction of captive mammals for conservation. *Bioscience* 39(3): 152–161.

Koestler, A. 1967. *The Ghost in the Machine.* New York: Macmillan.

Leopold, A. 1949. *A Sand County Almanac.* New York: Oxford University Press.

Lewin, R. 1992. *Complexity: Life at the Edge of Chaos.* New York: Macmillan.

Maturana, H. R., and F. J. Varela. 1980. Autopoiesis: The organization of the living. In *Autopoiesis and Cognition,* ed. H. R. Maturana and F. J. Varela. Boston: D. Reidel Publishing.

Midgley, M. 1983. *Animals and Why They Matter.* Athens: University of Georgia Press.

Norton, B. G. 1984. Environmental ethics and the rights of future generations. *Environmental Ethics* 4:319–337.

———. 1987. *Why Preserve Natural Variety?* Princeton, N.J.: Princeton University Press.

———. 1988. The constancy of Leopold's land ethic. *Conservation Biology* 2:92–103.

———. 1990. Context and hierarchy in Aldo Leopold's theory of environmental management. *Ecological Economics* 2:119–127.

———. 1991. *Toward Unity among Environmentalists.* New York: Oxford University Press.

Norton, B. G., and B. Hannon. In preparation. Toward a biogeographical theory of environmental value.

Norton, B. G., and R. Ulanowicz. 1992. Scale and biodiversity: A hierarchical approach. *Ambio* 21:244–249.

Rees, W. R. 1990. The ecology of sustainable development. *Ecologist* 20:18–23.

Regan, T. 1983. *The Case for Animal Rights.* Berkeley: University of California Press.

Regan, T., and G. Francione. 1992. A movement's means create its ends. *Animals' Agenda* 12(1): 40–43.

Thoreau, H. D. 1946. *Walden.* New York: Random House.

Ulanowicz, R. E. 1986. *Growth and Development.* New York: Springer-Verlag.

PART THREE

CAPTIVE BREEDING AND

WILD POPULATIONS

Most advocates of captive breeding programs consider their programs justified by the possibility of future reintroductions of animals into the wild, either to augment waning wild populations or to reestablish wild populations that have been extirpated from most or all of their traditional range. Some participants in this conference group (Robert Loftin, for example) favored a limited role for captive breeding programs, considering them justified only if they are directed at eventual reintroduction into the wild. Other participants would argue that reintroduction is only one reason for a captive breeding program and that a variety of benefits are adequate to justify captive breeding, even of some species for whom reintroduction is not a viable possibility. These benefits might include using animals as conservation ambassadors to protect habitats more broadly or displaying animals in educational or recreational exhibits. The likelihood of successful reintroduction of captive specimens into the wild will therefore remain a pivotal question in the assessment of the morality of captive breeding programs.

If in situ conservation is indeed the highest priority of zoos, it follows that activities such as procurement of breeding stock from the wild must be accomplished without negative effect on the survivability of wild populations. But this is hardly an all-or-nothing proposition, despite first impressions, because there are undoubtedly some cases in which the situation in the wild is so dire and the likelihood of habitat recovery and introduction are sufficiently high that most observers would consider captive breeding a good enough risk to justify temporary removal of animals from the wild in anticipation of a successful reintroduction later. The challenge is one of judging when risks to existing populations in the wild are so dire that the admittedly risky strategy of temporary removal is justified. It seems unlikely that these questions of probability are best decided according to inflexible or doctrinaire positions on either side.

The issues regarding the procurement of founder stock from the wild and the reliability of reintroduction plans are difficult to disentangle within the broader strategies of conservation practice. Similarly, disagreements regarding procurement and reintroduction lurk in the background of discussions of regulation of captive populations, but the moral issues in this case are even more difficult because, here, the spotlight is on specific animals who have already made their contribution to the genetic pool and in this sense have become surplus to the captive breeding program. If the main purpose of captive breeding programs is to increase the likelihood of survival of wild populations, careful regulation of the gene pool is essential to maintain the genetic diversity of the captive stock. But specimens of long-lived species can complete their genetically determined role in preserving the genetic pool early in their natural life; once again, the emphasis in captive breeding programs on maintaining genetic variability creates moral dilemmas in all those cases in which surplus animals cannot be returned to the wild.

In the three policy areas addressed in Part Three—procurement, reintroduction, and captive population regulation—in which concerns about the health and survivability of wild populations intersect with decisions affecting captive populations and individuals, the moral conflicts and paradoxes raised here are apparently unavoidable. Zoos, eager to contribute to protection of wild populations, have embraced goals and values that in some situations necessarily conflict with their obligations to do no harm to their captive charges, those individual, captive specimens of threatened species that are essential to their conservation goals. The complex and distressing moral questions that arise as a result of expanded human activities, dominance, and responsibilities appear in their most difficult forms in these chapters. Their authors struggle to state the conditions under which one moral commitment outweighs the other in the areas of interaction between wild and captive populations. This is clearly an arena where more discussion and debate are warranted.

PROCUREMENT OF ANIMALS FOR CAPTIVE BREEDING PROGRAMS

Captive breeding programs must be understood in their larger conservation context, including the effects of those programs on wild populations and in situ conservation. Proponents of zoos, however, insist that the occasional procurement of animals from the wild is justified. Thus it is necessary to assess the impact of procurement of breeding stock for captive populations on wild populations. In this exchange, Fred Koontz argues that captive breeding programs can be successful without harming wild populations and outlines a set of ethical considerations and procedures to ensure that taking from the wild for captive breeding programs benefits wild populations. Ardith Eudey questions whether all of the effects of captive breeding are beneficial to wild populations. Both authors agree that careful guidelines must be followed to ensure that such programs complement, rather than damage, in situ conservation efforts.

WILD ANIMAL ACQUISITION ETHICS FOR

ZOO BIOLOGISTS

Fred Koontz

Wildlife conservation has been designated the highest priority for zoos belonging to the American Zoo and Aquarium Association (Hutchins and Wiese 1991). Zoos contribute to the preservation of biodiversity by offering public education, safeguarding genetic resources, conducting scientific research, and raising sorely needed conservation funds (Conway 1988, McNeely et al. 1990, IUCN et al. 1991). Sadly, the natural world is in the midst of a species extinction crisis caused by human activities (Soulé 1991, Ryan 1992). The most that any conservation organization can reasonably expect to accomplish in the short term is to slow the rate of species extinctions and habitat loss, in hopes that humankind eventually will stabilize its population and find better ways to manage its impact on the environment. Ultimately, all conservation organizations, including zoos, must measure their success by assessing the health of both individual species and entire ecosystems in nature.

It is an ethical paradox that zoos must on occasion remove animals from the wild. After all, loss of animals from nature is the very thing that zoos are trying to prevent. A fundamental fact, however, is that to establish zoo-based breeding programs for endangered animals, a certain number of wild specimens must be obtained (Foose and Ballou 1988). The purpose of this essay is to attempt to resolve this apparent ethical paradox by discussing six ethical considerations, which when taken collectively can define situations that justify acquiring endangered wild animals for zoo programs. I will also propose an ethical checklist, that is, a set of procedural steps for zoo biologists to follow when engaged in such projects.

Any system of ethical standards must be defined in conjunction with the group that will abide by the rules. It is beyond the scope of this chapter to formulate a universal ethics of animal acquisition. Instead, I will limit my focus to one particular group and situation: zoo biologists acquiring endangered or threatened animals from nature for captive breeding programs. I will first describe the relevant moral attributes of zoo biologists as a group, because these beliefs will serve as the foundation for the ethical standards that I propose. The reader is cautioned from the onset that those persons with significantly different attitudes toward animals or wildlife conservation undoubtedly could argue for different ethical standards.

ATTITUDES AND VALUES OF ZOO BIOLOGISTS

For the purposes of this chapter, a zoo biologist is defined as an individual (1) working in any of the American Zoo and Aquarium Association (AZA) institutions (or in zoos and aquariums internationally that possess the characteristics that would allow them to be accredited by the same AZA standards) and (2) acquiring wild animals for organized captive breeding programs. In North America, there are 167 zoos and aquariums accredited by the AZA and more than 6,200 individual AZA members.

The AZA currently coordinates organized breeding programs for 117 Species Survival Plan (SSP) species, in which the populations are managed collectively among member institutions for genetic variability, demographic stability, behavioral compatibility, and long-term viability (for a technical review, see Foose 1989). The goal of the AZA is to have at least 200 SSP species by the year 2000 (Hutchins and Wiese 1991). Zoo associations similar to the AZA but representing professionals from other regions of the world (e.g., Australia, Europe, and Japan) have in recent years also initiated organized captive breeding programs for endangered species (for a directory, see Swengel 1993).

Like any professional group, zoo biologists hold to a variety of moral attitudes and values. There has been no formal study of their beliefs, but nevertheless, for the purposes of this chapter, I think that the trends are clear enough to allow me to describe the average member. It is important to note that attitudes of both individuals and collective groups are dynamic. This is certainly the case for zoo biologists. During recent years, the mission of modern zoos and aquariums has evolved rapidly, mostly as a direct response to the equally rapid deterioration of the earth's biota (Rawlins 1985, Wheater 1985). Below, I take an attitude snapshot of a typical zoo biologist.

Attitudes toward Animals

Many zoo biologists become interested in animals at a young age, and this early avocation often fosters a naturalistic attitude toward animals. The primary characteristic of this point of view is a strong interest and affection for the outdoors and wildlife (Kellert 1989). Observation and personal involvement with wildlife also are key to the naturalistic perspective. For many of these future zoo biologists, animals become living objects to identify, compare, and catalog. Often their curiosity leads them to collect live specimens to bring home for closer study, frequently to the dismay of their parents.

In the process of keeping animals, including domestic pets, it is not uncommon for the caretaker to develop strong affection and attachment to individual animals. The animal is the recipient of emotional projections somewhat analogous to those expressed toward other people. This may lead to anthropomorphic distortions, and the animal's intrinsic abilities may be idealized, leading to romanticized notions of animal innocence and virtue. Kellert calls this an extreme "humanistic attitude" (1989), while Burghardt and Herzog define it as "anthropomorphic empathy" (1989).

Most practicing zoo biologists consciously attempt to avoid anthropomorphic distortions in their work with captive animals, but from time to time and to various degrees, strong humanistic attitudes surface within the zoo community. This is especially likely for those persons who supply primary care to animals. In general, however, all zoo biologists agree that while severe anthropomorphic distortions are counterproductive, some kinds of anthropomorphism are beneficial to wildlife conservation and animal welfare goals (Lockwood 1989). The typical zoo biologist's attitude toward this positive anthropomorphism was described by Donnelley and Nolan when they defined "critical anthropomorphism" as empathy tempered by objective knowledge of the particular species' (or individual animal's) life history, behavior, and physiology (1990).

Naturalistic and humanistic attitudes, combined with professional training and experience, have led most zoo biologists to express their own moralistic attitudes toward the humane treatment of animals. The most basic and consistent element of this philosophy is strong opposition to inflicting pain, harm, or suffering on animals. Zoo biologists believe that they are professionals, with the technical skills and resources needed to provide humane care both to individual specimens and to populations of captive animals, including endangered species. They also generally state that amateurs, who have neither sufficient training nor the necessary institutional support, should not keep most types of wild animals, especially endangered species. In short, zoo biologists are strong advocates of animal welfare.

Zoo biologists take a scientific approach to determining animal welfare guidelines. Increasingly, ethological, ecological, genetic, and physiological data are being used to design captive care programs (Foose 1989, Koontz and Thomas 1989, Novak et al. 1994). This attitude recently has rekindled the idea that zoo animals are legitimate scientific subjects, especially for those studies dealing with conservation biology (Hutchins 1988, 1990; Soulé and Kohm 1989; Kleiman 1992).

The moralistic attitudes of zoo biologists differ from many animal moralists in regards to animal rights. Advocates of animal rights (e.g., Regan 1983) argue that each sentient individual, human or nonhuman, has morally inviolable rights; all human actions that violate individual rights should be abolished. While different animal rightists interpret these tenets with varying degrees of latitude, zoo biologists generally do not assign rights to animals.

For example, I believe only humans have rights, because only humans are responsible and participatory (or, in special cases like infants, potentially responsible and participatory) moral agents and can be held accountable for their actions within the context of a human-based ethical and legal system. At first glance, this view of rights is somewhat similar to the traditional Kantian perspective, but Kant's "rational animal" is replaced here with the responsible and participatory animal. Only humans can be responsible and participatory within a human-based moral system. The granting of rights (and the associated responsibilities) to individual humans implies participatory involvement and requires that the individuals be responsible for their actions. Indeed, those humans who do not accept their responsibilities—that is, those who do not participate by behaviorally abiding by the agreed-upon rules—can ethically and legally be denied certain rights and, for example, might be imprisoned.

It is, from my perspective of rights, terribly anthropocentric of animal rightists to assign unilaterally to animals a system of human-based rights. If a grizzly bear kills a human backpacker in Yellowstone Park, should the bear be held responsible by being put on trial and punished for his crime? Should we expect the bear, or any other nonhuman, to consult human moralistic principles when making choices? In my opinion, no. Interestingly, however, there have been many cases historically when animals were held accountable for their violent acts, tried, convicted, and even executed for their actions (Beach 1975).

Assigning rights only to humans does not in any way diminish the zoo biologist's moralistic commitment to animal welfare. It is a *human* right to live in a society that treats animals fairly, where animals are not made to suffer. This suggestion is similar to Carruthers's "contractualism" (1992), a philosophical argument for treating animals fairly, based largely on a strong respect for our humanistic feelings toward animals. Carruthers also contends that animals are not rational agents and thus should not be assigned rights.

Many zoo biologists and other conservationists believe it is also a human right to live in a world where other people are not destroying the very environment upon which we, and our descendants, will depend for survival. The critical question, the one that really counts (and that is often overlooked because of the more emotionally charged focus on animal rights) is, What ought to be the ethical relationship between humans, other living organisms, and ecological processes (regardless of whether other species have rights)?

The moral attitudes of zoo biologists have been shaped increasingly by an ecologically based conservation ethic (Leopold 1949, Burghardt and Herzog 1989, Kellert 1989, Ehrenfeld 1993). This emphasis is directed at a deep appreciation of the interrelationships of species and at protection of the ecological processes that maintain these relationships—and the resulting biodiversity of organisms. This approach has shifted the zoo biologist's moral focus of attention from individual zoo animals toward the health of entire populations, species, and ecosystems.

Thus, the moral fabric of zoo biologists is a unique blend of naturalistic, humanistic, moralistic, scientistic, and conservationistic attitudes. It is the ecologically based conservation ethic, however, that is increasingly coloring the fabric. This is a direct consequence of the rapid deterioration of the natural world and the strong desire of zoo biologists to be part of the solution to this critical problem.

It is interesting that the moral fabric of zoo biologists is considerably different from that of the typical zoo visitor. Kellert found that both the average American and the typical zoo visitor are guided primarily by a humanistic attitude toward animals (1989). I would hypothesize that most zoo visitors also are influenced in part by both naturalistic and moralistic attitudes. Of special importance here, however, is that Kellert found that only 7 percent of the American public is strongly oriented toward ecological thinking, the attitude that has become the primary driving force shaping zoo biologists' conservation ethic. I suggest that it is this incongruence between the attitudes of zoo biologists and zoo visitors that zoo education programs must address if visitorship is not to decline in the future. While this is a zoo director's financial problem, it is a zoo educator's golden opportunity to shape public attitudes toward an ecologically based conservation ethic.

Philosophical Values

Although the moral fabric of zoo biologists is composed of a diverse blend of attitudes toward animals, these opinions can be amalgamated into two relevant philosophical values: (1) appreciation of biodiversity and (2) respect for individual animals. The dilemma is that at times these two values come into conflict (Hutchins and Wemmer 1987).

"Biodiversity value" refers to the ecologically centered belief that naturally occurring species diversity should be appreciated not only for its aesthetic beauty but for its ecological function as well—and thus, it should be preserved. Unfortunately, this value is not an ideal one, and it gets muddled because in practice there is no clear consensus among biologists regarding exactly what is meant by the term or how biodiversity value can be measured and compared between species. For example, is there more biodiversity value in saving ten species of beetles or one species of rhinoceros? I would guess that most zoo biologists would choose the rhinoceros, despite ten's being greater than one. There have been several attempts to incorporate various weighting factors (e.g., taxonomic uniqueness and ecological function) when assigning biodiversity values, but there still is no quantifiable, cardinal value system for the concept. Biodiversity value remains largely an anthropocentric value (humans simply like rhinos more than beetles), despite attempts to define the term from an ecocentric position (what is best for the species and the ecosystem). Nevertheless, I believe the term is useful in helping us to choose among various actions.

Respect for the individual animal emerges as a value from the zoo biologist's humanistic and moralistic concern for individual animals and their welfare. Special emphasis is placed on the natural individual, that is, characteristics expressed by the animal when it is in its proper ecological niche and living in its natural habitat.

Attitudes toward Biological Conservation

Zoo biologists believe that zoo programs can contribute significantly toward wildlife conservation. More than three thousand vertebrate species are being bred in zoos and other captive breeding facilities (IUCN 1987). When a serious attempt is made, most species breed in captivity, and viable populations can be maintained over the long term (IUCN 1987). This does not mean that zoo biologists think that all species are suitable candidates for captive breeding; careful selection is essential.

Zoo biologists believe that modern zoos are only one component of a broader, holistic conservation strategy that strives ultimately to save wild animals in nature. As part of this comprehensive strategy, zoo biologists offer their collective expertise and resources for preserving endangered and threatened species. Most SSP-like breeding programs are established with the primary goal of being ready to return genetic stock to the wild, when needed. However, zoo biologists also are working hard to maximize other conservation and scientific agendas so that zoos can provide as much help to wildlife as possible (Hutchins et al. 1991). The important wildlife conservation role of zoos and aquariums is recognized today by most conservation scientists (Wilson and Peter 1988, McNeely et al. 1990, IUCN et al. 1991).

Why Procure Wild Specimens?

Zoos obtain animals from nature much less frequently than in earlier times. As a point of reference, about 93 percent of all the mammals and 75 percent of all the birds added to AZA collections in recent years were zoo bred (Conway, this volume). Zoos have become producers of wildlife rather than consumers. In addition, no species has ever become extinct, endangered, or threatened by capture for zoos (Conway 1968). There are three basic scenarios in which zoos might justify acquisition of animals from nature: (1) when founder stock is needed to establish an organized, long-term captive breeding program intended to contribute to conservation; (2) when genetic or demographic immigrants are needed to supplement an already established captive program; and (3) when the animals are needed to serve other compelling conservation goals or when they are obtained for reasons of compassion (as in cases of animal rescue).

Acquisition Scenario 1. The science of acquiring founder stock to establish captive breeding programs, as well as the whole field of small-population biology, has undergone rapid advancement in recent years (Foose 1989). The technically minded reader is urged to explore the literature; however, here it is necessary only to get some feeling for the magnitude of animal captures required for starting a long-term program for breeding endangered species in captivity.

For most long-term breeding programs for endangered vertebrates (e.g., where the hope is to maintain 90 percent of the original genetic diversity without additional imports of wild animals for 100 to 150 years), it is necessary to remove about 25 reproductively capable animals from nature. In most cases, the founders are bred rapidly to produce an expanded captive population, whose exact size depends on many program- and species-specific characteristics but usually varies between 200 and 500 animals, after which the population is carefully controlled to maintain it at the zoo carrying capacity for the species. For both biological and practical reasons the captive population is distributed among a number of zoos. This founding-growth-maintenance-management model has been used for most SSP programs to date in order to establish healthy, self-sustaining, long-term populations of endangered animals in captivity.

Acquisition Scenario 2. At times, captive breeding programs require periodic importation of unrelated wild-caught animals in order to continue to meet their genetic goals. If imports are needed at all, the exact number required per generation will vary, depending upon the specific goals of the program and the particular species' biology; however, the numbers are usually not large—e.g., one to ten immigrants per generation (for review see Lacy 1987, Willis and Wiese 1993). Also, on occasion an SSP-like program may develop a demographic problem (e.g., a very skewed sex ratio) that could threaten the captive population's survival. It is likely that wild immigrants for demographic reasons would be requested in these cases.

Acquisition Scenario 3. Sometimes zoo biologists obtain wild animals solely out of compassion (e.g., animal rescues) or for educational, scientific, or fund-raising programs. Zoo biologists often refer to such animals as conservation ambassadors. These animals usually are not members of an endangered species, but when they are, most zoo biologists believe that the acquisition must, first and foremost, do no harm to the health of the species, and second, the acquisition should contribute in some way to the in situ conservation of the species or its ecosystem.

ETHICAL CONSIDERATIONS WHEN ACQUIRING WILD ANIMALS

Below are six considerations that I believe should be discussed, and weighed collectively, when zoo biologists propose to collect endangered animals for zoo-based breeding programs.

The Conservation Impact

The zoo biologist's central attitude of an ecologically based conservation ethic requires that a comprehensive analysis of the conservation impact of removing the requested animals be made before any capture of wild animals is undertaken. The "no detriment" clause in the Endangered Species Act has a similar intent. This evaluation should be among the first tasks performed when undertaking projects to remove animals from nature. The environmental impact must be evaluated from both in situ and ex situ conservation perspectives. What will be the negative effects on the population, species, and ecosystem of removing animals from the in situ population? What will be the positive effects on the captive breeding program by adding animals to the ex situ population? Different analyses should be conducted for each of the possible removal strategies.

Evaluating the negative effects of removing animals from the wild population is often more difficult in practice than in theory. The most common problem is a lack of information on the exact animal census and spatial structure of the population. Unfortunately, in many endangered-species cases, waiting for an exhaustive data set would take too long (with the risk of extinction in the meantime), and also such surveys can cost considerable sums of money, which is often unavailable. There are several guidelines and methodological tools, however, that can be applied to guide action in the absence of full information.

First, captive breeding programs should be initiated before species numbers become critically low, when every animal becomes demographically significant. The IUCN suggests that ex situ populations be started before wild populations fall below one thousand individuals (IUCN 1987). The purpose of SSP-like captive breeding programs is to reinforce, not replace, wild populations. In recent years, new computer simulation programs can provide insights into the impacts that various management strategies and removals will have on the estimated probabilities of a population's viability and long-term survival (Lacy et al. 1992).

The most practical approach, and perhaps the most prudent ethical strategy when negative impacts cannot be reasonably estimated, is to remove the necessary wild animals for captive breeding programs from doomed populations. In a world that is logging so severely that less than 10 percent of its tropical forests will remain standing in twenty years (Brown et al. 1992), this approach should often be possible. Of course, the ability to collect some endangered animals from an area should never be used as an excuse to destroy the habitat, as the overall biological loss would undoubtedly be negative. Other sources of doomed animals that are worth considering are captive animals living in inhumane conditions (e.g., private collections) and openly traded market animals. Extreme caution, however, must be taken not to support accidentally a trade in captured animals. Of course, doomed animals must be evaluated on a case-by-case basis to determine if other options are possible (e.g., rehabilitation and translocation) that in specific cases might have greater conservation merit than inclusion in captive breeding programs.

Removing doomed animals for organized captive breeding programs might be justified, but how do we explain cases in which we know of doomed animals and habitats, but we choose not to remove them for captivity or for translocation? This ethical dilemma is one that all wildlife conservationists increasingly will be confronted with in the future.

When zoo biologists are measuring the predicted positive effects of establishing or maintaining a captive breeding program—including the value to in situ conservation if animals from that program are to be reintroduced back to nature—it is incumbent upon the collector to justify biologically the number of animals requested. The justification should reflect a number equal to the genetic, demographic, behavioral, scientific, and educational needs of the program. The genetic, demographic, and husbandry goals of the breeding program should be clearly defined in advance of the collection.

After the conservation impacts are evaluated, animal collection should proceed only if the overall impact is positive in terms of biological diversity. This ethical approach is based on the zoo biologist's strong moral attitude toward long-term, biological conservation at all ecological levels of organization. Of course, as noted above, biodiversity analyses often are not clear-cut. Nevertheless, I believe this approach gives us a starting point upon which to debate the ethical appropriateness of specific cases of wild animal acquisition. Note that with this approach, the primary argument is not whether it is ethically right or wrong to remove wild animals from nature (a moralist's perspective), but rather the discussion is focused on the biodiversity value to species and ecosystems of removing specific, individual animals.

As an aside, a common disagreement among field and zoo biologists centers on the relatively high financial costs of many captive breeding programs when compared to field-based conservation projects (for a discussion, see Hutchins and Wemmer 1991). Some conservationists argue that it is unethical to remove animals from nature because of the high cost, when so much needs to be spent in situ. I would respond that unless the money is clearly fluid (i.e., it can be readily transferred to field projects), then the financial cost of the breeding program has no real bearing on whether it is ethical to remove the animals. If the money is transferable, however, then this factor should definitely be included in the conservation impact.

Likelihood of Establishing a Successful Captive Population

Estimates for the likelihood of successfully establishing the proposed breeding program should be determined early in the planning process. Often experience with related species can be useful when making these predictions. The likelihood of success should be high in order for the project to proceed to the collection stage. Another important set of criteria to establish during the early planning stage is the program's measure of success. Are these success goals consistent with the number of animals being requested and the kind of program being established?

For many captive breeding programs, it is difficult to estimate either the eventual likelihood or the time frame associated with reintroducing the species back to nature. As a result, the assessment of the potential positive effects of the breeding program on the in situ population, species, and ecosystem levels is problematic. In such cases, it seems to me that the ethical argument in favor of removing founder animals from nature would be strengthened greatly by planning into these zoo programs multiple educational, scientific, and in situ fundraising options, so as to offset the unpredictable use of the zoo-based genetic resource for reintroduction. This is in contrast to the single reintroduction argument cited to justify collecting founder animals when reintroduction is likely to occur in a relatively short time frame. In general, however, I believe too many biologists are placing too great an emphasis on justifying (or not justifying) zoo-based breeding programs only on the value of a possible reintroduction. This narrow view has caused many zoos to miss great opportunities to contribute more fully to wildlife conservation efforts by expanding, for example, education programs, scientific research, and fund-raising efforts for in situ conservation projects.

Long-Term Commitment and Responsibility

Any proposal by zoo biologists to remove wild animals from nature, especially endangered ones, should be made only in conjunction with a good-faith, long-term commitment to the animals and their breeding program. This consideration is founded not only on the conservationistic perspective but also on the zoo biologist's moral and humanistic attitudes toward animals. Such long-term commitment and assumed responsibility are not only moral but financial as well. The message here is that removing animals from the wild should never be undertaken lightly.

The considerable obligations associated with establishing breeding programs for endangered animals normally will require institutional partnerships. In order to safeguard the moralistic concerns that the zoo biologist has for the captured animals and their descendants, it is necessary to clearly define the legal, financial, and moral responsibilities of each partner before any capture takes place. The AZA board of directors has approved a set of recommendations for forming such partnerships, which provides guidance in matters of finance, legal considerations, and management organization. Also, advances are taking place in management structures to oversee responsibilities and to coordinate captive breeding efforts (Seal et al. 1993). For our purposes here, we will assume that all the obligations and tasks needed to establish an SSP-like, organized breeding program can be met either before the actual capture of the founder animals or within a reasonable time after their acquisition.

Some conservationists have argued that it is better to establish captive breeding programs in the animals' countries of origin, rather than to export the animals to foreign countries. This view appears to suggest that the closer animals are to their native range, the more agreeable biologists should be to the idea of capturing founder animals. This perspective is often opposed by some zoo biologists because of their concern for the long-term care of the animals in less developed countries. The plain fact is that the animals would without a doubt receive better care in a modern zoo in a developed country than in any facility that could be built and serviced in a less developed area. For some species this might not be the case; for example, some delicate leaf-eating primates might fare better near their native habitat because of the availability of local browse for feed. The point here is that from the zoo biologist's perspective, the captive breeding site should be selected based on where the animals are most likely to receive the best long-term care, and hence have the best chance of survival, and where overall conservation impact can be maximized. To a zoo biologist, it is no more ethical if a captured animal is kept 10 miles or 10,000 miles from its capture site.

When assigning long-term responsibilities among captive breeding program participants, the question of ownership of the animals arises. Is it ethically more appropriate for one partner or the other to own the captured animals? Who should own the offspring? There is no clear consensus among today's practicing conservationists about this issue, other than the belief that endangered-species transactions should be noncommercial (IUCN 1987). Several different scenarios have been tried, but the growing trend seems to be to assign ownership of the founders and all future offspring to the country of origin. The animals are then placed on loan to the various participating institutions. But is this an ethically based decision or a politically based action? Putting politics aside, from at least this zoo biologist's conservationistic and moralistic perspective, wild animals do not belong to any one person or institution, they are part of the global commons. As such, I think that the best we can do is to assign them to be owned in trust by the entire managing partnership, hopefully for the collective good.

When discussing responsibility, many conservationists argue that ultimately the people of a political state have the task of protecting their wildlife. Unfortunately, the majority of wildlife is located in less developed regions of the world, and since the industrialized nations have contributed significantly to its demise, it makes sense for the more developed nations to assume the responsibility of assisting those poorer nations to manage their wildlife resources. Thus, captive breeding program partnerships should include the local wildlife management institutions, whenever possible, even if all the captive breeding facilities are located outside the country of origin. This type of management structure promotes holistic species management by bringing together wildlife managers who are responsible for free-ranging animals with those persons caring for captive animals of the same species.

Communication and Documentation

Proper communication and documentation of wild animal acquisition is an ethical issue. Obtaining proper legal permits and local permission to capture and transport wild animals is, of course, a prerequisite for any capture operation. There are numerous local, national, and international laws and treaties that must be addressed (for review, see Animal Welfare Institute 1990). However, I would like to stress that there are many other aspects of communication and documentation that must be considered.

After a group decides to acquire one or more wild-caught animals, it is necessary to reflect on the impact that this action may have on others with a stake in the species (e.g., field biologists working in the area). It seems reasonable to expect the zoo biologist to solicit the comments of all interested parties; however, four practical difficulties immediately arise from this suggestion. First, the zoo biologist might not know of all those who would want to be included in the decision-making process. Second, many field biologists are difficult to reach in a timely manner. Third, the zoo biologist might be unsure of the appropriate time to notify other parties. (Is it when you first get the idea? after a basic feasibility study? after you get the money? after a comprehensive conservation assessment?) And fourth, what if field and zoo biologists evaluate the conservation impact of a proposal differently?

Fortunately, there is a growing system of communication among all conservationists (e.g., Swengel 1993). I think that the point here is that zoo biologists must make a good-faith effort to communicate with the larger conservation community, but this is a two-way process. The most practical recommendation at this time is for the zoo biologist to initiate communication early on with the AZA, the Captive Breeding Specialist Group of the World Conservation Union (IUCN), and the appropriate taxonomic specialist groups of the IUCN's Species Survival Commission (SSC) and the International Council for Bird Preservation (ICBP). "Early on," however, will often mean after initial exploratory contact has been made with officials of the country of origin and other members of the proposed working group. Otherwise, the zoo biologist might be in the awkward position of announcing an acquisition project before discussing the idea with the local authorities or the possible local program partners. I would suggest that it is up to the particular SSC or ICBP chairperson to know who within their ranks should be contacted.

There is no easy solution to the problem of reaching people in remote places, nor is there a formula for deciding what to do if the field and zoo biologists evaluate the conservation impact differently. One approach that is helping in some cases is the Conservation Assessment and Management Plan, an IUCN-sponsored effort involving both field and zoo biologists, which provides a comprehensive means of assessing priorities for intensive wildlife management, sometimes including captive breeding, within the context of the broader conservation needs of threatened taxa (Seal et al. 1993).

Another practical problem with communication is that often opportunities to rescue doomed animals (which could be of vital importance to their species' conservation) arise suddenly. In these cases, speed is of the utmost importance, not just from a conservationistic perspective but, considering the animals' welfare, from a moralistic one as well. In these cases, communication with all interested parties often is not possible. It is difficult, for example, for me to be understanding of lengthy permit procedures in these kinds of situations, when the animals often die while waiting for their traveling documents to be processed. It seems obvious that all parties should try to be as flexible as possible in such rescue situations.

Proper documentation includes recording where the animals were obtained and tracking all animals closely throughout the breeding program's duration. Necessary documentation also includes researching and publishing as much scientific information as possible about the animals in the breeding program. After all, this philosophy is consistent with the zoo biologist's scientist attitude and, more important, with the idea of preserving biodiversity.

Animal Welfare

The zoo biologist's moralistic attitudes (and to a lesser degree humanistic beliefs) concerning animal welfare require that all animal acquisition projects be designed to assure that proper animal welfare standards can be met. This includes all phases of the project: capture, transport, quarantine, integration of the animals into any existing program, and long-term care. For example, it would be unethical to acquire animals from a dealer without first knowing that the capture methods were humane. I believe that in most cases it is necessary for the working group to conduct the animal capture themselves, in order to ensure that professional animal care is being provided. The norm today should be to include veterinarians with relevant experience on capture teams.

Before animals are obtained from the wild, the zoo biologist must be reasonably sure that proper, humane care can be provided to the animals (and their descendants) once they arrive. Facilities should be in order. The biologist should have some basis for predicting that the husbandry and care procedures will be successful. In practice, there will be times when this is not possible, because there may be no similar species already in captivity. This is an argument for not waiting until species are critically endangered to begin a captive breeding program. It is also why zoo biologists sometimes suggest capturing wild animals of nonendangered species to serve as husbandry models.

Risk Management

Obtaining wild animals for captive breeding programs is an intervention-type, conservation management tool. It is in many ways inherently risky. Accidents happen. Often field conditions are difficult and resources are limited. It is easier for others to criticize after the fact than to advise before. What kinds of precautions do our ethics demand? The ethical framework established in this chapter is for use by one professional group, namely, zoo biologists. Thus, it seems that the best guideline that can be offered here is, Do all the things that a reasonable zoo biologist would attempt if faced with the same situation and resources.

A CHECKLIST FOR WILD ANIMAL ACQUISITION

This chapter has presented six ethical concerns for zoo biologists to consider when contemplating wild acquisition of an endangered species: (1) conservation impact, (2) likelihood of captive breeding program success, (3) long-term commitment and responsibility, (4) communication and documentation, (5) animal welfare, and (6) risk management. Although these six topics, taken together, can help us to decide if specific acquisition projects are ethically appropriate, they do not provide a step-by-step method of initiating such projects. The procedural checklist presented here is based upon both the moral attitudes of zoo biologists and the six ethical considerations. Of course, any such list of procedural steps is just a starting point, and no doubt it will have to be modified on a case-by-case basis.

Step 1: Proposal Development

Identify the working group and its resources.

Make initial contact with other interested parties.

Define the goals of the breeding program.

Determine number of animals required.

Estimate likelihood of captive breeding success.

Define measures of success.

Tabulate results of working group's conservation assessment and overall program feasibility study.

Select a removal strategy.

Select a site for the captive breeding program.

Design animal care and welfare strategies.

Assign responsibilities to working-group members.

Assign future ownership of animals.

Establish a project schedule.

Step 2: Proposal Review

Solicit formal comments from interested parties.

Modify proposal, if appropriate.

Step 3: Permissions and Permits

Obtain local, national, and international permissions and legal permits.

Step 4: Capture, Transport, and Quarantine of Animals

Take steps to provide proper animal welfare.

Take reasonable precautions for accidents.

Document the capture operation properly.

Step 5: Program Establishment or Integration of New Animals

Document the captive breeding program properly.

Maximize scientific research on the animals.

Step 6: Long-Term Captive Care

Provide ongoing animal and population care.

Publish updates of program status.

Maintain program dialogue with wildlife officials from the animals' country of origin.

CONCLUSIONS

The underlying theme throughout this essay has been that while zoo biologists' primary guiding philosophy is to preserve biodiversity, zoo biologists also value respect for the individual. This blend is one that directs zoo biologists ethically to strive for ecosystem and species health but at the same time not to neglect any animal's welfare. This causes difficult choices at times and demands skillful balancing of ethical values and actions. Procuring endangered wild animals for captive breeding programs is one such situation that requires zoo biologists to evaluate their actions carefully with respect to their moral and ethical beliefs.

Removing animals from nature, at first evaluation, seems contradictory to the mission of modern zoos and aquariums and to the goals of conservation biology. By carefully examining the effects of the removal, however, with respect to conservation impact, likelihood of captive program success, long-term commitment and responsibility, communication and documentation, animal welfare, and risk management, it is plausible to obtain a summary evaluation that supports removal and that is both consistent with institutional goals (e.g., to save wildlife and to provide public education concerning wildlife conservation) and consistent with the moral values of zoo biologists. Thus, removing endangered animals from nature in appropriate cases is not an ethical paradox for zoo personnel but instead a situation that demands that they take morally directed action in pursuit of the goals of wildlife conservation.

ACKNOWLEDGMENTS

I thank the following persons who helped me to formulate my opinions expressed in this chapter: Ann Baker, Rick Barongi, Richard Block, William Conway, James Doherty, Susan Elbin, Ardith Eudey, Valerius Geist, Eugene Hargrove, Michael Hutchins, Don Lindburg, Anna Marie Lyles, Bryan Norton, George Rabb, Gay Reinartz, Geza Teleki, Chris Wemmer, and Wendy Westrom.

REFERENCES

Animal Welfare Institute. 1990. *Animals and Their Legal Rights: A Survey of American Laws from 1641 to 1990*. Washington, D.C.: Animal Welfare Institute.

Beach, F. A. 1975. Beasts before the bar. In *Ants, Indians, and Little Dinosaurs*, ed. A. Ternes, 48–53. New York: Charles Scribner.

Brown, L. R., C. Flavin, and H. Kane. 1992. *Vital Signs: The Trends That Are Shaping Our Future*. New York: W. W. Norton & Co.

Burghardt, G. M., and H. A. Herzog, Jr. 1989. Animals, Evolution, and Ethics. In *Perceptions of Animals in American Culture*, ed. R. J. Hoage, 129–151. Washington, D.C.: Smithsonian Institution Press.

Carruthers, P. 1992. *The Animals Issue: Moral Theory in Practice*. New York: Cambridge University Press.

Conway, W. G. 1968. The consumption of wildlife by man. *Animal Kingdom* 71(3): 18–23.

———. 1988. Can technology aid species preservation? In *Biodiversity,* ed. E. O. Wilson, 263–268. Washington, D.C.: National Academy Press.

Donnelley, S., and K. Nolan, eds. 1990. Animals, science, and ethics. *Hastings Center Report* 20, No. 3 (May-June): Special supplement, 1–32.

Ehrenfeld, D. 1993. *Beginning Again: People and Nature in the New Millennium.* New York: Oxford University Press.

Foose, T. J. 1989. Species survival plans: The role of captive propagation in conservation strategies. In *Conservation Biology and the Black-Footed Ferret,* ed. U. S. Seal, E. T. Thorne, M. A. Bogan, and S. H. Anderson, 210–222. New Haven, Conn.: Yale University Press.

Foose, T. J., and J. D. Ballou. 1988. Management of small populations. *International Zoo Yearbook* 27:26–41.

Hutchins, M. 1988. On the design of zoo research programs. *International Zoo Yearbook* 27:9–19.

———. 1990. Research. In *Proceedings: American Association of Zoological Parks and Aquariums Annual Conference,* 329–333. Wheeling, W.Va.: AAZPA.

Hutchins, M., and C. Wemmer. 1987. Wildlife conservation and animal rights: Are they compatible? In *Advances in Animal Welfare Science 1986–1987,* ed. M. W. Fox and L. D. Mickley, 111–137. Washington, D.C.: Humane Society of the United States.

———. 1991. Response: In defense of captive breeding. *Endangered Species Update* 8:5–6.

Hutchins, M., and R. J. Wiese. 1991. Beyond genetic and demographic management: The future of the Species Survival Plan and related AAZPA conservation efforts. *Zoo Biology* 10:285–292.

Hutchins, M., R. J. Wiese, K. Willis, and S. Becker, eds. 1991. *AAZPA Annual Report on Conservation and Science, 1990–1991.* Bethesda, Md.: AAZPA.

IUCN. 1987. *The IUCN Policy Statement of Captive Breeding.* Gland, Switzerland: World Conservation Union.

IUCN, UNEP, and WWF. 1991. *Caring for the Earth: A Strategy for Sustainable Living.* Gland, Switzerland: World Conservation Union.

Kellert, S. R. 1989. Perceptions of animals in America. In *Perceptions of Animals in American Culture,* ed. R. J. Hoage, 5–24. Washington, D.C.: Smithsonian Institution Press.

Kleiman, D. G. 1992. Behavior research in zoos: Past, present, and future. *Zoo Biology* 11:301–312.

Koontz, F. W., and P. Thomas. 1989. Applied ethology as a tool for improving animal care in zoos. In *Animal Care and Use in Behavioral Research: Regulations, Issues, and Applications,* ed. J. Driscoll, 69–80. Beltsville, Md.: National Agricultural Library.

Lacy, R. C. 1987. Loss of genetic diversity from managed populations: Interacting effects of drift, mutation, immigration, selection, and population subdivision. *Conservation Biology* 1(2): 143–158.

Lacy, R., T. Foose, J. Ballou, and J. Eldridge. 1992. Small population biology and population and habitat viability assessment. In *Population and Habitat Viability Analysis (PHVA): Briefing Book Core Materials.* Apple Valley, Minn.: World Conservation Union SSC Captive Breeding Specialist Group.

Leopold, A. 1949. *A Sand County Almanac.* New York: Oxford University Press.

Lockwood, R. 1989. Anthropomorphism is not a four-letter word. In *Perceptions of Animals in American Culture,* ed. R. J. Hoage, 41–56. Washington, D.C.: Smithsonian Institution Press.

McNeely, J. A., K. R. Miller, W. V. Reid, R. A. Mittermeier, and T. B. Werner. 1990. *Conserving the World's Biological Diversity.* Gland, Switzerland: IUCN (World Conservation Union).

Novak, M. A., P. O'Neill, S. A. Beckley, and S. J. Suomi. 1994. Naturalistic environments for captive primates. In *Naturalistic Environments in Captivity for Animal Behavior Research,* ed. E. F. Gibbons, E. J. Weyers, E. Waters, E. W. Menzel. Albany: State University of New York Press.

Rawlins, C. G. C. 1985. Zoos and conservation: The Last twenty years. In *Advances in Animal Conservation,* ed. J. P. Hearn and J. K. Hodges, 59–70. London: Zoological Society of London.

Regan, T. 1983. *The Case for Animal Rights.* Los Angeles: University of California Press.

Ryan, J. C. 1992. Conserving biological diversity. In *State of the World: 1992,* ed. R. Brown, 9–26. New York: W. W. Norton & Co.

Seal, U. S., S. A. Ellis, T. J. Foose, and A. P. Byers. 1993. Conservation assessment and management plans (CAMPs) and global captive action plans (GCAPs). *Captive Breeding Specialist Group News* 4(2): 5–10.

Soulé, M. E. 1991. Conservation: Tactics for a constant crisis. *Science* 253:744–749.

Soulé, M. E., and K. A. Kohm, eds. 1989. *Research Priorities for Conservation Biology.* Washington, D.C.: Island Press.

Swengel, F. 1993. *Global Zoo Directory.* Apple Valley, Minn.: World Conservation Union SSC Captive Breeding Specialist Group.

Wheater, R. 1985. Zoos of the future. In *Advances in Animal Conservation,* ed. P. Hearn and J. K. Hodges, 111–122. London: Zoological Society of London.

Willis, K., and R. J. Wiese. 1993. Effect of new founders on retention of gene diversity in captive populations: A formalization of the nucleus population concept. *Zoo Biology* 12:535–548.

Wilson, E. O., and F. M. Peter, eds. 1988. *Biodiversity.* Washington, D.C.: National Academy Press.

TO PROCURE OR NOT TO PROCURE

Ardith Eudey

U nprecedented environmental degradation and species loss make imperative the reassessment of conservation strategies, including an objective
evaluation of the benefits and costs of ex situ conservation, with an
apparent emphasis on the preservation of individual animals, and in situ
conservation, with a focus on species, communities, and ecological processes.
The major function of zoos in protecting biodiversity may prove to be conducting education programs designed to raise the public's ecological awareness. Conservation programs must be strengthened in habitat countries, and the input of
field biologists must figure in all assessments of population status and recommendations for conservation action.

In preparing to respond to the arguments raised by Fred Koontz in his development of a set of ethics governing the acquisition of wild animals by zoos, it is
appropriate to identify the constituency for which I am an advocate—indeed a
moral advocate—just as he found it necessary to describe the "relevant moral
attitudes" of zoo biologists affiliated with the AZA. Three groups immediately
come to mind, one being field biologists or field conservationists, especially
primatologists, of whose ranks I am a member. More frequently than not, primates are the charismatic or flagship species selected to be the focus of campaigns
to save tropical forest habitats. The number of participants in the conference and
the number of contributors to this publication who are primatologists, both
activists and zoo people, further underscore the importance of this mammalian
grouping. Another group that comes to mind is local forest rangers, whose
responsibility it is to protect endangered and threatened species and the habitats
essential for their survival. The more dedicated these people are, the more they

146

themselves may belong to the category of endangered species. In 1973, I began long-term field research in Huai Kha Khaeng Wildlife Sanctuary in west-central Thailand, in conjunction with the Royal Forest Department's Wildlife Conservation Division (see Eudey 1980). In 1990, Khun Seub Naksathien, then chief of the sanctuary, took his own life as a final gesture to dramatize the vulnerability of his rangers and the inadequacy of conservation efforts in Thailand. The third constituency for which I am an advocate is local people, including cultural minorities such as the hill-tribe peoples in Thailand, who may find their traditional behavior and standard of living threatened by both development schemes and conservation action. In 1986, I witnessed the involuntary relocation of Hmong hill folk from Huai Kha Khaeng Wildlife Sanctuary, which was done in the name of conservation (Eudey 1989).

BIODIVERSITY CRISIS

My overall response to the chapter on animal acquisition, and what must be considered my major disappointment with it, derives from its focus on individual animals, specifically the megavertebrates; species, communities, and ecological processes are virtually ignored. As a consequence, the decision to procure animals from the wild is accepted as given. The inherent risks of wild animal acquisition (and captive breeding of rare species), as acknowledged by zoo biologists, make this focus especially surprising. The relative benefits and costs of ex situ and in situ conservation, which can be measured in ethical as well as monetary terms, are not examined, even though it appears that such assessment has been assigned the highest priority. Thereby, the value of other vertebrates, invertebrates, and flora is ignored. Indeed, the merits of in situ conservation efforts may even be trivialized by approaching the problem only from the standpoint of assessing the impact of removing animals from wild populations (and then proceeding to claim that evaluating negative impacts may be difficult). In this context, Koontz proposes the concept of conservation utilitarianism to justify the removal of specific animals from the wild in terms of their use or value in promoting biological diversity. No provision is made to evaluate the overall environmental and societal benefits that may accrue to in situ conservation action.

Most programs modeled after AZA species survival programs appear to be established with the primary goal of being ready to return genetic stock to the wild, when needed. With candor, Koontz continues that it is difficult to estimate either the eventual likelihood or time frame of reintroduction of individuals; that is, the use of genetic resources for reintroduction is unpredictable. He would justify the acquisition of wild-caught animals by expanding scientific study on them. Carried a step further, captive breeding may be seen as an end in itself, however. Although it may be characterized as potentially preserving species (or at least megaspecies) diversity, it does not contribute to the well-being of evolving ecosystems and may result in their degradation.

As a corollary of the above, the potential to strengthen and enrich conservation action in habitats of developing countries through collaborative efforts is overlooked by Koontz, although such collaborations have been successfully undertaken by institutions such as the New York Zoological Society. Perhaps this omission may be attributed to the characterization of the problem as a species extinction crisis when it is actually a global environmental disaster (although it frequently is expressed in terms of what may be a more comprehensible species loss). Contrary to the description of the zoo biologist's morality, it is not a human *right* (or expectation) to live in a world where people are not destroying the very environment upon which they depend for survival. Rather, it is a human *responsibility* not to destroy the environment, and a responsibility that has been abrogated as technology renders the human population more distant from nature. The developed world, increasingly referred to as the North, may be seen as having a moral obligation to assume more financial obligation to the developing world, or South, as part of this responsibility. At the very least, financial support for in situ conservation should *always* precede procurement.

THE ROLE OF ZOOS IN CONSERVATION

Wildlife conservation has been identified as the highest priority of the AZA. The zoo biologist, the agent of preservation, supposedly is driven by an "ecologically based conservation ethic," although a concern with interrelationships and process, in reality, is not expressed in Koontz's essay. This mission is said to have had a rapid evolution, "mostly as a direct response to the equally rapid deterioration of the earth's biota." One may question whether this response is direct or indirect, however. Zoos have been hard-pressed to justify their existence in a changing world, and this justification may be merely expedient or politically correct, although the influence on policy of an increasing number of scientifically trained zoo biologists cannot be discounted.

The zoo community is described as contributing to the "preservation of biodiversity" in four ways: (1) offering public education, (2) safeguarding genetic resources, (3) conducting scientific research, and (4) raising conservation funds for in situ conservation. According to a 1989 study cited by Koontz, only 7 percent of the American public is strongly oriented toward ecological thinking (Kellert 1989). Zoo education programs must address the public's lack of ecological understanding and in so doing should raise public awareness in order to obtain support for conservation in the wild. This may well prove to be the most significant way in which zoos can contribute to protecting biological diversity. Support for in situ conservation should not be contingent upon the acquisition of animals from the wild, an ethical argument put forward by Koontz (this volume). Koontz appears to imply that financial support may be available within zoos for captive breeding but that it is not readily transferable to field projects. It also might be appropriate to educate the public (and zoo administrators) that money for conservation action should be more fluid between the former and the latter.

The contention that no species has ever become "extinct, endangered, or threatened by capture for zoos," based upon a 1968 communication by Conway, can be challenged. The Convention on International Trade in Endangered Species of Wild Fauna and Flora (CITES) was established to govern zoo acquisitions as well as all other traffic in wild animals and plants. CITES infraction reports continue to implicate zoos in illegal acquisitions. The rarer a species, the more zoos throughout the world may be willing to resort to any means necessary (or to pay any amount necessary) to obtain individual animals. Within the past several years, one zoo has offered $65,000 and another has offered as much as $125,000 for a wild-caught gorilla (IPPL 1987, 1989), a CITES Appendix I species listed as endangered by the U.S. government. Proper communication and documentation in the acquisition of animals is an ethical issue. Animal dealers have been known to fabricate the existence of zoos and produce bogus documentation so that wild-caught animals can be traded as captive-bred (see, e.g., IPPL 1991). The acquisition of animals from dealers may be highly unethical in that the method of obtaining some species almost invariably is inhumane. Both great apes and lesser apes, for example, typically are captured as infants by killing their mothers. Zoo biologists may well advise zoo curators always to check their sources carefully.

I would concur with zoo biologists that animals already held in captivity, such as openly traded market animals, are an example of a doomed population and may constitute a source for captive breeding, to augment in situ conservation efforts. A vigorous program of confiscation, without compensation and including the imposition of legal penalties, may contribute to stopping internal market trade.

HABITAT COUNTRIES

Zoo biologists may demonstrate a lack of sensitivity to the conservation needs of developing countries. In other contexts, what has been interpreted as the exploitation of the natural heritage or natural resources of developing states by industrialized nations has been the focus of criticisms raised by the South against the North, reaching a crescendo at the June 1992 United Nations Conference on Environmental Development, or Rio de Janeiro Earth Summit. By far the most questionable statement made by Koontz is the following: "The *plain fact* is that the animals would *without a doubt* receive better care in a modern zoo in a developed country than in *any* facility that could be built and serviced in a less developed area" (emphasis added). This statement stands in contradiction to the claim that a captive breeding project should be established where overall conservation impact can be maximized, which invariably would be in a habitat or source country. The development of, and support for, conservation action in habitat countries should be a mandatory component of any captive breeding program. Likewise, the presence of a proper facility in the country of origin would eliminate administrative delays in rescue operations, such as those attendant upon obtaining traveling documents, a situation deplored by zoo biologists.

The question of ownership of captured animals (founders) and any captive-born progeny is perhaps even more sensitive. Koontz questions whether the assignment of ownership of such animals to the country of origin is based upon ethical or political considerations. Claiming that wild animals are part of the "global commons," he advocates that they be "owned in trust by the entire managing partnership, hopefully for the collective good." The composition of the partnership and the obligations of the partners remain obscure. It is just this kind of argument that the South has attacked, contending that the North wants access to resources of the developing world without compensation for them. Acquisitiveness is creating a situation in which wild animals may be treated as commodities in developing states, to be exchanged for dollars for conservation or other purposes.

THE ROLE OF FIELD BIOLOGISTS

Zoo biologists and field biologists are not necessarily adversaries with respect to the capture of animals in the wild. Ultimately, the field biologist may feel morally obligated to protect a population or species and the habitat necessary for its survival, but we are led to believe that this also is the position of the zoo biologist.

In all probability the field biologist will be the first to know, either directly or indirectly, if a population is threatened by a crisis or doomed. Likewise, the field biologist is in the best position to prognosticate the survival of a population or species in the wild, and almost no one, including field biologists, is beyond communication at this time. The fact that there are no more remote places is symptomatic of environmental exploitation and degradation worldwide. I find it difficult to understand how the zoo biologist could propose the acquisition of any wild animals without first communicating with organizations such as those indicated by Koontz, including the AZA and the appropriate specialist groups of the IUCN Species Survival Commission (SSC), which comprises both taxonomic groups and special-interest groups such as the Captive Breeding Specialist Group and the Trade Monitoring Group. The action plans being compiled by the taxonomic groups provide a useful reference to the conservation status of species and the conservation action considered necessary for their survival. The SSC specialist groups also should be able to comment on the ethics and standards of local and international animal dealers. The acquisition of wild animals should never be discussed with government officials without first obtaining scientific input on the status of the target population.

CONCLUSIONS

The intention of this critique has been to focus on those aspects of zoo acquisition ethics as delineated by Koontz that directly impinge on field biologists and conservation action in habitat countries. The accelerating rate of species loss and accompanying reduction in biodiversity make it morally imperative that any further discussion of wild animal acquisition include an objective assessment of the benefits and costs of ex situ and in situ conservation.

REFERENCES

Conway, W. 1968. The consumption of wildlife by man. *Animal Kingdom* 71(3): 18–23.
Eudey, A. A. 1980. Pleistocene glacial phenomena and the evolution of Asian macaques. In *The Macaques: Studies in Ecology, Behavior, and Evolution,* ed. D. G. Lindburg, 52–83. New York: Van Nostrand Reinhold.
————. 1989. 14 April 1986: Eviction orders to the Hmong of Huai Yew Yee village, Huai Kha Khaeng Wildlife Sanctuary, Thailand. In *Hill Tribes Today,* ed. J. McKinnon and B. Vienne, 249–258. Bangkok: White Lotus & Orstom.

IPPL. 1987. Gorillas exported from Cameroun. *International Primate Protection League Newsletter* 14(1): 3.

————. 1989. Two gorillas arrive at Mexican zoo. *International Primate Protection League Newsletter* 16(2): 3–4.

————. 1991. Thai dealer's dirty scheme. *International Primate Protection League Newsletter* 18(1): 9–10.

Kellert, S. R. 1989. Perceptions of animals in America. In *Perceptions of Animals in American Culture,* ed. R. J. Hoage, 5–24. Washington, D.C.: Smithsonian Institution Press.

REINTRODUCTION, CONSERVATION, AND ANIMAL WELFARE

M any captive breeding programs have made reintroduction of specimens back into natural habitats an important goal. Reintroduction efforts have been successful in reestablishing species that were once lost in nature. Though the science of reintroduction is still in its infancy, great strides are being made, and as more species become threatened with extinction, reintroduction techniques may assume greater importance. However, reestablishing or augmenting wild populations from captive ones raises a number of complex moral issues, because captive-bred animals sometimes lack the behavioral skills necessary to survive in the wild. In this exchange, Benjamin Beck frankly assesses the successes and failures of reintroduction programs and stresses the importance of a scientific approach. Robert Loftin reviews the arguments for and against reintroduction and questions whether most current programs fulfill basic requirements for justification. David Hancocks cautions zoos against promising too much in the area of reintroduction and recommends that all reintroductions must take into account broad social issues as well as zoological problems.

REINTRODUCTION, ZOOS,

CONSERVATION, AND ANIMAL WELFARE

Benjamin Beck

Reintroduction of captive-born animals is being used increasingly as a conservation strategy. Zoos have provided animals to about half of the 128 known programs, with reintroduction being one rationale for zoo endangered-species breeding programs. Only about 12 percent of the programs can be documented as successful. Reintroduced animals, like wild animals, are subject to levels of risk, pain, and stress that are much higher than those experienced in good zoos, sometimes making it appear that reintroduction of zoo-born animals is not humane. Further, it seems that to improve reintroduction success by preparing zoo-born animals to cope in the wild, they must be subjected to risk and stress in the zoo, thereby decreasing their individual welfare before, as well as after, reintroduction. The use of experienced conspecific guides, intense postrelease management, and meticulous scientific documentation will optimize reintroduction success and individual welfare for many species.

IS REINTRODUCTION A COMMON STRATEGY?

From a database on reintroductions that my colleagues and I are compiling for the Reintroduction Specialist Group of the World Conservation Union's Species Survival Commission (Beck et al. 1992), we can currently document 145 projects that have released captive-bred animals in this century to reestablish or reinforce a wild population to promote conservation. Translocations of wild animals and reintroductions for primarily recreational or commercial purposes are not included. This figure strikes most as quite low given the tremendous variety and intensity of recent worldwide conservation activity, but many of the reintroductions have been recent and we know of more in planning.

More than 13 million captive-bred animals have been reintroduced. Most were fish, toads, and salamanders, but more than 70,000 captive-bred mammals, birds, and reptiles of 100 other species have been reintroduced. This figure subjectively strikes most as quite high, although again there is no clear way to calculate expected frequencies.

TO WHAT EXTENT ARE ZOOS INVOLVED?

Zoo-born animals were involved in 76 (59 percent) of the 129 reintroductions in which we could determine the source of the animals. More than 20,000 animals, about 10,000 of which were mammals and birds, were released in these 76 zoo-related projects. The term "zoo" in this calculation includes aquariums, meaning that on average only one zoo in about five in the developed world has been involved in reintroduction. The number of reintroduced mammals and birds is equivalent to the collections of only four or five major North American zoos.

Conversely, zoos participate in reintroduction in ways other than providing zoo-born animals for release. For example, the Gladys Porter Zoo provides husbandry expertise in the Kemp's ridley sea turtle reintroduction but does not provide zoo-born turtles. The Frankfurt Zoological Society is a major financial supporter of the golden lion tamarin reintroduction but has provided only two zoo-born tamarins.

Nevertheless, it does not appear that zoos are the primary proponents, animal providers, funders, or managers of reintroduction programs. State and federal wildlife agencies are involved in a vast majority of reintroductions, including many of those that release zoo-born animals, and appear to be the major driving force. This perhaps explains why the bulk of reintroductions have been sited in temperate North America, Europe, and Australia and New Zealand, where public wildlife agencies are relatively well funded and staffed. But reintroduction is a justification for zoo breeding programs, and reintroduction is one, though not the only, contribution that zoos make to conservation (Mallinson 1991).

IS REINTRODUCTION SUCCESSFUL?

We can find evidence that only 16 (11 percent) of the 145 reintroduction projects contributed to the establishment of a self-sustaining wild population. We made the judgment of success based on a stable or growing population of at least 500 individuals and independence from provisioning or other direct human support, or on smaller but growing population sizes accompanied by a formal genetic and demographic analysis of the sort advocated by Foose that predicts that the population is self-sustaining (1991). We used published reports, where available, and information provided directly by project managers. These 16 projects reintroduced captive-born wood bison, plains bison, Arabian oryx (Oman), Alpine ibex, bald eagle, Harris' hawk, peregrine falcon, Aleutian goose, bean goose, lesser white-fronted goose, wood duck, masked bobwhite quail, Galapagos iguana, pine snake, and Galapagos tortoise.

Obviously the other 89 percent are not all failures; indeed, in some projects there is encouraging progress toward a self-sustaining population. There are indirect benefits from reintroductions where a self-sustaining wild population may never be established, namely, increased public awareness and support for conservation, professional training, enhanced habitat protection, and increased scientific knowledge (Kleiman et al. 1991, Lindburg 1992). But although there are several different ways to measure success, we must acknowledge frankly at this point that there is not overwhelming evidence that reintroduction is successful.

IS REINTRODUCTION HUMANE?

The Universities Federation for Animal Welfare recently published a set of welfare guidelines for the reintroduction of captive-bred mammals (UFAW 1992). The introduction states, "It is clear that the well-being of an individual mammal may be compromised by releasing it into the wild, because it will be exposed to risks which are absent in captivity. The procedure of reintroduction, however, must be viewed in the broad context of the overall value of the operation. If there are good reasons for believing that a viable wild population can be established from the reintroduced animals then *the risk to an individual may be compensated for by the gain for conservation*" (emphasis in original). The guidelines of course strongly advocate all possible reduction of individual risk, and detail a number of steps to do so. But this statement reinforces the growing understanding of conflict between reintroduction success and the welfare of individual animals (e.g., Beck 1991). The conflict is certainly evident after reintroduction when animals are subjected to the rigors of the natural environment, but the conflict will also be evident before reintroduction as we strive to breed and keep animals under conditions that will enhance reintroduction success.

Those studying and working to conserve natural biological processes, ecosystems, species, and populations have made us aware of the unforgiving environmental harshness and relentless competition of the natural world. Wild animals suffer extremes of temperature and precipitation, fluctuations in abundance of food and water, attacks by predators and parasites, untreated illness and injury, and intra- and interspecific resource competition. Social primates are often ejected from groups and left to wander alone in unfamiliar areas. All such factors presumably cause pain or stress. Of course species have evolved mechanisms to cope, and coping with such stressors is woven into each species' biological heritage and nature. But this doubtlessly provides little comfort at the individual level, where well-being pertains.

We have recorded many such cases of risk during the reintroduction of golden lion tamarins (Kleiman et al. 1986, 1990, 1991; Beck 1991; Beck et al. 1991). We have seen a newly reintroduced golden lion tamarin sit immobile and shivering high in a tree in a cold (10°C) rain for twelve hours. We found her before dawn the next day on the ground, soaked, with a body temperature of 30°C. We were able to rescue and rerelease her, and she went on to survive for four years and reproduce. Not so fortunate was a young male, nearly self-sufficient after sixteen months in the wild, who was found dead, wedged head-down in a tree cavity; he was presumably feeding on insects and could not turn around or back out of the hole. A female was fatally bitten by a coral snake that she successfully ate. A young pair with an infant tried to enter their nest box, which had been taken over by Africanized bees; the female was found dead on the ground, and the infant disappeared; the male was found hideously swollen on the ground but we were able to save him despite his having been stung at least twenty-seven times. A pregnant female was attacked by an ocelot but escaped as our observer distracted the cat (our protocol calls for such intervention when pregnant females are attacked). The tamarin nonetheless aborted during the night and died two days later despite treatment. Most disturbing to contemplate is the stress on eight reintroduced tamarins that have been stolen from the forest and smuggled into the in-country pet trade. Poaching and habitat destruction have for decades decimated the wild population of golden lion tamarins; reintroduction tragically gives new life to this commerce. Cold rain, Africanized bees, coral snakes, and poachers, all threats to well-being endured by wild and reintroduced tamarins, are illustrative of the postreintroduction risks addressed in the UFAW guidelines and are risks that never would have accrued had the tamarins remained in captivity. Watching reintroduced tamarins nap at midday in a patch of sunshine in a tree and watching their wild-born young play, feed, and locomote effortlessly in the forest provide stressless moments for both monkey and observer, but it is clear that reintroduction is not entirely humane.

Another welfare dilemma concerns how we can breed and maintain animals in captivity *before* release so as to improve their probability of successful reintroduction. At the heart of the issue are naturalistic exhibits. As Robinson has noted (1988), these are mostly stage sets. We create natural-looking trees, vines, lichens, rocks, stream banks, waterfalls, pools, temperature, humidity, and background sound. Most of these features are unappreciated by the animals, but they do seem to delight and comfort zoo visitors and, we hope, make them more receptive to educational messages (Beck and Castro 1994). Parenthetically, they also distort our educational message, leaving visitors thinking that nature is a place devoid of stress, pain, and death. Most naturalistic exhibits do approximate the normal intragroup social environment. But what is unnatural about naturalistic exhibits is the absence of predators, food and water shortages, parasites, untreated illnesses and injuries, environmental extremes, and territorially aggressive conspecific neighbors, thereby compromising reintroduction success.

How can zoos prepare animals to cope with the predation, starvation, parasitism, climatic extremes, locomotor and orientation challenges, and social competition that they will encounter after reintroduction if we don't expose them beforehand? And how can we expose them beforehand without decreasing the welfare of individual animals in zoos? Many argue that we should create an environment for animals specifically slated for reintroduction that will allow them to learn essential information and skills, develop physical strength, develop stamina and agility, develop specific resistances, and make short- and long-term physiological coping adaptations. Much information and many skills, physical conformation, resistance, and physiological coping mechanisms have a prominent genetic component; genetic subsets underlying critical survival capacities might drift if selection is relaxed over generations of breeding and maintenance in zoos (Hart 1990, Vrijenhoek and Leberg 1991). Thus we must make life in zoos truly more naturalistic, not only for animals slated for reintroduction but also for the entire breeding stock. To do so will mean a dramatic decrease in the welfare of zoo animals. Not to do so will mean ongoing handicapping of reintroductees, and a depressed reintroduction success rate (Beck 1991). How do we resolve this dilemma and explain it to our many zoo constituents, including our own staffs, who for years have been able to optimize simultaneously the welfare of the individuals and the health of the populations that we manage?

ENHANCING REINTRODUCTION SUCCESS AND ANIMAL WELFARE

It is clear that reintroduction of captive-bred animals is being used as a conservation strategy, and that zoos are involved. It is also clear that the success of reintroduction must be improved, that reintroduction conflicts with animal welfare, and that improving the success of reintroduction will detract from the welfare of many animals in zoos. Our case for using truly naturalistic welfare-reducing zoo habitats will be stronger if we can show that we are developing and, where appropriate, using other techniques that enhance both reintroduction success and individual animal welfare.

Griffith, Scott, Carpenter, and Reed, using an analysis of projects involving the reintroduction of captive-bred animals and translocation of wild-born animals, and using project managers' judgments of success, estimated that 38 percent of projects reintroducing captive-bred animals were successful, and that 75 percent of projects translocating wild animals were successful (Griffith et al. 1989). Although the estimate of success of reintroduction of captive-bred animals is significantly higher in the Griffith et al. study than in the present study, let us focus here on the difference between reintroduction of captive-borns and translocations of wild-borns. Griffith et al. found the latter to be successful roughly twice as frequently. In the golden lion tamarin program we had reintroduced 86 captive-born tamarins and 7 wild-borns by 1990. The wild-borns were presumably taken for the pet trade, confiscated by Brazilian authorities, and turned over to us for release. Four of the 7 (57 percent) reintroduced wild-borns survived, against 22 of the 86 (26 percent) captive-borns, after comparable periods of comparable postrelease support and management. The reintroduced wild-borns produced more first-generation offspring per individual. We have also rescued and directly translocated some wild tamarins whose forest was being cut (Pinder 1986). Comparable data on the translocated wild-borns are not available since they are no longer closely monitored, but three large groups were seen in 1991 in the area into which two groups, totaling 13 tamarins, were translocated in 1986.

The Griffith et al. survey results and our results from the tamarin reintroduction point jointly to a strategy to improve reintroduction success: where there is a surviving wild population, it would seem more prudent to translocate excess or doomed individuals to suitable, unoccupied habitat. Translocated or confiscated wild-borns can also be socially integrated with captive-borns, and the mixed group reintroduced. The presence of a wild-born guide or tutor seems to compensate for some behavioral deficiencies in captive-bred reintroductees, presumably through some form of observational learning. In some cases, such as the reintroduction of masked bobwhite quail *(Colinus virginianus ridgwayi),* where wild conspecifics have not been available, closely related surrogates (Texas quail, *C. v. texanus*) have been used (Carpenter et al. 1991). Further, now that we have self-sufficient groups of reintroduced tamarins and their offspring, we can mix veterans with newly reintroduced greenhorn zoo-born tamarins without taking any members out of the truly wild population.

A long period of postrelease support may also be helpful. Support of reintroduced tamarins includes daily food supplementation for between six and eighteen months; veterinary care for injury, illness, parasitism, and exposure; artificial shelters; rescue and return of lost animals; and initial isolation from conspecific competitors; all while the tamarins adjust to life in the wild and begin to reproduce. Their wild-born offspring appear to be on a greatly accelerated trajectory to independence, and thus need far less such care (Beck et al. 1991).

The use of conspecific or surrogate tutors and intense postrelease support are strategies to compensate partially for deficiencies caused by captive breeding and management without compromising the welfare of animals in zoos. They improve reintroduction success and the welfare of individual reintroductees. The strategies create a vital demand for zoo personnel, who already have the skills required for these sorts of postrelease management techniques. This, in turn, provides an essential link between the zoo, animal welfare, and in situ conservation and wildlife management communities.

A third technique to enhance welfare and success is to document thoroughly any stress-producing regimens to which we expose reintroduction candidates before release and then to evaluate quantitatively their effect on postrelease survival and reproduction. In this way we can spare subsequent reintroductees the stressful experiences that would intuitively seem to enhance postreintroduction success but don't actually prove to do so. We found, for example, that our initial prerelease training for golden lion tamarins conferred only a transient advantage, and accordingly we revised the prerelease protocol (Kleiman et al. 1986, Beck et al. 1991). The prerelease experience that currently seems to hold the most promise for golden lion tamarins is to allow them to live at liberty in a forested area in the home zoo (Bronikowski et al. 1989); differences in postrelease behavior between zoo groups with and without such free-ranging experience are currently being analyzed.

The drawbacks are that the use of tutors, intensive postrelease management, and thorough scientific documentation all extend the duration of the reintroduction program and increase personnel costs. But concerns about the success and humaneness of reintroduction will inevitably reflect on zoos, and concerns about the value and humaneness of zoos will affect the use of reintroduction as a conservation strategy. It seems that all possible measures must be taken to optimize reintroduction success and the welfare of reintroduced animals.

ACKNOWLEDGMENTS

Funding specifically for golden lion tamarin reintroduction came from the Wildlife Preservation Trust International, Friends of the National Zoo, Frankfurt Zoological Society—Help for Threatened Wildlife, and the Woodland Park and Apenheul zoos. Other funding for the Golden Lion Tamarin Conservation Program came from the Smithsonian Institution (International Environmental Sciences Program, Educational Outreach Program, Office of Fellowships and Grants), National Zoological Park, Friends of the National Zoo, World Wildlife Fund, National Science Foundation, National Geographic Society, U.S. Fish and Wildlife Service, and Canadian Embassy, Brazil.

The Instituto Brasileiro do Meio Ambiente e dos Recursos Naturais Renovaveis (IBAMA, formerly IBDF) authorized and supported our work in Brazil. The staff of the Centro de Primatologia do Rio de Janeiro (FEEMA) provided medical care, necropsy service, and invaluable advice; we are especially grateful to Adelmar Coimbra-Filho and Alcides Pissinatti. The Fundacao para a Conservacao de Natureza (FBCN) has helped in many ways. Numerous zoo staff members, volunteers, and our Brazilian coordinators and observers have made the work possible. Many of the thoughts expressed here were stimulated by discussions with Devra Kleiman, James Dietz, Lou Ann Dietz, Jon Ballou, Andy Baker, Elizabeth Franke Stevens, and Beate Rettberg-Beck.

REFERENCES

Beck, B. B. 1991. Managing zoo environments for reintroduction. In *Proceedings: American Association of Zoological Parks and Aquariums Annual Conference,* 436–440. Wheeling, W.Va.: AAZPA.

Beck, B. B., and I. Castro. 1994. Environments for endangered primates. In *Naturalistic Environments in Captivity for Animal Behavior Research,* ed. E. F. Gibbons, E. J. Weyers, E. Waters, E. W. Menzel, 259–270. Albany: State University of New York Press.

Beck, B. B., D. G. Kleiman, J. M. Dietz, I. Castro, C. Carvalho, A. Martins, and B. Rettberg-Beck. 1991. Losses and reproduction in reintroduced golden lion tamarins *Leontopithecus rosalia. Dodo: Journal of the Jersey Wildlife Preservation Trust* 27:50–61.

Beck, B. B., L. G. Rapaport, and M. Stanley Price. 1992. Reintroduction of captive-born animals. Presented at the Sixth World Conference on Breeding Endangered Animals in Captivity, Jersey, Channel Islands, U.K., 3–5 May.

Bronikowski, E. J., B. B. Beck, and M. Power. 1989. Innovation, exhibition, and conservation: Free-ranging tamarins at the National Zoological Park. In *Proceedings: American Association of Zoological Parks and Aquariums Annual Conference,* 540–546. Wheeling, W.Va.: AAZPA.

Carpenter, J. W., R. R. Gabel, and J. G. Goodwin, Jr. 1991. Captive breeding and reintroduction of the endangered masked bobwhite. *Zoo Biology* 10:439–449.

Foose, T. J. 1991. Viable population strategies for reintroduction programmes. In *Beyond Captive Breeding: Re-introducing Endangered Mammals to the Wild,* ed. J. H. W. Gipps, 165–172. Oxford: Oxford University Press.

Griffith, B., J. M. Scott, J. W. Carpenter, and C. Reed. 1989. Translocation as a species conservation tool: Status and strategy. *Science* 245:477–480.

Hart, B. L. 1990. Behavioral adaptations to pathogens and parasites: Five strategies. *Neuroscience and Biobehavioral Reviews* 14:273–294.

Kleiman, D. G., B. B. Beck, J. M. Dietz, L. A. Dietz, J. D. Ballou, and A. F. Coimbra-Filho. 1986. Conservation program for the golden lion tamarin: Captive research and management, ecological studies, educational strategies, and reintroduction. In *Primates: The Road to Self-Sustaining Populations,* ed. K. Benirschke, 960–979. New York: Springer-Verlag.

Kleiman, D. G., B. B. Beck, A. J. Baker, J. D. Ballou, L. A. Dietz, and J. M. Dietz. 1990. The conservation program for the golden lion tamarin, *Leontopithecus rosalia. Endangered Species Update* 8:82–85.

Kleiman, D. G., B. B. Beck, J. M. Dietz, and L. A. Dietz. 1991. Costs of reintroduction and criteria for success: Accounting and accountability in the golden lion tamarin conservation program. In *Beyond Captive Breeding: Re-introducing Endangered Mammals to the Wild,* ed. J. H. W. Gipps, 125–142. Oxford: Oxford University Press.

Lindburg, D. G. 1992. Are wildlife reintroductions worth the cost? *Zoo Biology* 11:1–2.

Mallinson, J. J. C. 1991. Partnerships for conservation between zoos, local governments, and non-governmental organizations. In *Beyond Captive Breeding: Re-introducing Endangered Mammals to the Wild,* ed. J. H. W. Gipps, 57–74. Oxford: Oxford University Press.

Pinder, L. 1986. Translocacao como tecnica de conservacao em *Leontopithecus rosalia.* M.A. dissertation, Universidade do Rio de Janeiro.

Robinson, M. 1988. Building the biopark. *Zoogoer* 17:4–10.

UFAW. 1992. *Welfare Guidelines for the Re-introduction of Captive Bred Mammals to the Wild.* Hertfordshire, U.K.: Universities Federation for Animal Welfare.

Vrijenhoek, R. C., and P. L. Leberg. 1991. Let's not throw out the baby with the bathwater: A comment on management for MHC diversity in captive populations. *Conservation Biology* 5:252–254.

CAPTIVE BREEDING OF

ENDANGERED SPECIES

Robert Loftin

I t is an increasingly common practice throughout the world to take rare animals into captivity for the purpose of breeding them to augment the world population. This has been done for many purposes, including pro- ducing rare and economically valuable animals for commercial sale, pro- ducing game animals to be released and hunted, and saving animals that are threatened with extinction in the wild. This chapter will concentrate on the latter practice—captive breeding of species that are threatened with extinction in the wild. The goal of captive breeding in this sense is, or should be, the production of animals that can be reintroduced into the wild and survive there ultimately without human assistance. This practice will be referred to simply as "captive breeding" for the sake of brevity.

Many captive breeding programs have been highly publicized as well as sharply criticized in recent years. I will begin, in the first two sections, by surveying the arguments for and against captive breeding programs. Then several visible breed- ing and reintroduction programs will be examined in light of the arguments for and against captive breeding. Finally, I will consider the extent to which captive breeding and reintroduction programs can be justified on the philosophical foun- dations of Aldo Leopold's land ethic and discuss general rules to guide future reintroduction programs.

ARGUMENTS SUPPORTING CAPTIVE BREEDING

The first and perhaps the strongest argument for captive breeding is the argument that there is simply no alternative for some species. It seems clear that without human intervention, some animals are certainly doomed in the wild either through habitat loss, genetic inbreeding, overharvesting, environmental contamination, epidemic diseases, some combination of these, or perhaps other causes. This is indeed a powerful argument. If there is no alternative to extinction other than captive breeding, then few would oppose the practice, because few are willing to accept the extinction of a species without expending every effort possible to avoid this outcome; but there are always other alternatives. One could attempt to address the causes of endangerment rather than initiate desperate, stopgap captive breeding. Obviously, this becomes an exercise in line drawing: when and how does one decide that there is no alternative? There can be no formula for that. Informed humans must make a judgment call. Despite the frightful pragmatic difficulty of determining exactly when there is no feasible alternative to captive breeding, there is little difficulty conceptually. If it is the case that there is no alternative to a captive breeding program, if that is the only way a species can be preserved, then we ought to at least attempt it. Sincere and informed humans will differ on when that point is reached, but in those cases (if any) in which breeding in captivity is the sole hope of survival, that is a sufficient condition to justify such a program.

As I will argue below, sometimes captive breeders have apparently rushed to the judgment that there are no alternatives, but this is not to say that there are alternatives in every case. Every animal and every situation is different and it is hard to generalize. Cases in which there was probably no alternative to extinction in the wild include the Arabian oryx, the California condor, the red wolf, the black-footed ferret, and probably others. In cases with truly no alternative to extinction in the wild, taking the remnant into captivity for the purpose of augmenting the population through captive breeding is justified. The difficulty is to discern when this is and is not the case.

The second argument supporting captive breeding is that even if reintroduction proves to be unfeasible for whatever reason, it is better to have the species in captivity than not to have it at all. Some authors dispute this claim (Jamieson 1985), but advocates of captive breeding can make the telling point that extinction is irreversible. Once the species is gone there is nothing humans can do save lament its passing. So long as the species lives even in captivity there are at least some alternatives, few though they might be. Conditions could conceivably change, more habitat might become available, public attitudes might shift, or environmental contamination might decrease. Unlikely as these scenarios are for some animals, at the very least keeping the biological species in existence in some form, even in a cage, keeps some future alternatives open to some extent.

A third argument supporting captive breeding is somewhat more philosophical. According to this argument many species are near extinction because of human activities; therefore, humans have a moral duty to restore what we have damaged and degraded. According to this philosophical argument, principles of justice demand that we humans redress the imbalance by rebuilding what we have destroyed. While this is a compelling argument, it is not an argument for captive breeding specifically, unless it is used in conjunction with the first argument. Provided there are alternatives to captive breeding, the justice argument may support the much more general practice of environmental restoration. If captive breeding is the only (or the best available) means to accomplish this end, then it is justified. But it may or may not be the best means to this end. Some other program such as habitat preservation might be more effective. Hence, this argument supports captive breeding only in some cases.

Perhaps the major factor undermining this argument is the fact that the wild does not and cannot exist anymore. The idea of putting things back like they were, of restoration of the wild, is not feasible. The best that we might hope for is to manufacture some facsimile of the wild as it once was. As Jan DeBlieu has shrewdly observed, "The predominant goal of [captive breeding] programs is not to restore animal populations to their original conditions but to reshape them so they can exist in a thickly populated, heavily developed, economically expanding nation" (1991). There are two ways in which animals have been reshaped by captive breeding programs: genetically and behaviorally.

In some cases, captive breeding programs have reshaped animals genetically by deliberately introducing genes from populations that are alien to the region in question. It has been pointed out that the reintroduction of the European bison to Poland was accomplished with animals that were "hybrids of native and non-native subspecies mixed with a little blood from American plains bison . . . and domestic cattle." The same thing is true of the much publicized reintroduction of the peregrine falcon to eastern North America, which has been widely hailed as a good example of a highly successful captive breeding project.

In 1946, when DDT came into wide use, there were about 350 pairs of peregrines of the subspecies *anatum* nesting in eastern North America. The aeries were located almost exclusively in mountainous areas well inland. By 1965 the peregrine was gone from eastern North America—the subspecies *anatum* was extinct. After the extinction of the eastern subspecies, the captive breeding program began under the leadership of Tom Cade and the Peregrine Fund at Cornell. Some falcon chicks were taken from aeries in Alaska, Canada, and the western United States. Some adult falcons were contributed to the program by falconers, who took a strong interest in the program from the outset. Birds from Chile, Scotland, and Spain were used. According to Cade, "We purposely wanted to create a mishmash population. We felt that was the best way to guarantee enough genetic diversity for the birds to survive and reproduce in the eastern environment" (DeBlieu 1991).

Obviously Cade was right because the hybrid falcons bred very well in captivity, so well that in 1974 two chicks were hacked back into the wild from the top of a tall building (note the location) in New Paltz, New York. Plans were also made to release falcons from hacking towers placed in coastal marshes, a habitat where falcons had not bred historically. By 1991 more than sixty pairs were breeding along the Atlantic Coast on skyscrapers in cities and on towers in salt marshes far from the traditional mountain cliffs where the extinct subspecies *anatum* had had its aeries. Everyone seems pleased, but it is worth noting that the falcon flying free in the East today is not an example of restoration of what once was but rather of introduction of another bird, of the same species to be sure, but with a somewhat different genetic makeup and different behavior patterns. What we have are facsimile falcons. So long as most people, especially reporters, cannot tell the difference, it does not seem to matter.

A fourth argument is that captive breeding programs raise consciousness regarding the plight of whole natural systems. Captive breeding programs are highly dramatic and make good press. Proponents of this argument point to the educational value of captive breeding programs and conclude that these programs are justified by the attention they bring to the plight of the natural world (Durrell and Mallinson 1987). Although it is true that humans learn from captive breeding programs, exactly what they learn is more problematical. Captive breeding is an extreme example of single-species management. It is clear that single-species management directs attention away from entire systems and toward single species, usually the more conspicuous species preferred by humans. The publicity given to captive breeding also fosters the high-tech managerial approach to environmental problems. It creates the impression that omnipotent humans can cope with the global decrease in biological diversity through in vitro fertilization. In the midst of a recent controversy in Florida over speed limits for powerboats that were hitting manatees, one observer suggested that the Florida Game and Freshwater Fish Commission simply capture the manatees, place them in a large freshwater lake where they would be safe, and permit the powerboaters unrestricted use of the waterways. Although this suggestion is clearly absurd, the managerial attitude that lies behind it is typical of an unfortunate number of citizens—if some species of animals are making things inconvenient for humans, the problem can be solved by developing new techniques and by manipulation of populations. Do publicity campaigns for captive breeding programs encourage or discourage this unfortunate attitude?

A fifth argument is that captive breeding programs provide additional incentive for habitat preservation. Once captive populations reach a certain level there will be no place to release the animals unless additional habitat is set aside. In a sense, we might end up with a species all dressed up with no place to go. Since we clearly need some place to release captive-bred animals, more habitat will be preserved. The counterargument is that this has not happened to any great extent so far. Supporters of this argument can point to few projects where additional habitat has been obtained specifically to release animals bred in captivity. The approach has been to look around for some place to release animals that is already protected by public ownership or to strike an agreement with private landowners to permit release on their lands. Contrary to this argument, it could also be argued that captive breeding programs are frightfully expensive and these programs divert money from field studies and from habitat acquisition programs.

A sixth argument for captive breeding is that in the future, human attitudes will change; rather than less habitat for wild animals in the future, there may well be more. Captive breeding is a stopgap to buy time until the human social and political climate changes to the point that conditions will improve the prospects for reintroduction. Several discussions I have seen on this subject speak in terms of a fifty-year period. The general argument seems to be, if we can preserve a species for fifty years, conditions will be more favorable for the species at that point than they are now. This argument seems to fly directly in the face of current trends, especially in human population growth. While some habitat is being set aside, the net effect of human activities globally is a sharp decrease in available habitat for wildlife. In short, this argument seems to be whistling in the dark.

A final argument supporting captive breeding is that there simply is no wild anymore, no natural systems that are as they once were. Systems have been altered to the point that animals must change in order to survive. Change and adaptation are natural processes because nature is a dynamic system rather than the static system some seem to imagine it to be. Those animals that are adaptable are usually not the ones that are endangered. Since change is a natural process and the environment has changed to the point of irreversibility, management is fully justified in helping animals to adjust and adapt to changed ecosystems. By training animals to adapt to new conditions and extinguishing their traditional behaviors, humans are merely abetting natural processes and benefiting the animals as well as the systems.

The main objection to this argument is that it is based on human arrogance. A species is what it is only because of its traditions, its culture. Once these traditions are extinguished, the animal is no longer wild even if the morphological aspects of the species are unaffected. In essence, the wild animal is extinct, having been replaced by a quasiwild animal.

ARGUMENTS AGAINST CAPTIVE BREEDING

Having looked at some of the reasons that have been given to support captive breeding programs, let us briefly survey some of the arguments that can be made against the practice.

Captive breeding is an extreme example of single-species management, in which an attractive species is singled out for an enormous amount of attention and effort. This is biologically misleading in that it creates the impression that systems consist of mere collections of entities, each of which can stand alone. An equal amount of effort expended on the preservation of complete ecosystems through habitat preservation would be far more valuable in the long term.

Captive breeding represents an anthropocentric, high-tech approach of manipulative intervention in nature. This attitude of anthropocentric manipulation is what caused the decrease in biological diversity that the planet is experiencing at the present time; thus captive breeding reinforces and perpetuates the attitudes that make such desperate action necessary in the first place. As Michael Fox has put it, "With the high-tech innovations of operant training devices, behavioral monitoring, ova transplantation, and genetic engineering, the contemporary zoo is fast becoming an endorsement of capitalist industrial technology" (Fox 1990). Rather than fostering this managerial approach to nature, we should be following a philosophy of "Nature knows best." It is the attitude of manipulation that got us into the difficulty; we will escape the current situation only when we renounce that attitude.

Captive breeding and subsequent reintroduction are excessively cruel to individual animals that are born and reared in captivity and thus lack the skills needed to survive in the wild. Breeding animals in captivity is in some sense breeding the wild out of the animal. Those traits that make it likely that the animal will thrive in captivity are usually precisely the opposite of the traits needed to make it in the wild. Moreover, cultural traditions and social structures are disrupted in captivity, altering those behaviors that make a wild animal what it is. Mortality among captive-reared animals is high when they are released into the wild because they have lost the information and behavioral patterns necessary to get by out there. Hence, captive breeding is merely producing disposable animals, doomed to spend their lives in a cage somewhere in a breeding facility or to be exposed to a quick, cruel death in the wild.

In some cases, the genetics of the species or subspecies have been altered in captive breeding programs. Animals with different genetic backgrounds are bred into the population, sometimes deliberately, in order to increase the variation in the breeding population. The end result is an animal that may look like the ancestral form but is subtly different biologically. In these specific cases, critics of zoos and other captive breeding programs can argue that the end—of preserving a naturally occurring species—is not achieved by the program.

It has also been argued that by encouraging the man-manipulates-nature attitude, and by suggesting that alternatives to habitat protection exist, captive breeding gives corporate culture an excuse to continue its rapacious ways. Through the concept of mitigation in one form or another, critics argue, society feels that the present course of action, while not ideal, is at least acceptable. After all, we are making heroic efforts to save the animals, so it is permissible to proceed along the present path of plundering the planet.

Captive breeding is an impractical waste of effort, it is sometimes argued. Why produce animals in captivity to be put back out into the same situation that brought them to the brink of extinction in the first place? If conditions were such that the species could survive in the natural situation, it would not be endangered. On the other hand, unless something changes in the habitat, the same situation will merely repeat itself. Someone has defined stupidity as doing the same thing over again while expecting different results.

Captive breeding can actually be counterproductive, if taking animals out of the wild disturbs the remnant wild population and actually hastens the demise of the wild population. When the species is dangerously reduced to a remnant population, it becomes a self-fulfilling prophecy to take a significant portion of the remnant into captivity. This not only reduces the actual number of wild animals but the harassment and disruption involved also alter social structure and hasten the downward spiral toward extinction.

Captive breeding shifts resources away from field studies to lab work. While captive breeding may produce greater numbers of individuals, the knowledge of how the species fits into its environmental niche goes begging. Since knowledge of the causes of extinction and understanding of the factors that enhance the chances that the animal will make it in the wild are more important than actual numerical count, this argument suggests that captive breeding ought to be de-emphasized in favor of field studies of the organisms in their natural environment.

Captive breeding sacrifices the interests of individual animals to the interests of the species. Breeding facilities cannot keep all animals, so animals with common genes are often surplused for the sake of those with rare genes. Those whose bloodlines are well represented become dispensable (Luoma 1987). One recent writer on this subject favors killing the animals once they have replaced themselves in the captive population (Cherfas 1984).

Captive rearing lowers fitness by pampering animals with easily available food and protection from predators, parasites, and disease, so that the process of natural selection is disrupted. Animals that are produced have a lower level of fitness compared to animals reared in the wild, thus the fitness of the species is lowered overall. Hence, the species has a better chance if it is left alone in the first place. Captive breeding may produce a higher number of animals in terms of simple numerical count, but since those individuals are less fit to survive in the wild, the net effect may be harmful on the whole.

EVALUATION OF SOME VISIBLE REINTRODUCTION PROGRAMS

Having surveyed the arguments for and against captive breeding, I believe we can conclude that there are situations in which captive breeding and reintroduction programs are indeed justified. But the numerous arguments and concerns stated also suggest that each particular case must be examined carefully before a judgment can be made. In this section, I will present in detail several highly visible, and purportedly successful, reintroduction programs. We will see that these programs may not always fulfill the conditions implied by the arguments surveyed above.

It is important to point out that in many of the most publicized cases of captive breeding, the criterion of no alternative has not been met. Let us first consider the case of the golden lion tamarin, which is especially interesting since it has been so often put forward as a paradigm of a successful captive breeding program. In 1972, when the Wild Animal Propagation Trust became interested in captive propagation of this tamarin, deforestation and the capture of wild animals for medical research, the pet trade, and exhibition in zoos had reduced the species from its original range in eastern Brazil to a few small areas northeast of Rio de Janeiro. Only about 2 percent of the original rainforest habitat of the species remained intact (Luoma 1991). No one knew exactly how many tamarins were left in the wild. The trust located 70 tamarins in zoos throughout the world, and over the next decade scientists, led by Devra Kleiman, developed much improved methods of propagating tamarins in captivity. But in 1974 a much more significant event took place when the Brazilian government, at the urging of Brazilian primatologist Adelmar Coimbra-Filho, set aside 4,856.4 hectares (12,000 acres) of habitat at the Poco das Antas Reserve.

By 1984 tamarins were propagating so well in captivity that 20 were released into the wild at Poco das Antas. At that point there were about 100 wild tamarins on the reserve. The way for the release had been paved by an education program carried out by James and Lou Ann Dietz. This program educated the local people on the rarity and importance of tamarins. Despite the careful training given to released animals by Benjamin Beck to prepare them to survive in the wild, only 3 out of the 20 were alive (plus a pair of twins born in the wild to one pair of released tamarins) in June 1985. Releases have continued at Poco das Antas and on private ranches nearby up to the present time. Of the 91 tamarins freed between 1984 and 1990, only 32 were still alive in the summer of 1991, but the good news is that 38 young animals born to former captives in the wild were also a part of the total population of about 450 to 500 animals. My point is that about four-fifths of the total population are wild animals that continued to make it on their own in the wild without human assistance.

It therefore seems that the most important things done for the golden lion tamarin have been setting aside the Poco das Antas preserve and educating the local people, rather than breeding tamarins in captivity, although the wild population may have been enhanced genetically by the introduction of additional bloodlines from the captive population. Proponents of captive breeding will argue that the preserve and the educational program would not have been established without the attention focused on the captive breeding program, but my point is that captive breeding was not the only alternative available in the case of the golden lion tamarin. If the money spent on the captive breeding program had been expended on the acquisition of additional habitat and the educational program alone, the tamarin would no doubt be approximately as well off as it is today with the captive breeding program.

Another case that has been widely acclaimed as a victory for captive breeding is the case of the whooping crane. Here again, close scrutiny of the history of the program reveals that the no-alternative argument does not apply to this case. In the thirties this crane had been reduced to two flocks, one that wintered at Aransas on the coast of Texas and one nonmigratory flock of about 13 birds on the coast of Louisiana. In 1940 a severe storm wiped out the Louisiana flock, and by 1945, when Robert Porter Allen began to study the species, only 18 birds were left in the Aransas flock. It was not until 1954 that the breeding grounds of the Texas flock were discovered at Wood Buffalo Park in Canada. A decade later, when the captive breeding program for this species was established at Patuxent Wildlife Research Center, Maryland, the wild flock had increased to 32 birds, thanks primarily to an educational campaign carried out by National Audubon that discouraged shooting along the migratory route of the bird. In 1967 six eggs

CAPTIVE BREEDING OF ENDANGERED SPECIES · 173

were removed from wild nests in Canada and brought to Patuxent. Finally, in 1975 the first egg hatched at Patuxent from an artificially inseminated crane— after eight years only one pair had bred in captivity.

Partly because of the obvious difficulty of breeding whooping cranes in captivity, attention turned to cross-fostering of whooping cranes with sandhill cranes in 1975. Between 1977 and 1978, 75 eggs were taken to Wood Buffalo Park and placed in sandhill crane nests at Gray's Lake. Of those, 45 hatched and 11 chicks lived long enough to migrate with their foster parents to their wintering grounds at Bosque del Apache, New Mexico.

In the next five years breeding techniques at Patuxent improved, and eggs from the captive flock began to be taken to Gray's Lake for cross-fostering, but the cross-fostering experiment was not turning out well. Despite the placement of nearly 300 eggs at Gray's Lake for cross-fostering in sandhill crane nests, in 1991 there were only 13 whooping cranes there, nine males and four females, none of which gave any indications of breeding on their own.

In the meantime, the wild flock at Aransas had grown steadily. In the winter of 1990–1991 there were 136 whooping cranes at Aransas, all reared by their parents in the wild without human intervention. Obviously, the captive breeding program had done little to help the species, and the birds had improved the situation on their own. Clearly, this case is not one that fits the no-alternative scenario.

In other cases, in which the requirement of no alternative is met, reintroduction programs appear futile because the programs are undertaken without a plan for correcting the original problem that threatened the species in the wild. If we just do the same thing over, why do we expect the results to differ? In my opinion, the attempts to reintroduce the red wolf into the wild are almost certain to fail despite the enormous amount of effort expended on the program and even though it is often presented as a paradigm of a successful captive breeding program. The reason that the red wolf became endangered was only partly loss of habitat and persecution by humans. The major factor was hybridization with coyotes. In 1975 the decision was made to let the red wolf go extinct in the wild and capture all the pure wolves remaining in the wild for the captive breeding program. Of the 400 animals trapped from the wild as a result of this decision, only 40 were judged to be pure enough to admit to the captive breeding program. Even some of those were later judged to be hybrids, so in the end the future of the species rested on just 17 animals that were judged to be pure red wolves (DeBlieu 1991).

The program was successful and red wolves were introduced at five sites.

Generally about half of the wolves died within the first year after release, but with plenty of wolves in captivity to replace the casualties, a population can be sustained in the wild. That is, until they encounter coyotes.

You will recall that the reason the red wolves were brought into the captive breeding program in the first place was that they were breeding themselves out of existence by hybridizing with coyotes. The theory is that with so few pure wolves left in the wild, the social structure of the species was disrupted to the extent that the wolves began to mate with the coyotes, which had been invading their range in the east for some decades. It seems obvious that with far fewer animals in the wild today than when hybridization started, the red wolves will have to be watched day and night to guard against hybridization. In short, the program is premised on continual human intervention forever. On coyote-free islands, hybridization will not be a problem, but these islands are small and family groups will quickly become inbred. Therefore, genetic shuffling of new bloodlines on and off the islands will be necessary forever.

Another program that has been widely acclaimed as an example of a successful program is the case of the black-footed ferret. Yet when one looks at the causes of endangerment, it becomes difficult to see why. The immediate cause of the demise of the ferret was epidemic disease, in the ferrets and in the population they preyed upon. Breeding ferrets in captivity is unlikely to establish populations in perpetuity as long as the animals are susceptible to epidemics.

Despite the frightful difficulties plaguing captive breeding programs, some have been successful by any standards. Of the programs I am familiar with the most successful has been the reintroduction of the Arabian oryx. In the thirties Arabian nobility began hunting the animals from motor vehicles. By 1960 there were fewer than 100 in the wild, all at the southern end of the Arabian peninsula. Then a hunting party from Qatar killed 48 animals, about half the wild population. At that point the Faunal Preservation Society of Great Britain became interested in the species. The next year (1962) another 16 animals were killed by hunters, leaving fewer than 30 oryx remaining in the wild. Major Grimwood captured 3, which were sent to Kenya and then to Phoenix, Arizona, where they were joined by 7 more from the London Zoo and from the private collections of noblemen from Kuwait and Saudi Arabia. The first calf was born in captivity in 1963. By 1971 there were 30 animals in captivity in zoos, and the next year the last 6 wild oryx were killed by hunters. In 1974 Sultan Qaboos of Oman became interested in the project and offered to support it financially. In 1980 there were some 80 to 100 animals held by wealthy Arabs, approximately the same number in the captive breeding program in zoos. In 1981, 14 oryx were released at Yalooni in Oman, and the next year 10 more were released after a period of acclimatization in large pens. A key element in the success of this program was involving the local population, a nomadic group known as the Harasis, in the program, hiring them as rangers and cultivating good relations with them. In 1984 a second herd of 11 animals was released. The period 1984–1986 was very dry, so the oryx were fed supplementary rations. In 1987 and 1989 two more herds were released to enhance the genetic diversity of the population. By 1990, 99 animals were ranging in the wild, they were reproducing well in the wild, the survival rate for released animals was very good, and the hunting problem had been controlled. The target of the program is 300 animals in the wild.

The cause of the demise of the Arabian oryx was sport hunting. Human behavior is much easier to control than that of any other animal (if one has the necessary resolve). Therefore, once this cause of mortality was brought under control, the reintroduction became relatively easy. There was plenty of habitat, no pandemic diseases, and enough genetic diversity for the species to make it. Hoofed stock generally breeds well in captivity, so the program worked.

Given this pattern of successes mixed with failures, it is obviously difficult to generalize regarding the necessity and likely success of captive breeding programs. An observer might get the impression that the institution of a captive breeding program depends more on accidental features of a situation—the existence of a dedicated advocate for a particular species well placed in a modern zoo with captive breeding facilities, for example—than on a careful evaluation of the program based on its unavoidability and likelihood of success. An important priority for the future must therefore be to develop and apply natural criteria for determining when a captive breeding program is justified. Hopefully, this book and the various arguments listed above in this chapter will provide guidance for such an understanding.

REINTRODUCTION, TRAINING, AND THE WELFARE OF RELEASED ANIMALS

I represent the land-ethic position in environmental ethics, and in this role I will examine captive breeding programs from that viewpoint. In contrast to the animal-rights position defended by Tom Regan (1983, this volume), the land-ethic position emphasizes the well-being of ecosystems and stems from Aldo Leopold. J. Baird Callicott, the ablest defender of the land ethic, articulates Leopold's ethic as placing the locus of value in entire ecosystems rather than in individual nonhuman animals, or even in species (Callicott 1989). Individuals and entire species have a role to play in an ecosystem and are valued for that functional role, not for themselves. Thus the land ethic can tolerate actions that are detrimental to individual animals but draws the line at actions detrimental to the system as a whole. This philosophy is based in large part on a certain understanding of the science of ecology wherein individuals are dispensable so long as there are enough other individuals to play their role in the system. This is why our detractors, with deliberate abusiveness, refer to the land-ethic position as "environmental fascism."

Thus the land-ethic position in environmental ethics generally attributes more value to a member of an endangered species than to an individual of a common species, even if the endangered organisms are at the same or lower level of cognitive or sensory development. Thus the land ethic can value an endangered plant more than an introduced rabbit that is eating it, even though the rabbit is much more highly sentient and has an incomparably higher level of cognition. The rabbit may well be in the wrong place at the wrong time ecologically and is thus judged to be disrupting the system. If this is the case, then human action to the detriment of the rabbit in favor of the plant is justified.

According to the land ethic, each species has a role to play as an energy manager within the ecological system and therefore has a functional relationship within it, because other species in the system depend on efficient energy management for their continued flourishing. Accordingly, those particular members of a species that is threatened with extirpation take on special value as essential to the maintenance of energy flows in the larger system. The land-ethic position therefore differs sharply from the animal-rights position, because the latter cannot attribute more value to an individual of an endangered species than to a member of a common species, other things being equal. Thus a California condor individual is due exactly the same moral consideration as a European starling, no more and no less (Singer 1975). It is therefore difficult for animal rights advocates to justify any special efforts to save species by affording special treatment to members of remnant populations. This difficulty makes the animal-rights position unattractive to most committed conservationists.

It is also a widely recognized shortcoming of the animal-rights position, from the viewpoint of conservationists, that it cannot attribute intrinsic or inherent value to plants and those animals so low on the chain of cognitive and sensory development that they cannot be considered subjects of a life. Thus plants can have instrumental value at most. On the other hand, the land ethic has no difficulty extending moral consideration to plants, microorganisms, insects, or even abiotic components of an ecological system, not for their own sake but for the sake of the system as a whole—thus the holistic thrust of the land-ethic position.

This philosophical difference makes a profound practical difference when it comes to the question of management of wild animals. According to animal rights advocates, "With regard to wild animals, the general policy recommended by the rights view is let them be" (Regan 1983). In contrast to the hands-off posture intrinsic to the animal-rights position, the land ethic can accept management of natural systems if the goal is to restore the system to a facsimile of its original condition or to maintain the system in a semblance of its natural condition. Thus the land ethic can accept captive breeding if and only if the goal is restoration. It cannot accept captive breeding if the goal is to keep the species on ice, that is, to keep the species in captivity forever.

When I first began to think about captive breeding as a moral issue, I was under the naive impression that the goal of the Species Survival Plans (SSPs) was to reintroduce endangered species into the wild and phase them out in captivity. I soon realized that this is not what is being contemplated. The plan is to reintroduce some individuals back into the wild and perhaps build self-sustaining populations out there, but also to keep a breeding population of the species in captivity forever. Zoos are fond of calling themselves modern versions of Noah's ark, but there is one important difference—the animals got off and left the empty ark behind once it had fulfilled its purpose. Is that what the SSPs have in mind, or are SSPs to be perpetual arks? With the red wolf, to take just one example, the plan calls for 330 animals in captivity forever and 220 in the wild.

Most Species Survival Plans include no provision for reintroduction. In most cases it remains a pious hope. What is being contemplated in most cases is keeping the species alive in captivity for 200 years. Arguing from the land ethic, therefore, I assert that only those plans that include reintroductions are deserving of social support. The goal should be to get the species out of the cages and into the wild where they belong.

The reason that most SSPs do not include reintroduction as more than a pious hope is that there is simply no habitat available. Even if we can breed the species in captivity, we will end up with an animal all dressed up and no place to go. Zoos therefore ought to devote more attention to preserving whole intact eco-systems than they do. When the Cleveland Zoo expends some $24 million on a new rainforest exhibit, one has to wonder whether it would have been wiser, from a conservationist viewpoint, to preserve a significant amount of existing intact rain forest in Brazil or Belize rather than to build one from scratch in Cleveland.

Captive breeding is the extreme example of autecological (single-species) management. Single-species management is, quite simply, biologically unsound. It is also philosophically unsound, it can be argued, in that an individual or a species has little value apart from its context. As David Brower has put it, "A [California condor] is only five percent bone and feathers. Ninety-five per cent of condor is place." When an animal is removed from its context, the ecosystem, it is degraded immediately—most of its value is lost. Therefore, captive breeding ought to be only a temporary measure to perpetuate species until they can be restored to the wild.

For this reason, there is an important conceptual distinction between a zoo and a captive breeding facility. The difference is in purpose. The purpose of a captive breeding facility is, or ought to be, to restore ecosystems by reintroducing species into the ecological niches they once occupied. The purpose of a zoo is to retain animals in captivity for the purpose of display, including entertainment, education, and economic development (chiefly to promote tourism). Although these two activities are often carried on by the same corporate entity, thus blurring the distinction, conceptually the two are quite distinct—a zoo is not a captive breeding facility.

Reintroduction is frightfully difficult and can be very hard on individual animals. Survival rates are often low because pen-reared animals sometimes lack the skills needed to survive in the wild. Nevertheless, we must reject the argument that reintroduction is too cruel to be morally justified in any case. We must reject the outdated philosophy of Heini Hediger, who argued that humans are actually doing wild animals a favor by taking them into captivity where they will not be subjected to the hardships and hazards of the wild (1964). Hediger argued that wild animals are not really free, since they are bound by instincts into behavior patterns that are inflexible. He also pointed out that some animals live longer in zoos than in the wild as evidence that the ones in captivity are the lucky ones. That has been challenged empirically in recent decades, but even if Hediger was right, even if some animals do live longer in zoos, their lives, and ours, are correspondingly impoverished (Jamieson, this volume).

Therefore, we are morally obligated to attempt to condition individuals as thoroughly as we can before release so that their chances of survival in the wild are maximized, even if this entails inflicting suffering on individual animals. Beck has suggested that a part of the conditioning of golden lion tamarins for release should include the sudden introduction of predators into their cages to condition the monkeys to avoid predators in the wild (Beck et al., unpublished manuscript). Any monkeys killed by successful predators would be eaten by the predator in full view of the surviving members of the group. This is harsh, but it is no harsher than what these monkeys have to face daily in the wild. If it can be shown that this kind of conditioning actually works, if it improves predator avoidance in the wild, these harsh measures and others equally harsh are justified.

For how long should the zoos monitor the reintroduced animals? The answer, apparently, is "Forever." I'm not sure humans realize the implications of that. One of the most important aspects of intact natural systems is that they are self-sustaining—they take care of themselves. But once the system is degraded by human action to the point that we decide to take the species in out of the cold and put it on ice, we have made a permanent commitment to that wild species that is irrevocable. We have assumed the obligation to protect it from extinction, and extinction is forever. Hence, survival is forever as well. Not 200 years, not 50,000 years, not 100,000 years, but forever. If you are not awed by that prospect, you should be.

For eight years, I built nest boxes for a threatened subspecies of American kestrel in Florida. American kestrels are common and in no need of human assistance throughout most of their range, but the Florida breeding population is classified as threatened. Many scientists believe nest cavities are the limiting factor and have suggested that nest boxes erected and maintained by humans in suitable habitat would augment the population. And they do. The kestrels move right in and fledge young if the box is in the right place. But after eight years, I asked myself the question, Who is going to be doing this one thousand years from now? This brought me some clarity, because I realized my efforts were futile unless I was in a position to make a permanent commitment on the part of future humans to maintain boxes for kestrels. If I could not make that commitment, I was not doing the kestrels any favors. Once they become dependent on boxes provided by humans for their continued survival, they become clients.

We have done that with one wild North American bird, the purple martin. The practice originated in colonial times because martins will mob raptors, thus a flourishing martin colony on one's farm provided significant protection for one's chickens. Today the purple martin is an industry, and in the western United States, they seldom if ever nest anywhere except in structures provided by humans. I decided that I would prefer not to put the kestrel on the same path to dependency. Better to try to preserve some remnant of kestrel habitat and let the birds make it on their own.

Hence, I believe that animals should be taken out of the wild for captive breeding programs only if there is no alternative to extinction. Although there really is no alternative in at least some cases, the priorities should be on maintaining the species intact in the wild and correcting the causes of endangerment in the first place. Captive breeding ought to be only a last-ditch, desperation effort.

Many will disagree with this conclusion because once a species gets to the point of desperation it becomes very difficult to save. Beck has pointed out in the previous chapter that out of 145 reintroduction projects only 16, about 11 percent, have made any contribution to maintaining a self-sustaining wild population. But this rather underwhelming result does not necessarily lead to the conclusion that reintroduction is so difficult and unlikely to help that it ought not to be attempted. The alternatives to captive breeding and release are extinction or keeping the species on ice (that is, in zoos) forever. Neither alternative is acceptable from the point of view of the land ethic. We are morally obligated to attempt reintroduction and to do our best to make it work whether we succeed or not. In short, we must try.

REFERENCES

Callicott, J. B. 1989. *In Defense of the Land Ethic*. Albany: State University of New York Press.

Cherfas, J. 1984. *Zoo 2000: A Look behind the Bars*. London: British Broadcasting Corp.

DeBlieu, J. 1991. *Meant to Be Wild: The Struggle to Save Endangered Species through Captive Breeding*. Golden, Colo.: Fulcrum.

Durrell, L., and J. Mallinson. 1987. Reintroduction as a political and educational tool for conservation. *Dodo: The Journal of the Jersey Wildlife Preservation Trust* 24:6–19.

Fox, M. W. 1990. *Inhumane Society: The American Way of Exploiting Animals*. New York: St. Martin's Press.

Hediger, H. 1964. *Wild Animals in Captivity*. New York: Dover Publications.

Jamieson, D. 1985. Against zoos. In *In Defense of Animals*, ed. P. Singer, 108–117. New York: Harper & Row.

Luoma, J. R. 1987. *A Crowded Ark*. Boston: Houghton Mifflin.

———. 1991. A wealth of species on the forest floor. *New York Times*, 2 July.

Regan, T. 1983. *The Case for Animal Rights*. Berkeley: University of California Press.

Singer, P. 1975. *Animal Liberation*. New York: Avon Books.

AN INTRODUCTION TO

REINTRODUCTION

David Hancocks

T here is a commonly held misconception that zoos are not only saving wild animals from extinction but also reintroducing them to their wild habitats. The confusion stems from many sources, all of them zoo-based. Zoos spend vast sums of money on promotion and public relations, quite possibly more than on conservation. The language of the promoter is always suspect, often disingenuous. The word "habitat," for example, has replaced "cage." People hear about zoos building new habitats and putting animals from their collections into the new habitats, and draw the wrong conclusions when they hear zoos also openly boast that they are arks destined to save the earth's wildlife. One resulting serious problem is the sense of complacency this engenders among the public, who are led to believe that zoos are taking care of the problem of endangered wildlife (Tudge 1992). In reality, most zoos have had no contact of any kind with any reintroduction program. Furthermore, the zoo world's cavalier attitude to habitat seems to suppose that wild habitats are in abundance, merely waiting for animals bred in the zoo to be released there.

Lee Durrell and Jeremy Mallinson claim that "reintroduction is a primary goal of keeping animals in captivity" (1987). This may be true only at their institution, on the Island of Jersey. But for a great number of other zoological parks, any such claim is based on flimsy foundations. As a practical matter, the conditions under which zoos keep animals are not at all conducive to reintroduction. The spaces in which animals are displayed in zoos are rarely tolerable for sustaining natural behaviors. The regime under which zoos maintain their animals in no way prepares their skills for survival. The conditions in which almost all zoo animals spend the majority of their lives, the holding quarters where they typically spend about sixteen of every twenty-four hours, are usually woefully inadequate for anything other than basic containment. There have been some encouraging developments in zoo exhibit design over the past two decades, although many of the improvements are only superficial and aimed more at creating better images than satisfying the animals' behavioral and psychological needs, but the design of off-exhibit holding areas has made no substantial improvements since the days of the nineteenth-century menageries. They still tend to be barren, cramped, and sterile spaces, full of harsh and reverberating noises, with no opportunity for social or any other natural interactions.

The topic of reintroduction, training, and welfare of released animals is still in its infancy in the zoo profession. It is appropriate that it has been included in this volume. This is a time of rapidly gathering crisis. The problems of wildlife conservation are immensely difficult, and they are new to us. Reintroduction of captive-bred animals is already talked about as if it were inevitable, as if it were a fait accompli. At present I think it can be said to be no more than a potentially useful, maybe vital, tool.

Reintroduction is an extraordinarily difficult procedure. It is far too complex for zoos to carry out on their own, and it is far more than just an exercise in zoology. It involves politics, social welfare, education programs, long-term funding stability, public support, many changes in zoo facilities and programs, and, especially, it involves restoration and protection of wild habitats.

This is a time of immense changes on our planet, with massive damages to natural environments. Concurrently we have seen many changes in attitude in recent years. In the twenty years that I have been living in the United States, I have seen significant shifts in attitudes to animals, to animal welfare, and to preservation of wild animals. There are new concerns about wilderness and about the health of the planet. Unfortunately those concerns do not yet rest heavily in the minds of most of our politicians and business leaders, but even they, I believe, indulge in wishful thinking. They would like to see zoos succeed. They would like to see wild animals saved by captive breeding programs. And they would like to see wild animals reintroduced to their natural habitats.

There is, unfortunately, no guarantee that this will happen. Certainly, if it does, to any degree, it will be the result of hard work, foresight, and dedication by a great number of people, both within and without the zoo profession. They will have to work together and involve a wide range of disciplines and backgrounds and cultures. The self-congratulatory attitude of zoos will have to be replaced with a little more humility. The public relations departments will have to learn to promote conservation, not just the zoo. Most important, we must recognize that successful reintroduction, if it happens, is going to depend as much upon what people are willing to lose as upon what they are hoping to gain.

REFERENCES

Durrell, L., and J. Mallison. 1987. Reintroduction as a political and educational tool for conservation. *Dodo: The Journal of the Jersey Wildlife Preservation Trust* 24:6–19.

Tudge, D. 1992. *Last Animals at the Zoo*. Washington, D.C.: Island Press.

CAPTIVE BREEDING, SURPLUS ANIMALS, AND POPULATION REGULATION

One of the goals of an organized captive breeding program is to maintain a genetically diverse and demographically stable, self-sustaining population of living organisms. In order to accomplish this goal, animals must be allowed to breed. However, one potential result of genetic and demographic management is the production of surplus animals. What is the ethical responsibility of zoos toward such animals? In this exchange, Robert Lacy argues that space is limited in zoos and that population management, including culling, is a potentially critical aspect of zoo breeding programs focused on the preservation of endangered species. Donald and Linda Lindburg suggest that welfare considerations and public relations are important as well and that certain species, such as great apes, may never be considered for culling. Frederic Wagner discusses captive breeding in light of a broad range of ethical issues, including the fate of individual animals that have been designated as surplus.

CULLING SURPLUS ANIMALS FOR

POPULATION MANAGEMENT

Robert Lacy

In a world of finite resources, it is impossible to maintain the lives of some animals without sacrificing the lives of others. Thus, the decision of whether to cull animals when they become surplus to the needs of a program (propagation, education, or research) is not a consideration of whether animals should be killed but rather which animals will be killed. Zoos often do not cull favored animals because of the pain that it causes to those humans that have long cared for the animals. Decisions about which animals to kill are based on considerations of human feelings, although the justifications are often couched in terms of animal welfare or animal rights. The continued maintenance of favored animals, after no further progeny are desired and no other uses for the animals can be identified, uses resources that are needed to prevent the suffering and death of other animals, both currently living and yet to be born. Because of unwise use of resources, evolutionary lineages and ecological communities of organisms, which otherwise might be conserved, will die.

Let me preface my remarks with a brief admission of my philosophical leanings. My judgments about right and wrong are primarily consequentialist or utilitarian. I care more about the effects of an action than the motivations that led to it or some other measure of intrinsic goodness of the action. I view most actions as having both good and bad effects, and I consider most ethical those actions that lead to the most good while causing the least harm. Perhaps because of my training in evolutionary biology, I usually look at the long-term, even multigenerational, impacts of actions. I am distressed when I witness the suffering of an animal, but I find abhorrent the view that benefits to nearby and present individuals should outweigh the quality and even existence of distant or future life.

In the terminology of many animal rights advocates, I am an environmental fascist. I value living, interacting, evolving communities of organisms more than the lives of the individuals that temporarily exist as elements of those communities. My views arise, in part, from the value I place on the system (environmental holism); in part, from a judgment that benefits to the broader ecological communities will necessarily lead to more benefit for more of the individuals within those communities (animal welfare collectivism). In spite of the overriding importance I assign to evolving communities of living organisms, much of what I will argue below will focus on the welfare of individual animals. I intend to show that even considerations of individual animal welfare and rights do not necessarily lead to the conclusions and actions usually ascribed to animal welfare and rights views.

SOME BIOLOGICAL REALITIES AND DEFINITIONS

Animals that are born will die. Except for those that are killed quickly by an efficient predator (perhaps human), most sentient animals will suffer pain when dying. Death from disease, starvation, injury, or failure of organ systems due to age usually occurs slowly and with suffering. Because of competition, predation, and avoidance of predation, many animals must die to sustain the life of those that survive long. Thus, the reality that many more animals are born than can live to reproduce and get old is not just an incidental consequence of the fact that species have evolved high reproductive rates. Deaths would occur even if each pair produced only two progeny, and generally, those species that are the most fecund also suffer the highest natural mortality. For some animals to enjoy more than a marginal existence—having good nutrition, lots of space, and freedom from competitors, predators, and disease—even more other animals must be sacrificed directly (e.g., by being eaten) or indirectly (e.g., by forgoing a meal or by being excluded from adequate habitat). The only way to eliminate death is to eliminate life. The only way to end all suffering of wildlife is to eliminate the wild.

Surplus animals are those that are no longer needed for the goals of a program. The concept implies a utilitarian view, with the goals defined by and perhaps serving humans. Animals that are surplus to one purpose may still be of considerable value for other goals, for example, breeding programs elsewhere, research, or exhibition.

Culling is the termination of the life of an animal before it would have died from unavoidable disease or failures of organ systems (old age or natural causes). Contrary to the definitions supplied to participants of the Atlanta conference, culling is not by definition painless and humane. Culling could be achieved by painful slaughter, although such would unnecessarily sacrifice individual welfare with no benefit to others and, thus, would show unethical disrespect for sentient animals.

CONSEQUENCES OF CULLING SURPLUS ANIMALS

Culling kills animals. Culling of a surplus animal also frees resources that would otherwise be used to maintain that animal, allowing those resources to be used for the benefit of other animals. In zoos, the resources required for maintenance of animals are usually described as "cage space," although this term encompasses funds for feed, keeper time, and facility maintenance, as well as actual physical space. When an animal without further value to the program (Species Survival Plan, other propagation program, research, education) is culled, the cage space made available allows for the production of another animal with greater value to the program, or the acceptance of an already living animal that may be surplus to the needs of another facility. In a world of finite resources for captive breeding, culling one animal directly allows another to live.

For example, because subspecies–hybrid orangutans occupy roughly one-third of spaces for the species in U.S. zoos, breeding of pure-subspecies orangutans has been severely curtailed. It is questionable whether adequate spaces are available for maintaining healthy populations of the two subspecies in zoos, if hybrids continue to occupy spaces for the twenty to thirty years more that most existing hybrids could live, and if other species are not displaced to release more spaces for orangutans. (There are two subspecies of orangutans, one from Sumatra and one from Borneo. Before the considerable genetic differences between the two forms were recognized, zoos frequently interbred the subspecies. Subsequently, the SSP has declared a moratorium on the further production or breeding of the hybrids, in the hope that the propagation program could sustain healthy populations of the pure subspecies.)

The orangutan example also points out another consequence of culling: more populations of wildlife can be maintained secure from extinction if animals not desirable for further breeding are culled. The cost of maintaining hybrid orangutans could be calculated in dollars (Lacy 1991, Lindburg 1991), but the cost can also be expressed in terms of species survival. The presence of surplus orangutans in zoos effectively prevents the protection of one great ape population, or another species with similarly large resource requirements. The survival of a population of animals in captivity or in the wild is directly related to the number of breeding animals (Soulé et al. 1986, Soulé 1987). Except in those few cases in which animals care for or otherwise benefit nonoffspring in a social group, postreproductive animals contribute nothing to future generations. The number of species maintained with finite resources, however, is usually directly related to the total numbers (breeding or not) of animals maintained per species. Thus, a maximization of the proportion breeding in each population will maximize the number of species protected. Because natural habitats have been destroyed, and resources to redress the damage at least partly are limited, many species will go extinct even though we have the knowledge to protect them. Extra resources spent protecting favored, but surplus, animals means that even more species will go extinct.

Placing such a value on species protection is clearly antithetical to many concepts of animal rights. I view, however, the loss of a species and, therefore, the prevention of all future lives of that life form and of any that could have arisen from it to be far worse than the loss of any individual life. Individuals are always mortal; species may continue through evolutionary time. (Although the fossil record shows that species typically persist for a few million years, many extinctions result from evolution into a different form, rather than the actual termination of an evolutionary lineage.) An individual can suffer an unnecessarily premature death; a species can suffer an unnecessary death.

CONSEQUENCES OF NOT CULLING SURPLUS ANIMALS

Not culling kills animals. If surplus animals are not culled, animals may not be bred, because the resources needed to sustain them are used by the surplus animal. One, perhaps undesirable, means of preventing deaths is to prevent the procreation of life. Yet existing animals and species are dying because of limited resources for their sustenance. Thus, even if additional animals are not to be bred in a captive program, the refusal to cull surplus animals expends resources that could be used to prevent deaths of other presently existing animals and extinctions of species. The millions of dollars spent on maintaining surplus animals in zoos could go far to advance animal welfare and species conservation, if we choose to redirect those funds.

Even if monetary resources cannot be redirected (i.e., society chooses not to do so), simply housing and feeding surplus animals causes the deaths of other animals. For a carnivore, the connection is obvious. Even for an herbivore (or a carnivore cruelly forced to live on a vegetarian diet), the provisioning of food to one animal is damaging to others. Probably the most destructive human activity is agriculture. Vast areas of the globe have been denuded of the natural biota in order to support food production for humans and for animals provisioned by humans. The deaths that occur because of this habitat destruction are not always painless. Poisons, traps, and farmers' cats do not always kill humanely. The pest control measures often incidentally kill many individuals of nontarget species. Even the setting aside of land to house comfortably the zoo's surplus (and nonsurplus) animals takes a toll, as predators, competitors, and many incidental resident species are cleared to make way for the managed population. For example, the San Diego Wild Animal Park is a magnificent, spacious habitat for many endangered species, but its maintenance requires the replacement of native vegetation by alien grasses, the exclusion of many native herbivores, and the control of coyotes. The "retirement homes" suggested by the Lindburgs will similarly displace and kill wildlife (Lindburg 1991; Lindburg and Lindburg, this volume).

It is inescapable that decisions to cull animals and decisions not to cull animals always involve a judgment that some animals will die and others will live. Often the deaths occur to animals about which we have no or only vague awareness, and often the deaths are not without suffering (especially in those cases in which the deaths are indirect and hidden).

WHY DO ZOOS (USUALLY) NOT CULL SURPLUS ANIMALS?

I would argue that zoos usually avoid culling surplus animals in order to minimize human discomfort. The decisions are based on the rights of humans to lead happy lives, not on considerations of nonhuman animal rights. When we care for an animal for many years, whether that animal is in a zoo or is a family pet, we learn a lot about its individuality, we empathize with it, and we receive emotional rewards from our interactions with it. Generally, the warmth we feel toward an animal relates in part to how similar its features and expressions are to our own, even if we misread those expressions. For example, bottlenose dolphins, with their upturned, smiling mouths, engender great sympathy. Animals with forward-pointing eyes are often perceived as more responsive to us than those with laterally placed eyes, even though most mammals with forward eyes are carnivores with a need for accurate depth perception (for pouncing on prey) and most mammals with lateral eyes are herbivorous, nonaggressive prey species. Infant animals elicit greater emotional response from humans, perhaps because we are neotenic and infant animals look more like us. Large animals are also easier for us to empathize with; small mammals get little consideration, even if (like bats) they possess remarkable sensory and maybe even cognitive capabilities (Griffin 1984).

It should be clear from the way in which we rank the rights of animals to be spared culling that our judgments are based on the feelings elicited in humans, not the feelings of the animals themselves. We usually do not cull animals in our care because we feel that we would suffer.

Our decisions about which animals can morally be culled are not based on assessments of the mental capabilities of the animals. We accept culling of pigs on farms, but not (usually) culling of peccaries in zoos. We accept harvest of cattle for meat, but would abhor killing of surplus gazelles to feed the families of curators. We shoot, trap, and poison coyotes, but find it unacceptable to euthanize a tiger until it is suffering from incurable disease or the effects of age. Within zoos, many of us cull small rodents, but not large ones. We cull bats, but not cats.

Nor can our decisions about culling be justified by assigning blame for animal deaths (pigs, cattle, coyotes, rodents, bats) to someone else. If we eat meat, wear leather products, cultivate artificial lawns, or own pets, we know that our actions are directly or indirectly causing destruction of life, often of sentient higher mammals. In fact, we are responsible for all of life because we have chosen, by our numbers and our life-styles, to disrupt the global ecosystem and put all forms of life at risk.

I must emphasize and reiterate: Decisions regarding the culling of animals in zoos have been based on maximizing human pleasure. They rarely are based on animal welfare or animal rights and are often counter to those goals.

SHOULD ALL SURPLUS ANIMALS BE CULLED?

No. Our feelings do matter. The emotional bonds we feel to animals near us, with which we feel that we share experiences, are essential motivators behind the care we give to animals. The attachments we form with other animals are probably as important as any economic arguments in instilling in people a sense of responsibility for preventing the destruction of the natural world. Perhaps we do need to protect, for inherently selfish reasons, some of our favorite animals, in order to nurture feelings of compassion, stewardship, and responsibility for life.

I have three cats in my house. Although I do not let them outside to prey upon songbirds, they do utilize resources. They serve no purpose but to increase my happiness. Maybe zoos can keep a favorite tiger or gorilla long past the end of its breeding lifetime for the same reason. I hope, however, that the reasons behind our reluctance to cull our favorite animals are openly admitted to ourselves and to the public. We are restrained by concern for our happiness, not by moral imperative.

CAN ZOOS AVOID PRODUCING SURPLUS?

Not always, unless we are willing to subjugate the futures of many animals and many species to the interests of a select subset of individuals in our zoos today, so that we can define all animals in our zoos to be nonsurplus. Often, an animal identified as surplus in one context (say, a Species Survival Plan) can still fill an important role elsewhere. Respect for that animal as an individual and as a living resource requires that we not needlessly discard it. Some exhibits are not designed and are not appropriate for breeding. Those exhibits should be inhabited by animals that have been declared undesirable for breeding.

As discussed by Lindburg and Lindburg in the next chapter, for many species an appropriate allocation of zoo space has been filled, or overfilled. As soon as a pair produces two offspring (or some other number that might be calculated to account for juvenile mortality or to allow for genetic management), the pair becomes surplus to the breeding program. Ideally, from a genetic and demographic standpoint, we would refrain from breeding a pair of animals until there was just enough time to obtain the desired number of progeny before the breeders become physiologically postreproductive or die (Lacy in press). Such a scheme would limit the surplus animals to those relatively few at any one time that were postreproductive but not yet dead. Unfortunately, because we cannot know when an animal will become postreproductive, and because some animals will not breed unless they are paired when still young, we often must obtain the desired progeny from a pair much earlier in their potential lifetime.

Zoos can, by careful, cooperative planning of breeding programs, keep the production of surplus animals to a minimum, perhaps well below the level currently being produced. Lindburg and Lindburg also describe well the possibility of using the spaces and other resources better, to produce more of the animals needed for conservation and fewer of those having little direct or indirect benefit to conservation and the welfare of the animals themselves.

Is it unethical for zoos to produce abundant births, perhaps out of a lack of effort to contracept breeding or perhaps to entertain the public, and then to cull those not desired? Clearly, this is the approach of much of the pet industry, of some livestock producers, of many research laboratories, and of much of wildlife management. This practice shows minimal regard for the welfare of individual animals (except in that the killing is often done in a painless way) and can waste considerable resources. If zoos have a mission to cultivate respect for wildlife, then respect for individuals must be part of that, even if it does not take precedence over the preservation of species. If zoos also have a mission of furthering conservation, then the unnecessary production of surplus animals is an inexcusable dereliction of our duty to use resources wisely for the protection of the natural world.

I have some difficulty with, but can accept, that often we will not cull surplus animals because we value our sensitivities over the preservation of species. I cannot condone the production of surplus simply to satisfy the short-term and shallow goals of convenience (e.g., not contracepting) or light entertainment of a few people. The costs in terms of individual animals, of species, and of desensitizing people to the value of animals are too great.

Every time a zoo produces an animal, that zoo takes on the responsibility not only for that animal's welfare during its life but also for its death. For me, the purposes for which zoos produce many animals are sufficiently compelling to justify the creation, management, and termination of the lives of animals. For each prospective animal, however, competing interests must be weighed, and they should be weighed prior to the birth of the animal, because that is when we make the irrevocable decision to be responsible for the life and death of that animal.

REFERENCES

Griffin, D. R. 1984. *Animal Thinking.* Cambridge, Mass.: Harvard University Press.

Lacy, R. C. 1991. Zoos and the surplus problem: An alternative solution. *Zoo Biology* 10:293–297.

————. In press. Managing genetic diversity in captive populations of animals. In *Restoration and Recovery of Endangered Plants and Animals,* ed. M. L. Bowles and C. J. Whelan. Cambridge: Cambridge University Press.

Lindburg, D. G. 1991. Zoos and the "surplus" problem. *Zoo Biology* 10:1–2.

Soulé, M. E., ed. 1987. *Viable Populations for Conservation.* Cambridge: Cambridge University Press.

Soulé, M. E., M. Gilpin, W. Conway, and T. J. Foose. 1986. The millennium ark: How long a voyage, how many staterooms, how many passengers? *Zoo Biology* 5:101–113.

SUCCESS BREEDS A QUANDARY

To Cull or Not to Cull

Donald Lindburg and Linda Lindburg

A dvocates of the moral worth of animals commonly argue the case on the grounds of utilitarian or intrinsic value, or from beliefs that possession of sentience places an animal's individual welfare above any form of human utilization. These concerns may ultimately be rooted in intellectual and emotional experiences of nature or even of displaced individuals (as in zoos) or pets, but in each case they foster strong preservationist sentiments. Zoos are philosophically aligned with environmentalists in attaching higher value to aggregates such as taxa or ecosystems than to individual animals. However, in attempting to contribute to conservation of highly endangered species through captive breeding, zoos face a moral dilemma in advocating euthanasia for unneeded individuals. The rationale for managerial culling is largely based on economic considerations—i.e., the sacrificing of individuals to save species—but this advocacy constitutes a trade-off in values that places zoos' credibility as conservators of wildlife at risk. Expansion of programs that would control nonessential breeding and reduction in holdings of species not in critical need of propagation or essential to educational programs are examples of underutilized options in conserving resources that could be directed at alleviating this problem.

The vulnerability of life forms to extirpation, especially those least adaptable to environmental change (Myers 1979, 1989; Ehrlich and Ehrlich 1981), is frequently invoked as a compelling rationale for the establishment of captive breeding programs. In a more perfect world, it is argued, conservation resources could be devoted wholly to preserving wild habitat. Faced, however, with the imminent prospect of extinctions on an unparalleled scale, there is concern that for some taxa it may already be too late for single-option approaches. Captive propagation is advocated as a way of preserving options in light of future uncertainties (Hutchins and Wemmer 1991).

Although such projects are much debated and as yet have demonstrated limited success, we believe carefully structured captive breeding programs are potentially important adjuncts to in situ conservation efforts (for a critical account, see DeBlieu 1991). We find sobering the observation of Reid and Miller that "even if all human impacts on the biosphere were to cease immediately, species extinctions due to the impacts that have already taken place would continue for decades" (1989, 1). In light of this unhappy reality, a hands-off approach in our dealings with wildlife as advocated by some moral philosophers is an unacceptable option (e.g., Singer 1990). We cannot roll back history to Edenic conditions and start anew. Nor can we avoid, regardless of the milieu in which we toil, decisions that will discriminate in favor of some at the expense of others. Our best hope is that in playing God, we are both benevolent and rational.

If, as seems likely, an enhanced role for zoos as arks is seen as a necessary aspect of larger conservation strategies (Foose 1983, Seal 1985, Soulé et al. 1986), the means employed in filling this role should be ethically defensible. Here, we address one of the more troubling aspects of captive breeding efforts, namely, the management of those that are surplus to breeding endeavors. Is there an entitlement on the part of surplus animals to live beyond their useful years? Is there a moral dispensation allowing the sacrifice of some for the greater good of others—i.e., a higher valuation of species than of individuals? Do zoos have options other than euthanasia for unneeded animals?

THE INEVITABILITY OF A CAPTIVE SURPLUS

Most zoo authorities would agree that the fate of surplus individuals ranks among the paramount issues facing zoos at the present time. The AZA has convened working groups to formulate and refine policy in this area for its member institutions (Diebold 1992) but is disadvantaged by lack of information on who these animals are and their number. There is, accordingly, a paucity of information on the costs of maintaining surplus animals and the cost-reduction measures employed by zoos. The AZA is in the difficult position of advancing guidelines for dealing with a problem whose dimensions are as yet only dimly discerned.

To those outside of zoos, a widely shared perception is that the number of surplus animals is large and that it results from unregulated breeding. Grandy, for example, wrote in the 1989 Humane Society News that he believed zoos were producing in excess of 8,000 surplus animals yearly, and that most of these were from unplanned births. Those working in zoos, however, recognize that numerous factors other than unplanned breedings contribute to the generation of a surplus. Lacking natural culling processes, captive populations contain individuals that have suffered reproductive incapacitation from advanced age, disease, or inherited disorders. Although some of these individuals may retain a social value to conspecifics, nonreproductive adults in carefully managed captive populations are usually regarded as surplus to those programs and as dispensable.

With the implementation of programs designed to maintain genetic diversity over long time periods (Flesness 1986, Foose et al. 1986, Hutchins and Wiese 1991), a new reality for zoos is the generation of a genetic surplus. In an address to the AZA in 1976, Conway was among the first to call attention to this point, stating that "restricting breeding is only a part answer to the problem, for surplus is unavoidable in biologically sound propagation programs" (Conway 1976, 20). One factor used to classify individuals as genetically surplus is overrepresentation in the captive gene pool, a status which usually derives from a bygone era when animals were bred without sufficient concern for individual differences in fertility or survivorship. Another factor is the limited space available in zoos and the consequent requirement that the population size of any given taxon must be restricted. The genetics of captive breeding is the genetics of small population demes, with all the attendant concerns for retaining as much diversity as possible.

We use as an example of a genetically generated surplus the lion-tailed macaque *(Macaca silenus),* an endangered primate for which survival planning and a studbook were initiated in 1982 (Gledhill 1983, 1985). A master plan finalized in 1989 revealed that the numbers living in North American zoos already exceeded by 20 percent the available quality spaces, determined to be about 200. In addition, this plan pointed out the necessity of a drastic curtailment in recruitment through birth in order to maintain a steady state. Practically speaking, individual females with the capacity of producing eight to ten offspring during their reproductive lives would henceforth be limited to producing two or three (Lindburg et al. 1989). Those that met their genetic quota while still relatively young would become surplus to propagation efforts long before their natural lives were over. A recent revision of the plan (Gledhill 1991), designed to improve further the genetic well-being of lion-tails, would reduce the population to 100 individuals (i.e., 56 percent of those alive at the time).

What options are open to zoos in dealing with this surplus population? One is to transfer unneeded individuals to zoos abroad, where spaces are available. But experience informs us that it would be only a matter of time until the North American problem of today would occur globally. Returning lion-tails to wild habitat is deterred by declining availability of suitable habitat and by the existence of a wild population of 2,000 to 5,000 whose immediate future is best insured by protecting it (Karanth 1992). Transferring surplus animals to research laboratories, animal dealers, or roadside menageries are unacceptable options for zoos. Thus, in bringing this population under controlled breeding and genetic management, zoos find themselves facing the hard choice, namely, either to maintain surplus individuals in perpetuity or to cull by way of euthanasia.

ETHICAL CONSIDERATIONS

Captive breeding programs for endangered species are not ends in themselves. Conceptually, they are intended to provide a safe environment for short-term increases in population numbers to a size permitting reintroduction to wild habitat. A sine qua non of these programs is that there be protected habitat to receive them. Although they differ in approaches, captive breeding and in situ conservation ultimately come together, therefore, in giving highest value to the protection of natural ecosystems and their component populations.

Some environmental philosophers stress the utilitarian value of natural phenomena (e.g., Passmore 1974, Myers 1979, Ehrlich and Ehrlich 1981), while others argue for an intrinsic value for natural wholes or systems as the products of *"a continuing historical process of immense antiquity and majesty"* (Ehrenfeld 1978, 208, emphasis in original). Rooted in the views of early North American conservationists such as Muir and Leopold that species or ecosystems have a right to exist, or at least not to be extirpated (Muir 1901, Leopold 1949), holists such as Ehrenfeld and Callicott attribute intrinsic value to the natural world in order to give conservationist arguments stronger moral standing than if they were based only on serving human ends (Ehrenfeld 1978; Callicott 1984, 1986). For our purposes, it is sufficient to note that environmental ethicists invariably see individuals as but the carriers of a small (though perhaps significant) fraction of the species' genome, momentary actors in the long evolutionary drama, and therefore only of utilitarian value to the higher cause of conserving the collective welfare. From his studies of the foundations of environmental ethics, Hargrove notes, for example, that in contrast to evolving entities (i.e., species), individual animals "do not endure long enough in terms of preservationist time scales for efforts at this level to be of much consequence," and admonishes that to be concerned with the welfare, rights, and interests of individuals is to "abandon proper attitudes toward wildlife in favor of improper sentimentalism" (1989, 128). In a section entitled "The Bankruptcy of Individualism in Species Preservationism," Norton concludes that "concern for welfare of individuals seems, in general, to be the wrong direction to look if the goal is to preserve species" (1987, 167). Clearly, the application of this ethic to surplus captives would render them expendable whenever they impede efforts to conserve their kind.

In contrast, animal rights philosophers value the individual animal above aggregates such as species or communities (e.g., Singer 1979, Regan 1983). Since an aggregation cannot be said to have a brain, reasoning power, sentience, or interests, it is difficult to ascribe rights in the sense that this is done for individuals. The standing of humans as moral agents requires that they respect the interests of sentient animals, including an interest in being exempted from human-caused suffering and death. Whereas the valuing of animals intrinsically or for food, sport, or aesthetic enjoyment engenders protection at the level of systems or taxa, animal rightists part company with ecologists, environmentalists, and naturalists in extending protection on the basis of individual welfare.

It is nevertheless consistent with animal rights philosophy to support efforts to save endangered species, not because their numbers are reduced but because their individual members "have valid claims, and thus rights against those who would destroy their natural habitat" (Regan 1983, 360). If humans would respect the interests of individuals, it is argued, the end result would be the preservation of biotic wholes, as desired by environmentalists. As is often pointed out, however, when concern for welfare or interests is extended only to organisms having sentience, many animals and all plants as well as inorganic aspects of nature do not qualify for protection (Norton 1987, Hargrove 1989).

Animal rights philosophy also distinguishes between killing to end suffering and killing healthy animals that, could they give expression to their interests, would prefer to go on living. This latter form, usually justified as in the animals' best interests, is referred to as paternalistic euthanasia and is contrasted with humane or preference-respecting euthanasia (Regan 1983). According to this ethic, the euthanizing of healthy zoo surplus animals would disregard their interests and is therefore immoral.

From the foregoing, it is clear that in valuing species above their individual members and in giving priority to rare over more common taxa, zoos and aquariums are philosophically aligned with environmental conservationists. A representative statement of the position of zoos is provided by Lacy (1991), who advocates euthanasia for all individuals classified as surplus, irrespective of their state of health. Lacy's position, while not the official policy of the AZA, is widely embraced by its membership. This view, that it is permissible to kill individuals to save species, is a paradox of the captive endeavor that has an analogy with predation and other forms of culling under natural conditions. Intuitively, however, there is an uneasiness in going beyond what ecology and evolution would allow, as happens in equating human culling with the predatory acts of morally unaccountable animals. These concerns for individuals usually arise from encounters with wildlife that inspire Darwinian feelings of "grandeur in this view of life" (Darwin 1936, 374), not from beliefs in individual rights, disdained by most zoo professionals. Environmental philosophers in fact often stress the role of aesthetic appreciation and intuition in shaping these values (e.g., Norton 1986, 275–277; Hargrove 1989, 129–132), and such deep personal experiences would account for the moral discomfort often expressed by advocates of managerial euthanasia (Lacy 1991).

A SITUATIONAL ETHIC

The reality is that while we await guidance from those best qualified to resolve questions of rights or intrinsic worth, we employ a strategy in decision making that has many of the elements of a situational ethic. Perhaps the clearest formulation of this ethic is found in the writings of theologians, in particular Joseph Fletcher (1966). Fletcher drew a contrast between three approaches that could be employed in making moral decisions: (1) the legalistic, by which every decision ensues from rigid adherence to prefabricated rules and regulations; (2) the antinomian, which disdains all rules and acts spontaneously in each case from special knowledge written on the heart; and (3) situationism, in which one comes to the decision-making situation armed with the ethical maxims of one's community and heritage. Fletcher illustrates the situational ethic with the story of a cab driver who volunteered his views during an election campaign: "I and my father and grandfather before me, and their fathers, have always been straight-ticket Republicans." His passenger, being of the same party, responded by saying, "I take it that means you will vote for Senator So-and-So." "No," came the reply. "There are times when a man has to push his principles aside and do the right thing." A situational ethic is not locked into what Fletcher terms "punt on fourth down" rules, nor is it devoid of any guiding principles. It is an ethic that rests ultimately on value, but sees value as situation-dependent.

For anyone concerned with the loss of biological diversity, values of another time may not be adequate. Hargrove's point that ethics do shape legislation and public policy is well taken (1989), and whether arguments from individual rights or from intrinsic value will carry the day is yet to be determined. The one certainty is that policies and their ethical underpinnings will change with the times. In the sections that follow, we analyze the bases for the zoo professional's discomfiting options in dealing with the surplus problem and suggest alternatives to the notion that the only course in saving species is to destroy some individuals.

ECONOMICALLY DRIVEN DECISIONS

Much of the ethical justification for culling healthy surplus animals appears ultimately to rest on the need to allocate finite resources responsibly. Primarily, these resources consist of the limited space available in zoos for breeding programs and the very high costs of caring for animals in captivity (Conway 1986, 1989). Space allocations appear to reflect primarily the interests of individual institutions rather than global planning that would address the breeding requirements of various taxa. Given the system in place today, its relevance to the surplus question is pointed up by Lacy when he asks, "Is it more ethical to care for one animal for a healthy life-span of 20 years, terminated by poor health, than to care for two animals, each with life-spans of 10 years, terminated by painless euthanasia?" (1991, 294). Posing the issue in this way suggests a trade-off among values, and Lacy votes in favor of the two-for-one value. If the basis for these decisions could be reduced to purely economic considerations, most zoo professionals would probably vote with him. However, public concerns with individual welfare indicate a strong shift in values whose course is unlikely to be altered by decisions that are economically motivated.

THE POLITICAL COST OF EUTHANASIA

Cherfas, in writing about contemporary zoos, observes that "very often those who describe themselves as conservationists find zoos anathema, while those who visit zoos have anything but conservation on their minds" (1984, 42). In an article entitled "The Trouble with Zoos" (1986), Fox discounts zoos' conservation goals, seeing rather an exploitation of wildlife for exhibition, entertainment, and profit. Those who give highest priority to concerns for the pain and suffering of animals will be inclined to accept Varner and Monroe's characterization of captive breeding as a "moral atrocity" (1991, 28). Zoos are having only measured success in convincing society and their publics that they are truly contributing to the effort to save some wild species from extinction. It is against this reality that the cost in terms of public approbation must be considered if financial relief is sought through the killing of surplus individuals.

Further consideration of the lion-tailed macaque case will highlight this concern. Since 49 individuals were declared surplus nearly three years ago, fewer than a dozen have been sent to zoos abroad (North American zoos are at capacity). Recent action of the species management group raised the number of surplus individuals to 143 (L. G. Gledhill, personal communication 1994). Although some of these may be successfully placed in zoos abroad in the years ahead, the economics of the situation would require that perhaps as many as 100 relatively healthy, relatively young individuals be euthanized in the immediate future. Justification for this action could be found, let us say, in allocating the resulting spaces to a highly endangered sister species, the Mentawai Island macaque, which, according to Tenaza's investigations (1989), is being rapidly decimated by hunting, by poisoning, and by loss of habitat.

The orangutan provides yet another example of a perplexing case. This species was once distributed over a vast region in Southeast Asia but today exists as remnant populations on the islands of Kalimantan and Sumatra. Chromosomal studies suggest the two forms have existed in isolation from one another for at least a thousand generations (Seuanez et al. 1979). Karyotyping of the existing North American zoo population revealed that 29 percent (88 individuals) are offspring resulting from mixed breeding and therefore undesirable for further propagation (Perkins and Maple 1990). Although zoos will encumber enormous costs over the next thirty years in managing these hybrids to extinction, neither the species management group nor any zoological institution has yet proposed the euthanizing of orangutan hybrids. Failure to proceed with euthanasia in this case provides a good illustration of advocacy from an economic perspective that is curbed by political considerations.

Although the numbers today classified as surplus by zoos across the nation are unknown, they are surely in the thousands if not tens of thousands. Managerial culling through euthanasia is indeed an option. It is the law of the land. But its practice is politically expedient on a limited, perhaps hierarchically defined, scale at best. The disadvantages of a hierarchical approach are similar to those encountered by animal rights philosophers in attempting to define the phylogenetic level at which sentience begins (Singer 1990). As a standard for deciding which animals shall live and which shall die, it holds the prospect of engendering endless conflict. More seriously, it leaves zoos open to a variable standard that would surely lead, unpredictably, to their actions being challenged in the courts, their institutional property damaged, and their reputations forever tarnished. Managerial euthanasia, even for the noble cause of saving other species, is not widely available as an option because society will not allow it.

In addition, it would be erroneous to conclude from this concern with political costs that zoos would necessarily embrace euthanasia if it were to be more widely sanctioned. The differing points of view held by zoo professionals themselves are inadequately characterized if said to rest either on unbridled sentimentalism on the one hand or on callous immorality on the other.

There is a view attending the killing of zoo animals that holds that proper education of the public on this subject will foster its acceptance as necessary to the conservation enterprise. On the other hand, such thinking may be but another example of what Robinson has called the "enlightenment fallacy." which, as he observes, is negated on almost every issue (Robinson 1990; see also Rodd 1990, 112–113, for discussion of this point). A message emanating from zoos that advocates both the saving and the taking of animal lives is likely to be, at best, a confused one, perhaps analogous to the claim by military authorities during the Vietnam War that villages like My Lai had to be destroyed in order to save them. The point of the argument is that the goodwill and support of society for zoos' efforts is a resource that may not be worth risking in favor of economically defined ones. It is unlikely that society at large will be much impressed or swayed by economically driven actions, and we suggest that if captive breeding has value for wildlife conservation, then the economic burden of maintaining surplus individuals in zoos should be eased by means other than killing them.

ALTERNATIVES FOR ADDRESSING THE SURPLUS ISSUE

A starting point in addressing the surplus problem is for zoos to place curbs on breeding where it is now unrestricted. Again, the dimensions of the problem are unknown. It is known, however, that in 1991 formal management plans that control breeding existed in North American zoos for only 56 taxa (Hutchins and Wiese 1991). Breeding in other cases seems to be loosely constrained only by what the market will bear, if at all. Since a surplus is an unavoidable by-product of even the most prudently managed programs, and since zoos increasingly use captive breeding of endangered wildlife to justify their existence, it could be argued that spaces now claimed by unrestricted breeding of common forms should be utilized for holding the surplus from more essential programs.

A related step is to undertake some reordering of current holdings in favor of species most at risk of disappearing from wild habitat. The primary contribution of zoos to conservation is in providing a temporary refuge for species whose survival in the natural state is imperiled, and in using their presence in this displaced milieu to elicit preservationist concern for wild habitat. Yet, many of the resources of zoos are today consumed by species not at risk. Although hard data on this point are scarce, an example is found in an analysis of primate birth records from the *International Zoo Yearbook* (Lindburg et al. 1986). This analysis revealed that only 20 percent of the more than 24,000 infants born to zoo-living dams for the decade 1971–1980 were classified as endangered, vulnerable, or rare (Table 1). If the space occupied by other taxa is similarly biased in favor of the more common species, zoos do in fact have some options in the diversion of resources to conservation efforts without, at the same time, significantly undermining education about diversity. Some may find it difficult to accept the proposition that resources now occupied by species not at risk should go to the support of an unneeded surplus, for it somehow seems less noble. But as suggested earlier, changing societal values have made culling by euthanasia less tenable as an option.

Our third advocacy is for the development of low-cost facilities for maintenance of surplus individuals. This concept, as earlier described hypothetically for orangutans by Lindburg (1991, 2), is as follows:

Consider again the case of the 20 year-old female orangutan, mentioned earlier. What if a zoo could pay a one-time fee, predicated on demographic determination of average life expectancy, to an AZA-sponsored facility that would give her a home for the rest of her natural life? If she and others like her were to be kept in functional but low cost facilities, preferably in a warm climate, perhaps even outside North America, could pooled funds from such fee schedules generate enough revenue to make retirement sanctuaries a viable alternative? Could alternate but acceptable research uses of these retirees help defer the costs of their keep? If retirement for even a modest number of species would work, should zoos have to bear the entire cost themselves? Perhaps of most importance, can this alternative make any appreciable dent in the surplus problem?

In his response to this proposal, Lacy described the cost, if applied to all species, as "staggering" (1991). He is, of course, correct. But exhibit space or even space

TABLE 1. PRIMATE BIRTHS IN ZOOS, 1971–1980

IUCN Status	Number of Taxa	Births	
		Number	Percentage of Total
Endangered	14	2,522	10.4
Vulnerable	19	2,493	10.2
Rare	3	158	0.7
Indeterminate	5	130	0.5
Unclassified	127	19,333	79.3
All	168	24,366	100.0

Source: *International Zoo Yearbook*

behind the scenes, more often than not situated in urban locales and in temperate climates, is also very costly space. It remains to be seen whether low-cost space, developed as an alternative to behind-the-scenes warehousing of surplus individuals, is in fact a viable alternative. It also remains to be seen whether breeding of the highest-priority cases might be better pursued outside zoos, with zoos becoming the repository for the surpluses generated by these programs.

Finally, it seems propitious to propose that resource consortia be developed to alleviate the costs of preserving surplus individuals. This leads ultimately to questions of how highly society values the captive approach to conservation, and whether zoos should be expected to bear this burden alone. In reviewing some of the conflicts between conservationists and the advocates of animal rights, Ehrenfeld asked whether their different agendas meant that the two movements must go their separate ways in animosity. "That," he states, "would be a tragedy, because they desperately need each other. Conservation needs the passion, the numbers, and the power of the animal rights movement. The animal rights movement needs the understanding of how the natural world works and the informed foresight that conservation science and management can provide. Can a *modus vivendi* be evolved without violating fundamental beliefs?" (1991, 2). A strong yes seems evident in Ehrenfeld's account of an accommodation between animal rights groups and the pet industry on a number of common goals. Zoos would do well to examine this model of cooperation between unfriendly entities, and the surplus question seems ideally suited to such an endeavor.

The role of government in providing resources for the alleviation of captive-breeding costs also should not be prematurely discounted. A retirement fund for chimpanzees used in nationally mandated biomedical research programs (Eichberg and Speck 1988) and discussions initiated in 1989 to establish "National Institutes for the Environment" (Announcements 1990) portray a vision of a greater role for government in conservation if not in broad-scale preservation of biological diversity. Obtaining the cooperation of either public or private entities in the furtherance of their goals would undoubtedly require that zoos embark on a further reexamination of their priorities and their methods of operation, and publication of this book could represent such an opportunity.

SUMMARY

Action steps relevant to zoos' future dealings with the surplus issue can be summarized as follows:

1. Provide continuing implementation at the national level for defining the scope of the surplus problem.
2. Begin the process of bringing all species in zoos under management in order to avoid the occurrence of nonessential births.
3. Begin defining a process for phasing out through attrition at least some of the species that are not essential to conservation goals.
4. Explore the feasibility of developing alternate uses and low-cost facilities for animals classified as surplus.
5. Develop liaisons with other entities, both public and private, that would constitute resource consortia for dealing with issues of common interest.

These steps are advocated from the certainty that many members of society, including many zoo partisans, object to managerial euthanasia on moral grounds, whether those objections reside in beliefs or attitudes about rights of, reverence for, contractual obligations to, or the aesthetic value of things natural. Humans stand apart from the rest of the biological world in terms of conscience and moral responsibility. As Rolston has put it, "Humans are in the world *ethically* as nothing else is" (1989, 238, emphasis in original). It follows that in exercising our judgments we must act honestly and responsibly, and in the present case this means that we must pursue the unrealized options that are available to zoos in dealing with healthy surplus individuals.

REFERENCES

Announcements. 1990. *Conservation Biology* 4:225–226.

Callicott, J. B. 1984. Non-anthropocentric value theory and environmental ethics. *American Philosophy Quarterly* 21:299–309.

———. 1986. On the intrinsic value of nonhuman species. In *The Preservation of Species,* ed. B. G. Norton, 138–172. Princeton, N.J.: Princeton University Press.

Cherfas, J. 1984. *Zoo 2000: A Look beyond the Bars.* London: British Broadcasting Corp.

Conway, W. G. 1976. The surplus problem. In *Proceedings: American Association of Zoological Parks and Aquariums Annual Conference,* 20–24. Wheeling, W.Va.: AAZPA.

———. 1986. The practical difficulties and financial implications of endangered species breeding programs. *International Zoo Yearbook* 24/25:210–219.

———. 1989. The prospects for sustaining species and their evolution. In *Conservation for the Twenty-First Century,* ed. D. Western and M. C. Pearl, 199–209. New York: Oxford University Press.

Darwin, C. 1936. *The Origin of Species by Means of Natural Selection.* Modern Library ed. New York: Random House.

DeBlieu, J. 1991. *Meant to Be Wild: The Struggle to Save Endangered Species through Captive Breeding.* Golden, Colo.: Fulcrum.

Diebold, E. 1992. Memorandum. AAZPA Committee on Surplus SSP Animals, Wheeling, W.Va.

Ehrenfeld, D. 1978. *The Arrogance of Humanism.* New York: Oxford University Press.

————. 1991. Conservation and the rights of animals. *Conservation Biology* 5:1–3.

Ehrlich, P. R., and A. H. Ehrlich. 1981. *Extinction.* New York: Random House.

Eichberg, J. W., and J. T. Speck, Jr. 1988. Establishment of a chimpanzee retirement fund: Maintenance after experimentation. *Journal of Medical Primatology* 17:71–76.

Flesness, N. R. 1986. Captive status and genetic considerations. In *Primates: The Road to Self-Sustaining Populations,* ed. K. Benirschke, 845–856. New York: Springer-Verlag.

Fletcher, J. 1966. *Situation Ethics.* Philadelphia, Pa.: Westminster Press.

Foose, T. 1983. The relevance of captive populations to the conservation of biotic diversity. In *Genetics and Conservation: A Reference for Managing Wild Animal and Plant Populations,* ed. C. M. Schonewald-Cox, S. M. Chambers, B. McBryde, and W. L. Thomas, 374–401. Menlo Park, Calif.: Benjamin/Cummings.

Foose, T. J., R. Lande, N. R. Flesness, G. Rabb, and B. Read. 1986. Propagation plans. *Zoo Biology* 5:139–146.

Fox, M. W. 1986. The trouble with zoos. *Animals' Agenda* (June): 8–12.

Gledhill, L. G. 1983. *North American Regional Studbook for the Lion-Tailed Macaque.* Seattle: Woodland Park Zoo.

————. 1985. Progress toward a master plan of population management for the lion-tailed macaque. In *The Lion-Tailed Macaque: Status and Conservation,* ed. P. G. Heltne, 379–383. New York: Alan R. Liss.

————. 1991. Minutes. Lion-Tailed Macaque Species Survival Plan meeting, September. Library, Woodland Park Zoo, Seattle.

Grandy, J. W. 1989. Captive breeding in zoos: Destructive programs in need of a change. *Humane Society News* Summer:8–11.

Hargrove, E. C. 1989. *Foundations of Environmental Ethics.* Englewood Cliffs, N.J.: Prentice-Hall.

Hutchins, M., and C. Wemmer. 1991. Response: In defense of captive breeding. *Endangered Species Update* 8:5–6.

Hutchins, M., and R. J. Wiese. 1991. Beyond genetic and demographic management: The future of the Species Survival Plan and related AAZPA conservation efforts. *Zoo Biology* 10:285–292.

Karanth, K. U. 1992. Conservation prospects for lion-tailed macaques in Karnataka, India. *Zoo Biology* 11:33–41.

Lacy, R. C. 1991. Zoos and the surplus problem: An alternate solution. *Zoo Biology* 10:293–297.

Leopold, A. 1949. *A Sand County Almanac.* London: Oxford University Press.

Lindburg, D. G. 1991. Zoos and the "surplus" problem. *Zoo Biology* 10:1–2.

Lindburg, D. G., J. Berkson, and L. Nightenhelser. 1986. The contribution of zoos to primate conservation. In *Primate Ecology and Conservation,* ed. J. G. Else and P. C. Lee, 295–300. Cambridge: Cambridge University Press.

Lindburg, D. G., A. M. Lyles, and N. M. Czekala. 1989. Status and reproductive potential of lion-tailed macaques in captivity. *Zoo Biology Supplement* 1:5–16.

Muir, J. 1901. *Our National Parks.* Boston: Houghton Mifflin.

Myers, N. 1979. *The Sinking Ark.* Oxford: Pergamon Press.

————. 1989. A major extinction spasm: Predictable and inevitable? In *Conservation for the Twenty-First Century,* ed. D. Western and M. C. Pearl, 42–49. New York: Oxford University Press.

Norton, B. G. 1986. Epilogue. In *The Preservation of Species,* ed. B. G. Norton, 268–283. Princeton, N.J.: Princeton University Press.

————. 1987. *Why Preserve Natural Variety?* Princeton, N.J.: Princeton University Press.

Passmore, J. 1974. *Man's Responsibility for Nature.* New York: Charles Scribner's Sons.

Perkins, L. A., and T. M. Maple. 1990. North American orangutan Species Survival Plan: Current status and progress in the 1980s. *Zoo Biology* 9:135–139.

Regan, T. 1983. *The Case for Animal Rights.* Berkeley: University of California Press.

Reid, W. V., and K. R. Miller. 1989. *Keeping Options Alive: The Scientific Basis for Conserving Biodiversity.* Publications Brief. Washington, D.C.: World Resources Institute.

Robinson, M. H. 1990. Setting straight our plundered planet. *Zoo Biology* 9:247–250.

Rodd, R. 1990. *Biology, Ethics, and Animals.* Oxford: Clarendon Press.

Rolston, Holmes III. 1989. Biology without conservation: An environmental misfit and contradiction in terms. In *Conservation for the Twenty-First Century,* ed. D. Western and M. C. Pearl, 232–240. New York: Oxford University Press.

Seal, U. S. 1985. The realities of preserving species in captivity. In *Animal Extinctions,* ed. R. J. Hoage, 71–95. Washington, D.C.: Smithsonian Institution Press.

Seuanez, H. M., H. J. Evans, D. E. Martin, and J. Fletcher. 1979. An inversion in chromosome 2 that distinguishes between Bornean and Sumatran orangutans. *Cytogenetics and Cell Genetics* 23:137–140.

Singer, P. 1979. Not for humans only: The place of nonhumans in environmental issues. In *Ethics and Problems of the Twenty-First Century,* ed. K. E. Goodpaster and K. M. Sayre, 191–206. Notre Dame, Ind.: University of Notre Dame Press.

————. 1990. *Animal Liberation.* Rev. ed. New York: Avon Books.

Soulé, M. E., M. Gilpin, W. Conway, and T. J. Foose. 1986. The millennium ark: How long a voyage, how many staterooms, how many passengers? *Zoo Biology* 5:101–113.

Tenaza, R. W. 1989. Primates on a precarious limb. *Animal Kingdom* 92(6): 26–37.

Varner, G. E., and M. C. Monroe. 1991. Ethical perspectives on captive breeding: Is it for the birds? *Endangered Species Update* 8(1): 27–29.

THE SHOULD OR SHOULD NOT OF

CAPTIVE BREEDING

Whose Ethic?

Frederic Wagner

The world is, at the hands of a burgeoning, technological human population, losing species at an unprecedented rate from a biota that evolved over a 3-billion-year period largely in the absence of humans. That loss is occurring from a combination of habitat destruction, direct killing, and environmental pollution at the hands of our recently arrived species. There are attempts at countermeasures. Many countries of the world have population-control policies, although the American government has faltered badly in its support for population control during the past decade. There are efforts at environmental protection and sustainable use of natural resources. But population growth, environmental degradation, and species extinction continue, and it is not at all clear when or if these trends will be reversed.

It is clear that zoos and aquariums have become a last potential refuge for many of the world's species. They are serving, and will continue to serve, as the only means for preserving some species in a museum and for educational purposes. They have provided, and will continue to provide, the only source of animals for reestablishing otherwise extinct, wild, and free-living species in areas where habitat protection and restoration, and control of human exploitation, make survival in the wild possible once again.

There have already been successes (DeBlieu 1991, Luoma 1992). The Arabian oryx, European wisent, Przewalski's horse, California condor, and North American red wolf were at one time all reduced to captive populations and extinct in the wild. All now exist in nucleus or restored, wild populations established from captive releases. Near-extinct black-footed ferrets, golden lion tamarins, and peregrine falcons have all had their numbers augmented or their species ranges extended by captive-bred releases.

I do not suggest that captive breeding is the only answer for species restoration in the wild. It unquestionably depends on habitat restoration, and if species can be saved by such means, it goes without saying that habitat restoration is preferable. I know of no one who seriously considers captive breeding a moral or practical equivalent to habitat preservation and restoration as Varner and Monroe imply (1991).

And I do not suggest that captive breeding and release are infallible means for saving species. There are likely to be failures due to habitat insufficiencies, inability to raise populations above the viability level, excessive inbreeding depression, and the difficulties of behaviorally reshaping captive-bred animals for survival in the wild. But it is one tool that can and does contribute to the total effort at maintaining global biodiversity.

Clearly the practice incurs the attendant problems of surplus numbers being addressed here. While, as the Lindburgs and Lacy point out, there are techniques for minimizing the problem—exchange with other zoos, reproductive inhibition, maintenance in low-cost nondisplay facilities, and euthanasia—it can never be fully eliminated. It is a consequence of maintaining living populations of species over several or many generations. If a population is to survive, it must reproduce. In polygynous species there will always be a surplus of males. There may also be contraceptive failures and mismanagement.

But these purely technical and objective issues are almost irrelevant to the real concerns of this book, which are ethical. The fact that captive breeding can be used as one tool for reestablishing free-living species poised on the brink of extinction says nothing about whether society should engage in the practice, or should be concerned about preserving global biodiversity. The science of ecology can suggest how it can be done, and something about what the consequences to society are of doing it or not doing it, but it cannot say whether society should do it.

I am not a philosopher by profession, although I have had some formal education in the discipline and have written on ethical questions surrounding environmental and animal welfare issues (Wagner 1988). It may be presumptuous of me to venture my own views on these matters, but I will make so bold as to do so and welcome the comments of professionals on my arguments.

As I read the literature on the morality of captive breeding, it seems to me that the authors expend a great deal of effort, perhaps subconsciously, in searching for ethical absolutes or airtight rationales that will provide firm direction through our moral dilemmas. There is frequent mention of intrinsic value, fundamental rights, ideal moral judgments, and ultimate obligations (Donnelley 1990). In many cases these are asserted as moral absolutes. My basic message is that this search is a vain one.

What I do not hear is any concerted effort at examining the basis for these assumptions and assertions, for going back to the ages-old questions of how or on what authority we determine what is a valid or ideal moral position. By whom, or on what authority, are we enjoined to behave in certain ways toward animals?

Historically there have been only two answers to these questions. One is an appeal to religious authority accepted on faith, in which case the injunctions become absolutes. The problem with this is that different religions hold contradictory views on some issues and, in asserting absolutes, put themselves in the position of pointing the finger of immorality at each other. If they are to be consistent, those who assert an absolute rightist position must call the Muslims immoral for the zebh slaughter of food animals. In doing so, they will condemn something that is prescribed in the Koran, the revealed word of God for faithful Muslims. A close analog to the assertion of absolutes is what Norton in Part Two calls "monistic" ethics.

The other answer, and the one to which I subscribe, is that moral codes are secular constructs of the human mind. Hence, ethical precepts are human artifacts, and when adopted by societies become the normative morality of the forms to which the Lindburgs refer in the previous chapter. The same perspective applies to the idea of rights and, for our purposes here, animal rights. My dictionary defines a right as "any power or privilege vested in a person by law, custom, etc." Any ecologist knows that apart from humans, animals have no rights in a hostile nature filled with predators, parasites, competitors, aggressive conspecifics, and insensitive physical factors in the environment (Wagner 1988). They have rights if, when, and to the extent that humans accord them.

Because they are human constructs, ethical ideas vary and there are numerous moral codes, as Rawls discusses (1971) and Russow reviews (1990). For this reason there is, as Russow states, no "adequate and generally accepted moral theory that will generate unassailable, defensible solutions. For every ethical theory advanced in philosophic discussion, several oppose it, and consensus among moral philosophers is not to be found."

The important point here is that rightness and wrongness are what people and societies decide upon and assert. Since ethical codes are prescriptions for human behavior, they are injunctions by some individuals and some societies on how others should behave. Until such codes are generally accepted by a society, and better yet codified into law, it is ethnocentric for one component of society to try to coerce another into following the first's notion of morality, or similarly one society vis-à-vis another.

Such is the case with animal rights issues. Donnelley has termed the assertion of animal rights absolutes "ethical dogmatism" (1990). Wenzel has called the efforts of European Canadian rightists to halt seal hunting by the Inuit (Eskimos) "colonialism" (1991, 4). These native people protest that wildlife harvesting—their subsistence and an integral part of their societal structure and art—long predates European philosophy and law. They take the practice as a right.

Regan has termed the holistic act of subjugating individual animals to the welfare of species and ecosystems "environmental fascism" (1983). But a sizable component of society values species and ecosystems. Indeed, Windle speaks of the "grief" over loss of species and biotas (1992). Some 80 percent of the American public now consider themselves environmentalists. The prevalence of these values in society is indicated by their codification into law by our elected representatives in the form of such legislation as the Endangered Species Act, the Marine Mammal Protection Act, and the National Forest Management Act, with its mandate to manage for biodiversity.

Given this concern in society, it seems to me pure ethnocentrism to try to coerce society into discounting those values in preference to individualism. It is nothing less than anthropocentrism to shrug off whole species into extinction in deference to the comfort of individual members of nonthreatened species. Lacy considers this attitude speciesism (1991), an interesting paradox since speciesism is anathema to animal welfare advocates.

Regan states that the "rights view" does favor species and ecosystem preservation (1983). Hence both sides in the debate desire at least this common goal. How much more progress we could make toward its attainment, if both sides would concede the relativity of their positions and, as Norton and the Lindburgs advocate in previous chapters, yield some ground toward a compromise that would enable them to join hands in achieving that common end.

There is an entire suite of techniques for addressing the captive surplus problem. One is holding animals in nondisplay facilities, which can range from well-designed environments simulating the animals' habitat to promote their sense of well-being, to low-cost, simple quarters, depending on the species' needs. Another is reproductive inhibition. One approach is chemosterilization, which has now been developed for a number of species (Seal et al. 1976, Matsche 1977, Johnson and Tait 1983, Wagner et al. 1991). Another approach is penning the sexes separately, again depending on the species. Still another technique is exchange with other zoos, with the stipulation that they be AZA-sponsored.

The most controversial solution is euthanasia. It has its opponents, but there are those who consider it more humane than keeping aged or unhealthy animals alive. No less committed an animal rights advocate than Ingrid Newkirk, national director of People for the Ethical Treatment of Animals, concedes the practical necessity of euthanasia (1992).

No one, or combination, of these techniques will be suitable in all situations. Each decision will be based on some mix of consideration for the welfare of the animal and of the species, social acceptability of the technique in a given situation or locale, as well as considerations of cost, space, personnel, and other resources. As Lindburg comments, "Resolving the zoo surplus issue will undoubtedly require a multiplicity of approaches" (1991).

The decisions become real moral dilemmas when such resources as funds, personnel, and space are limited. Should a zoo continue to house excess animals from a few species purely on animal rights or humane considerations when the resources committed to them could be used to maintain individuals from other species poised on the brink of extinction?

Clearly these are difficult questions for society to address. It would appear that answers can come by only one or the other of two routes. One is an acknowledgment that the moral bases for positions taken are relative and can yield the compromises that Norton and the Lindburgs call for. The other is assertion that the moral positions taken are absolutes. In this case there will be continued confrontation and coercion, and whatever the result, it will come through balance of power.

REFERENCES

DeBlieu, J. 1991. *Meant to Be Wild: The Struggle to Save Endangered Species through Captive Breeding*. Golden, Colo.: Fulcrum.

Donnelley, S. 1990. The troubled middle *in medias res*. In Animals, science, and ethics, ed. S. Donnelley, and K. Nolan, 2–4. *Hastings Center Report* 20, No. 3 (May-June): Special supplement, 1–32.

Johnson, E., and A. J. Tait. 1983. Prospects for the chemical control of reproduction in the grey squirrel. *Mammal Review* 13:167–172.

Lacy, R. C. 1991. Zoos and the surplus problem: An alternative solution. *Zoo Biology* 10:293–297.

Lindburg, D. G. 1991. Zoos and the "surplus" problem. *Zoo Biology* 10:1–2.

Luoma, J. R. 1992. Born to be wild. *Audubon* 94:50–59.

Matsche, G. H. 1977. Fertility control in white-tailed deer by steroid implants. *Journal of Wildlife Management* 41:731.

Newkirk, I. 1992. Total victory, like checkmate, cannot be achieved in one move. *Animals' Agenda* 12:43–45.

Rawls, J. 1971. *A Theory of Justice.* Cambridge, Mass.: Belknap Press, Harvard University Press.

Regan, T. 1983. *The Case for Animal Rights.* Berkeley: University of California Press.

Russow, L. M. 1990. Ethical theory and the moral status of animals. In Animals, science, and ethics, ed. S. Donnelley, and K. Nolan, 4–8. *Hastings Center Report* 20, No. 3 (May-June): Special supplement, 1–32.

Seal, U. S., R. Barton, L. Mather, K. Olberding, B. D. Plotka, and C. W. Gray. 1976. Hormonal contraception in captive lions *(Panthera leo). Journal of Zoo Animal Medicine* 7:12–20.

Varner, G. E., and M. C. Monroe. 1991. Ethical perspectives on captive breeding: Is it for the birds? *Endangered Species Update* 8:27–29.

Wagner, F. H. 1988. *Predator Control and the Sheep Industry: The Role of Science in Policy Formation.* Claremont, Calif.: Regina Books.

Wagner, F. H., J. Berger, D. R. McCullough, J. W. Menke, E. S. Murray, B. W. Pickett, U. S. Seal, and M. Sharpe. 1991. *Wild Horse Populations: Field Studies in Genetics and Fertility.* Washington, D.C.: National Academy Press.

Wenzel, G. 1991. *Animal Rights, Human Rights: Ecology, Economy, and Ideology in the Canadian Arctic.* Toronto: University of Toronto Press.

Windle, P. 1992. The ecology of grief. *Bioscience* 42:363–366.

PART FOUR

I f it is agreed that there are some situations in which the obligation to protect species from extinction dictates actions that are invasive in the lives of individual animals and perhaps severely detrimental to their welfare, zoos must accept the challenge of articulating policies to guide care for individual animals in captive breeding programs. This final section presents viewpoints on three aspects of the treatment of animals in these programs. The first point-counterpoint exchange explores the rules of good husbandry for animals enrolled in captive breeding programs, the second addresses the difficult problems surrounding experimentation on captive wild animals, and the third explores problems in public relations—how should zoos communicate their concern for conservation and how can zoos use individual animals to contribute to that goal without threatening their well-being or dignity?

Once animals are enlisted in captive breeding programs, those humans who control their fate accept, all participants agree, a heavy responsibility for their care and for protecting the legitimate interests of the animals. In most cases there is considerable agreement regarding acceptable care of captive animals, especially regarding health and comfort of animals. But there is considerable disagreement regarding what interests animals in fact have, and this leads to disagreements regarding what is owed the animals, especially in the area of authentic experience. In the areas of basic care, use of animals in research, and concerns regarding the exhibiting of animals there emerge considerable agreements regarding steps that can and must be taken now to improve the plight of animals and to protect them against unnecessary suffering, but once these changes are accomplished, it appears that there will remain also in this area deep-lying disagreements about how much farther zoo professionals must go to protect the legitimate interests of captive animals.

CAPTIVE CARE AND MAINTENANCE

A deeper understanding of the behavior and ecology of animals in the wild is making it possible for modern zoos to construct better exhibits and holding facilities for captive wildlife. Indeed, design technology has improved considerably in the last decade, resulting in larger and more naturalistic zoo exhibits. Such displays are not only better for the animals, but they also are more educational because they make a connection between an animal and its habitat. The keeping of animals under captive conditions, however, raises some important ethical issues, and some opponents have even portrayed zoos as prisons for animals. Can modern zoos meet the basic psychological and physical needs of wild animals? In this exchange, Terry Maple, Rita McManamon, and Elizabeth Stevens explore the foundations of modern professionally managed zoos and their philosophy regarding the care and maintenance of wild animals in captivity. Environmental philosophers John Wuichet and Bryan Norton raise several related concerns in response. They suggest that animal welfare can mean very different things to different people, making it difficult for zoos to get ahead of their critics. After positing two conceptions of animal welfare by which zoos are often judged, they explore the impacts of these different conceptions on the type of care given to animals in captivity and consider the implications for the practices of care-giving.

DEFINING THE GOOD ZOO

Animal Care, Maintenance, and Welfare

Terry Maple, Rita McManamon, and Elizabeth Stevens

Conservation, education, science, and recreation are the four basic directives adopted by zoological park and aquarium members of the AZA. Institutions may prioritize these elements differently; the order herein is ours. For all of us, however, the care and welfare of the animals in our institutions have become our highest operational priorities and enable us to achieve our basic directives. Good zoos mobilize public attention and commitment to global conservation by focusing on individuals—and entire collections—as representatives of their native habitats. Within the zoo's unique recreational niche, we have a powerful opportunity to present complex ecological concepts in a pleasant and more controllable milieu. We also have the opportunity to study these special creatures and to apply this knowledge to improving the health and welfare of both captive and wild animals.

The concept of a good zoo requires definition. We believe that good zoos exceed the highest standards of the profession and, furthermore, engage in the consistent pursuit of excellence in conservation, education, science, and recreation. By this definition, good zoos are leaders, not followers. Good zoos eschew mediocrity. To maintain their high standards, good zoos must recruit superior talent to the zoo workforce. By their deeds, good zoos are regarded highly by their peers and within their communities. To succeed in the pursuit of our ambitious industry goals, we must endeavor to make good zoos the rule, not the exception.

In this chapter we will argue that science is still the most underrealized of the four directives, despite its critical role in the care, maintenance, and welfare of captive animals. Second, since no zoo wants to be considered inhumane, some of the issues and decision-making processes that surround the welfare of zoo animals will be discussed. Finally, the motivational role that zoos play in educating their visitors about conservation is one of the most important roles and is inextricably tied to the care and welfare of the zoo collection.

THE SCIENTIFIC ZOO

Throughout its history, the AZA has expressed some discomfort with science, as the word "research" appeared, disappeared, and then reappeared in the masthead of AZA. Most recently it has been resurrected as "scientific studies." Science has never been more important to AZA than it is today. The term "science" has been used interchangeably with "research." The 1987 edition of *The Random House Dictionary* simply defines science as "systematic knowledge of the physical or material world gained through observation and experimentation."

As a practical matter, observation is our more common strategy, as experimental studies are much more difficult to accomplish in the traditional zoo setting. Zoo biologists recognize that they must deal with the limitations of small sample size, unavoidable extraneous variables, interruptions, and a lack of baseline data. In this respect, zoo biology is more like fieldwork than laboratory work. It is a challenging new science.

We strongly believe that the world's zoos must increase and improve research output in order to advance their conservation objectives. Output depends on the creation of research departments in zoos. In 1975, Dr. John Eisenberg concluded, "Research staff become pivotal if a zoo is going to develop a consistent in-house conservation program and become important in efforts within other countries undertaken in the cause of conservation."

Productivity is directly linked to the presence of dedicated scientific staff in the zoo. To maintain our credibility as conservation leaders, zoo directors must elevate the standing of science in their institutions. Our conservation and educational efforts are knowledge-based. To sustain a respectable population of talented and well-trained scientists, zoos must become more serious in gathering data—which means providing a sufficient scientific budget.

Unfortunately, as zoo budgets have grown, research budgets have remained quite modest. Zoo leaders report that they want to do research (Finlay and Maple 1986), but they cannot afford to pay for it. We have learned that effective conservation and research programs depend on surplus revenues. While most zoos have committed to the idea of conservation, only financially successful zoos have been able to sponsor effective research programs. This trend is likely to continue throughout the nineties. Although we would like to see more zoos doing research, it is likely that the bigger zoos will simply shoulder even more of the load.

One of the unspoken deficiencies in zoo research is a scarcity of dedicated talent. The problems that we must solve in the next decade are enormously challenging. Bright people are needed. The modern zoo must recruit top-quality individuals who will be accountable for advancing the conservation agenda. In far too many zoos, research is relegated to the middle levels of management or to veterinarians not trained as scientists. Too often, research personnel with marginal skills are recruited, or are compensated at levels well below the competitive rate for academic positions. Some zoos have hired master's-level staff to lead scientific programs; this is a dead-end strategy for building bridges to universities. The correct path to recruitment should be clear. Entry-level zoo salaries must be competitive with those of universities. We cannot settle for second-rate research talent in the zoo.

While we clearly need more help from the outside (hence zoo-university collaborations), we must employ staff in the zoo whose assigned responsibilities are science and science alone. Research cannot be delegated to busy animal managers fully occupied with day-to-day administrative responsibilities. As the technology of species survival becomes more sophisticated, new skills will be needed to master the tools of the trade. Local universities can be the answer, but we must do a better job of recruitment. Even Hediger expressed surprise at the failure of zoos and universities to develop complementary relationships: "Every university which boasts a botanical institute will also have—in Europe at any rate—a botanical garden. Yet I know of no zoological institute which is properly complemented by a zoological garden. To the best of my knowledge there is, in fact, not one single university zoo" (1969).

Hediger conceded that the history of zoos, namely their links to the circus industry, and the high cost of managing exotic (especially large and dangerous) creatures held little appeal for academicians. However, these relationships are finally developing, and bridges to academia have been built. In a recent survey of research in American zoos, Finlay and Maple found that research productivity in zoos was highly correlated with university affiliations (1986). Every zoo can develop meaningful relationships with universities and should be encouraged to do so.

An expanding network of zoo biologists and collaborating scientists from universities may be the salvation of our species survival programs (SSPs), but every zoo needs to contribute more human and financial resources. In the short run, we may need to recruit technical volunteers just to catch up. Zoo directors must also consider the appropriate breadth of the research niche. Perhaps science, in some way, should be connected to everything we do.

CARE, MAINTENANCE, AND WELFARE

The goal of care and maintenance should be to provide every animal in the collection with an environment that optimizes health and welfare. In addition to satisfying physiological needs, this should include a living environment that offers species-appropriate stimulation and a social environment that provides a species-representative social group.

We must acknowledge that to achieve this goal in the zoo we must overcome spatial and economic constraints, as well as disagreements about how animal welfare or well-being should be defined (Tannenbaum 1991). Indeed, some common ground of agreement between captive animal managers and our clientele (zoo visitors and financial supporters) must be reached in order to retain credibility on animal issues and to mobilize the financial resources necessary to achieve local and global conservation goals (McManamon 1993). Science must be engaged to provide objective guidance and feedback on the success of management for each taxon for which we have accepted responsibility. Zoo education must communicate the message that animals in our zoos are well cared for and must motivate the public to act on behalf of the ecosystem. At the same time, communication is a two-way street. In order to be successful, we must also understand the attitudes and values of our clients and realistically communicate with them within the context of their own expectations of animal-human interactions and issues.

The basics of captive care can be realized through a philosophy of husbandry and exhibition that is based on the natural history of the animal (Maple 1979). However, a fine-tuning of care requires a scientific approach to evaluating the zoo environment. In a successful zoo environment, the animals will experience well-being, which would include physical health that is equal to or better than that experienced in the wild, with corresponding longevity and quality of life; reproductive success (if intended); and species-typical levels of behavior. Abnormal behavior should be absent or rare (Clarke et al. 1982).

Although Novak and Drewson recognized the difficulty of constructing a definition of well-being applicable to all taxa, the scope of their construct is similar to our own: "An animal may experience well-being if it is free from distress most of the time (some distress may be unavoidable), is in good physical health, exhibits a substantial range of the species-typical repertoire, and is able to deal effectively with environmental stimuli" (Novak and Drewson 1989).

Managers of captive animals should never fool themselves with the belief that they can replicate nature in a captive setting (Hediger 1950). To expect this outcome would demonstrate an ignorance of the intricacies and complexities that characterize natural ecosystems. Zoos should instead direct their efforts toward providing their inhabitants with as many biologically and ecologically relevant stimuli as possible. Additional scientific resources must be directed toward the taxa for which our knowledge of these stimuli is primitive.

The new wave of zoo exhibitry is naturalistic (Coe and Maple 1984, Hutchins et al. 1984). Its most basic features include exposure to fresh air and natural sunlight; a soft substrate, such as dirt and grass, instead of concrete; trees and bushes for climbing, leaning, foraging, nesting, and visual cover; rocks for perching and hiding; and pools of water for swimming, bathing, and drinking. More subtle features include seasonal variations in climate, varying the feeding regime in time and location to simulate the natural condition, and creating more three-dimensional space and varied topography for exercise and exploration. Critics have argued that naturalistic exhibits have been designed to make ourselves happier about keeping animals in captivity and that "the animals aren't any happier in the new natural habitats" (Siebert 1991). We disagree.

It is true that zoo visitors' impressions are clearly influenced by how animals are exhibited (Maple and Finlay 1986). Thus, there is a practical educational reason for zoo exhibit aesthetics. The AZA accreditation process implicitly acknowledges this influence by noting the appearance of exhibited animals on inspection forms. However, scientific evidence also indicates that animals thrive in these naturalistic habitats. In several primate species studied, movement from cage-style to naturalistic exhibits has been associated with an increase in species-typical behavior, a decrease in abnormal behavior, and an increase in reproductive behavior (Clarke et al. 1982, Maple and Finlay 1986, Snowdon 1991).

This suggests that providing an outdoor, basically naturalistic exhibit is a necessary first step toward creating a better quality of life for animals in the zoo. However, a naturalistic habitat is beneficial only if it provides the biologically relevant stimuli for its residents. The next step is to focus on the finer details of the habitat, such as the particular stimuli that the animals perceive and that shape their behavior. Scientific research is the best means by which to determine these species-specific environmental stimuli. For example, in a study of lowland gorilla (G. g. gorilla) habitats, Ogden, Finlay, and Maple measured several environmental variables to determine which ones affected the gorillas' behavior (1990). The authors found that the components of the outdoor environment that elicited natural behavior (and that were preferred by the gorillas in all groups) were flat areas that contained an object in addition to grass substrate, such as rocks, trees, and logs. A study of common marmosets determined that one critical component of the marmosets' environment, more important than just outdoor space, was the availability of dense visual cover in the form of thick bushes (Chamove and Rohrhuber 1989).

We feel comfortable with our description of well-being as the exhibition of species-typical behavior, including breeding behavior and a lack of abnormal behavior, and good physical health. Still, there are many unanswered questions. Is the well-being of wild animals greater than that of captive animals? What constitutes well-being in the wild? The ability to find food and shelter? The ability to successfully avoid predators? The freedom to go in a desired direction and for any desired distance?

In many ways, zoo animals live healthier lives than their wild counterparts. They are freed from the tasks of searching for food and shelter, and they do not have to be constantly vigilant in avoiding predators. They are provided with medical surveillance and treatment for infectious diseases, parasites, and occasional injuries. At the same time, it is unrealistic to expect that the natural processes of aging, or the ability of parasitic and infectious organisms to change in virulence in response to their host's defenses, will be eliminated by medical techniques available in the zoo. The rapid and continuous progress in veterinary and human medicine will, for the foreseeable future, provide increasingly effective means to relieve pain, treat disease, and improve the quality of animals' lives in captivity and in the wild. For example, artificial hip replacement techniques have been used in a gorilla (Ott-Joslin et al. 1987) and a snow leopard, *Panthera uncia* (Paul et al. 1985), and advanced diagnostic imaging techniques, such as CAT scans, can be applied to captive wild animals (Miller 1993).

Nor are such advances limited to endangered species. Current veterinary publications make it clear that the expected level of medical care for nondomestic pets now includes the use of laparoscopy, ultrasound, microsurgery, expensive antibiotics, and other treatment modalities (e.g., Frye 1991, Ritchie et al. 1993). The availability and application of such advanced techniques are communicated by zoo public relations departments and private practitioners to demonstrate the quality of medical care provided to individual animals.

It is true that captive animals have less freedom of choice in terms of movement. However, even migrating birds, who could go anywhere in the country, overwinter in exactly the same spots (i.e., the same bushes) every year. In fact, although animals have the freedom to move as they please in the wild, many species enforce their own limitations through territorial behavior. Territorial boundaries (marked by scents, food availability, or open aggression from neighbors) are no less real simply because they are imperceptible to humans. Many other species experience limitations to their movement forced on them by shrinking habitats and the disappearance of essential resources. Whether wild animals experience greater well-being than captive animals is a subject of debate that will always be susceptible to the imposition of human values. More objectively, it can be said that wild animals face more potential dangers and are definitely more vulnerable to health problems. Through a scientifically based plan of naturalistic habitat design, it is possible to provide the opportunity for the display of species-typical behavior, and this constitutes an essential component of well-being in zoo animals.

An additional, essential tool for encouraging species-typical behavior in any exhibit, naturalistic or otherwise, is enrichment. Enrichment involves adding one or more items (food or nonfood) to an animal's enclosure to render it richer in stimulation. The goal of the enrichment plan, again, is to encourage normal levels of general activity and foraging. Many wild animals spend a significant portion of their time (often much greater than 50 percent) feeding and foraging (Maple and Hoff 1982). Basic principles of feeding enrichment attempt to mimic the natural condition by manipulating the availability of food over time and by providing foods in a manner that requires greater handling time than do meals served in bite-sized pieces. Instead of being delivered in one or two feedings, the day's ration is spread out over the day into more frequent smaller feedings. The location of feeding should also vary. Food could be scattered or hidden in different areas to encourage exploration and searching behavior. Fruits can be fed without peeling them and chopping them into bite-sized pieces, feeding of browse results in a greater amount of handling time, and finally, foods can be presented in various unnatural puzzle boxes or other constructs to encourage more handling and manipulation prior to consumption (Bloomsmith 1989).

These basic principles, labeled as enrichment, should actually be labeled "routine feeding management." The development of enrichment techniques relies heavily upon scientific research to determine the optimum methods. Although enrichment, in general, is universally successful, the specific forms that enrichment must take vary according to the species and to the housing of the specimen. Enrichment should become a standard component of captive animal husbandry. It should no longer be thought of as an extra—as something that can be skipped on a busy day. It should be the goal of captive animal managers to make richer the lives of their animals on a daily basis. Enrichment should be the standard.

The exhibition of species-typical behavior should be the goal of all zoo husbandry programs. There are, however, limitations on the extent to which this can

Figure 4. Willie B., a gorilla, in his zoo environment. Naturalistic zoo habitats should be realistic enough to remind people of their obligation to protect wild habitat. (Photograph by J. Sebo)

be achieved. Some forms of predatory behavior, such as chasing and ultimately killing prey, cannot realistically or humanely be promoted in captivity. Similarly, some forms of predator-avoidance behavior in prey animals, such as vigilance, may not be exhibited at typical levels in captivity, in the absence of predators. The necessity to be constantly vigilant may actually be distressful to the animals. And so, captive managers are faced with the dilemma of whether normal stressful behavior should be facilitated in captivity because it represents a part of the species' natural behavioral repertoire, or whether it should be discouraged because it is distressful. Scientific measurements of stress in both captive and wild animals would aid in addressing this dilemma.

Noting that the absence of stress is not necessarily good, Snowdon discusses the construct as it applies to the human primate: "A moderate level of stress that we can regulate ourselves appears to be preferable. Given that human beings seek constant intellectual, physical, and social challenges, one might argue that psychological well-being could be defined in terms of our ability to adapt to changes in our environment" (1991).

Clearly, primate environments in the zoo should introduce elements of change, and to the extent that it is practical, environmental stimuli should be controlled by the animals' behavior. This idea is derived from the pioneering experiments of Markowitz (1978). Good zoos provide opportunities for animals to express their autonomy.

For many creatures, high social density would present problems in captivity, but for others, such as the Chilean flamingo *(Phoenicopterus chilensis),* it represents a necessary condition for successful breeding (Stevens 1991). The pursuit of well-being in the zoo requires a thorough understanding of the animal's life in the wild. In this way, good zoo biologists depend upon the work of productive field biologists.

THE HUMANE AND OPEN ZOO

Determining the degree to which each of these elements of captive care (exhibit and holding-area design elements, level of medical care, behavioral enrichment) will be provided to various taxa (and to individuals within the same species) necessitates complex decision-making regarding resource allocation for the present and into the foreseeable future.

Presently, there is no clear or consistent ethical philosophy espoused by zoological institutions. Zoo education programs, for instance, frequently house animals in off-exhibit facilities that would be deemed unacceptable for exhibit purposes, and the exercise opportunities afforded these animals are limited to being held or transported. Behavioral enrichment opportunities, when provided, are frequently limited to exhibit space where the public may enjoy observing active animals. However, many animals spend the majority of the twenty-four-hour day in a deprived holding area, and this may actually encompass the most active hours of the day for many species, for example, elephants, hippos, and rhinos (Maple and Archibald 1993).

While the above situations are often seen but not discussed or justified, similar institutional decisions about resource allocation for medical care are more frequently verbalized, at least within the institution. Positive resource allocation is frequently communicated openly to zoo clientele, but negative decision-making is much more problematic and is rarely acknowledged publicly. With the same prognosis for successful treatment, is the same expensive medical therapy as likely to be provided to an endangered, genetically underrepresented reptile as it is to a locally famous but genetically surplus mammal?

Within domestic animal medicine, decision-making of this type is most frequently within the purview of the individual owner. The veterinarian is expected to present reasonable options, costs, and prognoses, but it is the owner who ultimately decides his or her willingness and ability to allocate money for treatment, based upon the perceived value of the animal to the client. For example, euthanasia for a racing stallion with a fractured leg might be selected if the animal could no longer perform competitively. Alternatively, however, the same fractured leg in a breeding bull or a pet goat might be repaired. Animals treated in a general practice might be euthanatized because of the cost of surgery, but a teaching hospital might elect to subsidize the cost of treatment for a similar case in order to teach many students how to save other patients when funds are available. Although some critics might object to this approach as utilitarian, it is generally accepted by veterinary clientele and the general public. Practically speaking, it occurs because the owner provides the funds to support the decision being made. In the zoo, however, management generally preselects the outcome without proactive consultation with the financial provider—the community or clientele.

While the role of the owner is thus usurped by the caretaker, the rationale is rarely consistently based on our advertised valuation of animals. Resources are still selectively allocated to mammals, or to exhibition rather than to holding facilities, while fund-raising and publicity efforts are frequently focused selectively on individual animals or breeding success. Conversely, zoo professionals consistently emphasize that charismatic animals are merely flagship species for ecosystems and that even the most uncharismatic species of animals and plants are equally valuable and worthy of preservation. The inherent contradiction in philosophy is obvious, and our credibility as animal managers is severely strained by such ethical inconsistency. In fact, if we do not investigate and provide the highest quality, natural history–based husbandry and medical care possible to all of the animals in our care, we deserve the criticism. Otherwise, we shall be demonstrating the same irresponsibility as exotic pet owners who cannot support or grow tired of their pet monkeys.

Returning to the concept of medicine, it is well accepted that effective treatment and communication can occur only within a clinical partnership in which the client perceives that he or she is treated with compassion and empathy, not just competent professionalism. For our purposes, it is critical that zoo managers be perceived as caring, compassionate individuals with regard to animal issues, as we endeavor to engage the public in an active partnership to achieve our mutual conservation goals. Instead of preselecting options for them, such as whether publicly owned rabbits in the children's zoo should be provided with the same level of medical care that any pet owner could expect and obtain, we need to entertain seriously the possibility of proactively soliciting input on resource allocation. Otherwise, we risk winning the battle but losing the war—creating serious miscalculations of public sentiment and trust. Rather than limiting the results to a single zoo manager's loss of a job due to public scandal, the entire profession risks loss of credibility and trust on animal issues (Maple and Archibald 1993).

We would propose, therefore, that the most rational approach to these difficult and complex issues is to view the solicitation of community input as sharing responsibility rather than as abandoning it. Rather than seeing the community, including animal welfare activists, as outsiders, the good zoo should solicit its input in establishing what the institution stands for. Zoo managers should constantly play devil's advocate with themselves, questioning their chosen option for any animal (space allotment, euthanasia, social grouping, etc.) by testing it against the various perspectives represented in the community. The zoo should be viewed as the center of a network of participating partners in the community. The expectations of a rural community may be different from those in an urban environment, as their needs and backgrounds may vary.

Conversely, it is also the zoo's responsibility to ensure that this input and decision-making are rationally based, not simply reflective of a loud or well-financed minority. The benefits of resource investment in in-situ projects, as opposed to holding large numbers of animals in the zoo, should be clearly articulated to the community. With guidance, a well-informed community is more likely to appreciate and make those difficult global decisions. Zoo managers, however, should not denigrate concern for individual animals as emotionalism or irrationality, particularly when it is our marketing of individual animals and promotions through adopt-an-animal programs that exploit such attachment and that help to pay our operating expenses.

It is not reasonable to expect political and financial support for a zoo, and to market individual animals, without also expecting to provide accountability for management decisions. Even if providing differing levels of care for hoofstock, primates, reptiles, or birds may not be opposed by animal welfare and rights activists, they should be opposed by our internal professional standards. We should be able to articulate clearly the reasons that resources are not allocated toward certain animals. If an institution opposes placing wild mammals in the pet trade, yet supplies birds or reptiles to such outlets, how can this be justified professionally and philosophically?

The zoo therefore should be significantly ahead of the animal rights and welfare lobbies, through self-examination, anticipation of challenges, and proactive innovation. These are, rightfully, the reasonable expectations of our clientele—the right to ask questions and to be provided thoughtful, complete, and consistent answers.

THE MOTIVATIONAL ZOO

The motivational zoo realizes that its mission is to conserve wildlife and natural habitats through changing the attitudes of its visitors. It motivates them to think about and to act by promoting conservation. In fact, public education plays at least as large a role, if not a greater role, in conservation as does captive propagation. Captive propagation should not be upheld as the cornerstone of our identity; existing zoos have only enough space to sustain adequate populations of no more than 900 species (Conway 1986). As a result, although the role of zoos in captive propagation is important in terms of the overall effectiveness and scope of zoos' potential impact on preservation, conservation education is the undisputed cornerstone of zoo programs. Yes, zoos contribute to conservation through their captive breeding programs, but why is conservation necessary in the first place? It is because of human actions. So why not deal with human actions by trying to influence human actions? Educating and motivating humans may be the best method for contributing to conservation.

Critics of zoos claim that people do not go to the zoo to learn about wildlife conservation; instead, they go as part of a social outing and for amusement. Furthermore, early critics of the zoo suggested that zoos are misguided in believing that they are serving an educational function (Sommer 1974). These critics argue that not only are zoos not educational, but visitors may actually be coming away from the zoo with the wrong messages, for example, that animals pace back and forth or seem lethargic and detached, indicating boredom and psychopathology.

Captive care and maintenance, and their underlying philosophy, are directly related to whether a zoo will succeed as a motivational zoo. If zoo managers apply their scientific approach to maintaining animals and exercise the charge to run a humane zoo, the zoo will become a motivational zoo. If captive environments are managed as we have outlined herein, visitors will not see lethargic and psychotic animals. Instead, they will see active animals in a naturalistic habitat behaving much the same as their wild counterparts. Such good zoos are truly inspirational.

The general public may not come to the zoo looking for an education, but they may leave having been educated or at least strongly affected by their experience without even realizing it. Seeing two animals groom each other, hearing a lion's roar, observing a flamingo roll its egg, or having a close-up encounter with an elephant or a gorilla can leave a lasting impression on a visitor. It can be so stimulating that the visitor will want to learn more; maybe the visitor will be

Figure 5. Children in close contact with zoo animals. This type of contact is an emotional bridge to conservation awareness. (Photograph by J. Sebo)

motivated simply to read some of the graphic panels at the zoo, or perhaps the motivation will extend to reading related books and magazines. In the best-case scenario, the visitor will be moved to the point of taking some tangible form of conservation action. The more informed the visitor is, the more likely it is that a conservation consciousness will be instilled. The zoo experience can be that initial spark that motivates a member of the general public to become more responsible about conserving the environment.

We recognize a need to confirm that indeed the zoo experience is having the impact we anticipate. The scientific zoo must evaluate its progress in just this way. Scientific research contributes to conservation not only through animal research but also through studies on the public impact of our education programs (Swanagan 1993). Since few published studies actually measure the effect of zoo educational programs, we have identified this as a pressing agenda item for good zoos everywhere.

Structured educational programs at the zoo can be very motivating experiences, especially where zoo scientists participate in the development of the programs and share their specialized knowledge with those who teach. In a perfect zoo world the work of conservation biologists is the curriculum for conservation educators.

CONCLUSION

It is clear that elevating and expanding the role of science in zoos is a key factor in defining the good zoo. The application of science to each of the areas we have reviewed strengthens the zoo's ability to achieve its directives according to the highest standards of our profession. Through the endeavors of zoo biologists, we have learned how to house animals in ways that meet their basic metabolic needs, while providing appropriate (i.e., ecologically relevant) environmental and social stimulation. We recognize that zoos cannot duplicate the entire menu of variables found in nature, but they can and must apply scientific principles to provide more effective and stimulating environments for captive wildlife. From a scientific foundation, we can implement our conservation philosophy and thereby become more ethically accountable for our management decisions. Finally, when we apply the findings of zoo biology to animal management, we discover new pathways for communicating our conservation message to zoo visitors. The message can be delivered actively, through empirically based education programs, or passively, through exhibitry that encourages species-appropriate behavior patterns. Visitor experiences such as these have proven to be effective motivators of zoo patrons. Ultimately, this must be the mission to which the modern zoo aspires: motivating zoo visitors to take action to preserve and protect the diversity of life inhabiting our fragile earth.

REFERENCES

Bloomsmith, M. A. 1989. Feeding enrichment for captive great apes. In *Housing, Care, and Psychological Well-Being of Captive and Laboratory Primates,* ed. E. F. Segal, 336–356. Park Ridge, N.J.: Noyes Publications.

Chamove, A. S., and B. Rohrhuber. 1989. Moving callitrichid monkeys from cages to outside areas. *Zoo Biology* 8:151–163.

Clarke, A. S., C. J. Juno, and T. L. Maple. 1982. Behavioral effects of a change in the physical environment: A pilot study of captive chimpanzees. *Zoo Biology* 1:371–380.

Coe, J. C., and T. L. Maple. 1984. Approaching Eden: A behavioral basis for great apes exhibits. In *Proceedings: American Association of Zoological Parks and Aquariums,* 117–128. Wheeling, W.Va.: AAZPA.

Conway, W. G. 1986. The practical difficulties and financial implications of endangered species breeding programmes. *International Zoo Yearbook* 24/25:210–219.

Eisenberg, J. F. 1975. Design and administration of zoological research programs. In *Research in Zoos and Aquariums: A Symposium,* 12–18. Washington, D.C.: National Academy of Sciences.

Finlay, T. W., and T. L. Maple. 1986. A survey of research in American zoos and aquariums. *Zoo Biology* 5:261–268.

Frye, F. L. 1991. *Biomedical and Surgical Aspects of Captive Reptile Husbandry.* 2d ed. Malabar, Fla.: Krieger Publishing.

Hediger, H. 1950. *Wild Animals in Captivity.* London: Butterworth & Co.

———. 1969. *Man and Animal in the Zoo.* London: Routledge & Kegan Paul.

Hutchins, M., D. Hancocks, and C. Crockett. 1984. Naturalistic solutions to the behavioral problems of captive animals. *Zoologische Garten* 54:28–42.

Maple, T. L. 1979. Great apes in captivity: The good, the bad, and the ugly. In *Captivity and Behavior,* ed. J. Erwin, T. L. Maple, and G. Mitchell. New York: Van Nostrand Reinhold.

Maple, T. L., and E. Archibald. 1993. *Zoo Man: Inside the Zoo Revolution.* Atlanta, Ga.: Longstreet Press.

Maple, T. L., and T. W. Finlay. 1986. Evaluating the environment of captive nonhuman primates. In *Primates: The Road to Self-Sustaining Populations,* ed. K. Benirschke. New York: Springer-Verlag.

Maple, T. L., and M. P. Hoff. 1982. *Gorilla Behavior.* New York: Van Nostrand Reinhold.

Markowitz, H. 1978. Engineering environments for behavioral opportunities in the zoo. *Behavior Analyst* Spring: 34–47.

McManamon, R. 1993. The humane care of captive wild animals. In *Zoo and Wild Animal Medicine: Current Therapy,* ed. M. E. Fowler, 3d ed., 61–63. Philadelphia, Pa.: W. B. Saunders.

Miller, R. E. 1993. X-rays for exotics: Diagnostic radiology at the zoo. *Zoo Life* 4:89–90.

Novak, M. A., and K. Drewson. 1989. Enriching the lives of captive primates: Issues and problems. In *Housing, Care, and Psychological Well-Being of Captive and Laboratory Primates*, ed. E. F. Segal. Park Ridge, N.J.: Noyes Publications.

Ogden, J. J., T. W. Finlay, and T. L. Maple. 1990. Gorilla adaptations to naturalistic environments. *Zoo Biology* 9(2): 107–121.

Ott-Joslin, J., T. Turner, J. Galante, and B. Torgerson. 1987. Bilateral total hip replacement in a 26-year-old lowland gorilla *(Gorilla gorilla)*. *American Association of Zoo Veterinarians Proceedings:* 516–518.

Paul, H. A., W. L. Bargar, and R. Leininger. 1985. Total hip replacement in a snow leopard. *Journal of the American Veterinarian Medical Association* 187(11): 1262–1263.

Ritchie, B. W., G. J. Harrison, and L. R. Harrison, eds. 1993. *Avian Medicine: Principles and Applications*. Lake Worth, Fla.: Wingers Publishing.

Siebert, C. 1991. Where have all the animals gone? The lamentable extinction of zoos. *Harper's,* May, 49–58.

Snowdon, C. T. 1991. Naturalistic environments and psychological well-being. In *Through the Looking Glass: Issues of Psychological Well-Being in Captive Nonhuman Primates*, ed. M. A. Novak and A. J. Petts. Washington, D.C.: American Psychological Association.

Sommer, R. 1974. *Tight Spaces*. Englewood Cliffs, N.J.: Prentice-Hall.

Stevens, E. S. 1991. Flamingos breeding: The role of group displays. *Zoo Biology* 10:53–63.

Swanagan, J. S. 1993. An assessment of factors influencing zoo visitors' conservation attitudes and behavior. Master's thesis, School of Public Policy, Georgia Institute of Technology.

Tannenbaum, J. 1991. Ethics and welfare: The inextricable connection. *Journal of the American Veterinarian Medical Association* 198(8): 1360–1376.

DIFFERING CONCEPTIONS OF

ANIMAL WELFARE

John Wuichet and Bryan Norton

How shall we conserve species when other moral obligations—especially obligations to treat individual animals with care and respect—conflict? In the previous chapter, Maple, McManamon, and Stevens have proposed a general framework for setting priorities among the competing goals of zoo management. While we agree with their general approach to achieving a balance among management goals, we question their optimism in suggesting that a balancing of conservation activities with individual animal protection—however judicious—can result from consensus-building among and between zoos, the community, and activists with varied viewpoints. Rather than attempting to avoid conflict by keeping ahead of animal rights advocates, we recommend that zoos and zoo professionals, more realistically, accept the inevitability of conflict and take steps to understand, coexist, and manage within a contentious social context.

Maple et al. also state that member zoos of the AZA pursue four directives—conservation, education, science, and recreation—and they ask, What exactly is the role of care-giving in this complex of goals and objectives? The authors see the four AZA directives as an interconnected web of goals that depend on one another for mutual success: recreation is the stimulus that makes educating the public possible, education is a vehicle for achieving the mission of wildlife conservation, and the educated and entertained public in turn provides the funds for employing science to aid in the conservation of animals in the wild. While lamenting the present lack of funding in zoos for both science (which they all agree "is still the most underrealized of the four directives") and conservation (which they say sadly depends to a large extent on surplus revenues), they are satisfied with the role education currently plays as "the undisputed cornerstone" of zoo missions. Insofar as it is a means to education, they also weigh recreation heavily.

In this context, they see concern for the welfare of captive animals as a foundation that underlies all four of the directives—a sine qua non for the efforts to achieve other goals. The authors begin by asserting that "the care and welfare of the animals in our institutions have become our highest operational priorities and *enable* us to achieve our basic directives," and they conclude by saying science "*strengthens* the zoo's ability to achieve its directives" (emphasis added). These assertions point to an underlying sense of the interconnectedness of the various zoo missions, with science in the service of conservation, and humane treatment as a unifying mission for the zoo community. While we agree that the four directives can complement each other, we also believe they can sometimes conflict. We also doubt whether there is anything inherent in these directives that guarantees the welfare of captives.

Although we agree with their general approach to setting zoo goals and echo their call for a stronger commitment to science in all zoos, we also believe it is important—as the authors realize—to qualify this defense of zoo aspirations with the clear recognition that it stands mainly as a promissory note of better things to come than as a defense of past practices. It will remain so until most zoos—and not just the good zoos, as is true today—put some real dollars where their professed goals are.

While we accept the authors' basic approach, we are less optimistic that zoos and zoo professionals, as well as other wildlife managers, can avoid the coming conflicts with animal rights advocates—a small but growing minority among zoos' main clientele, the public. When Maple et al. say the zoo should be "significantly ahead of the animal rights and welfare lobbies, through self-examination, anticipation of challenges, and proactive innovation," we fear they may underestimate the dangers and difficulties ahead.

In response to Maple et al., therefore, we raise several somewhat related concerns. First, we question whether zoos and animal rights activists are going in the same direction. In other words, it may be that zoos and animal rights activists are both interested in the welfare of captive animals, but the two have very different ideas regarding precisely what "animal welfare" means. If so, it may be difficult for zoos to get ahead of their critics. Second, we question whether all animal rights activists are themselves going in the same direction; if some zoo critics employ one conception of animal welfare while others mean something entirely different, then soliciting input from the community may yield only more heated and frequent controversies. But more fundamentally, we also are concerned that zoos have more than animal rights activists as clients, which may place additional conflicting demands on zoo professionals seeking to increase community input. After positing two different conceptions of animal welfare by which zoos are often judged, we explore the impacts of these different conceptions of welfare on the type of care given to animals in captivity and consider the implications for the practices of care-giving.

DIFFERING DIRECTIONS AMONG ZOOS AND ANIMAL RIGHTS ACTIVISTS

We agree with the proactive approach of Maple et al. but are concerned whether zoos can achieve an idyllic state of agreement and camaraderie with the conflicting demands placed upon them by modern society. Moreover, reducing the horizon of managerial conflicts to a division between zoos and animal rights activists in the first place may be an oversimplification. The modern zoo must deal with a wide array of clients, including but not limited to the animals as individuals, the institution itself, zoogoers, activists, those who fund zoos, those who legitimate or regulate zoos, and the various species the individual animals represent. Since the zoo professional often must consider the demands of each of these clients in the course of making a management decision, we question whether any policy can simultaneously satisfy all of them. The best strategy may be to recognize that some conflict is inherent in zoo administration and to focus on ensuring productive avenues for conflict resolution rather than attempting to avoid conflicts before they arise by trying to predict and assuage the future concerns of myriad and divergent interests.

We are also concerned whether some of the zoo's many clients enjoy enough consensus within their own camps to make prediction of their future demands possible. Are all animal rights activists, that is, going in the same direction? The zoo can incorporate public opinion into managerial decision-making only to the extent that (1) consensus generally exists in the whole public community or (2) zoos can create or foster that consensus despite serious differences with and among their critics. We doubt both (1) and (2). Even today, critics of zoos do not agree on what is wrong with zoos, much less how to fix these problems. What if there is consensus on the problem but not on the solution? What if the public takes a position for a particular zoo management strategy but is not willing to allocate the funds necessary to execute such a program, or is not willing to give up other benefits in order to achieve the desired goal? What if the desires of the public are so varied that a general opinion cannot be ascertained?

TWO CONCEPTIONS OF ANIMAL WELFARE

Opinion polls show that the public exhibits a wide variety of attitudes toward wildlife (Kellert 1989, Donnelley 1990). Some of the variation in positions taken by members of the public on issues surrounding captive care can be accounted for by variation in the underlying conceptions of what it means for a captive animal to have well-being. These variations give rise to at least two quite different criteria of animal well-being—what we shall call the criterion of welfare and the criterion of authenticity. We do not pretend that zoos categorically subscribe to one and animal rights activists subscribe to the other. In fact, we recognize that some may tend toward one in some situations and toward the other under altogether different circumstances. Yet, a careful description of these two conceptions will draw attention to the hidden complexity of the task of zoo management and pave the way for the suggested improvements to the Maple et al. approach that follow. The zoo professional attempting to grapple with the divergent interests of myriad clients will be better prepared to mitigate conflicts and facilitate consensus if the ideological differences—the underlying interests rather than the more tactical positions (Carpenter and Kennedy 1988)—are clearly understood.

The Welfare Criterion

According to the welfare criterion (which we believe is the implicit basis of the arguments of most zoo professionals, some zoo critics, and Maple et al.), treatment of captive animals must achieve a level of well-being comparable to, or better than, the life they could be expected to live in a wild context. The welfare of an animal, in this view, can be measured first by standard physical criteria, such as longevity and freedom from disease, and second by psychological criteria, such as the exhibition of species-typical behavior. Proponents of the welfare criterion see zoos as capable of providing adequate levels of welfare as measured by these criteria. If it was discovered that zoos were not, in fact, capable of providing acceptable levels of these criteria in captivity, it would be arguably appropriate to return animals to the wild. From this perspective there is no inherent difficulty in the idea of individual animals in captivity per se, or if there is, the difficulty is outweighed by the benefits their captivity provides for some other client, such as the larger population, species, or ecosystem.

The perspective of the welfare criterion of animal well-being is evident in the previous chapter. The criteria Maple et al. see as relevant to the evaluation of captive animal welfare include species-appropriate stimulation, a species-

representative social group, and the satisfaction of physiological needs. By reference to Clarke et al. (1982), they add the criteria of longevity, quality of life,
reproductive success (if intended), and species-typical behavior. By concurring
with Novak and Drewson (1989), they add freedom from avoidable distress.

Zoos, zoo biologists, and zoo professionals are not the only ones to take the
perspective of the welfare criterion. It is evident, for example, in the Brambell
Report of 1965, the findings of a committee appointed by the British government to investigate claims of unfavorable conditions in intensive livestock husbandry:

We must draw the line at conditions which completely suppress all or nearly all the
natural, instinctive urges and behavior patterns characteristic of actions appropriate to the
high degree of social organization as found in the ancestral wild species. . . . In principle
we disapprove of a degree of confinement of an animal which necessarily frustrates most
of the major activities which make up its natural behavior. (Singer 1975, 135)

Although animal liberationists have been severely criticized for failing to note the
apparent differences between our moral relations with domestic animals and
those with wild animals (Callicott 1989; Norton 1991, 241), we are concerned
here not with the treatment of wild animals generally but with the treatment of
wild animals in captivity, making the Brambell Report more relevant than it
might otherwise be. Even though it focuses on the treatment of domesticated
livestock rather than the care of captive wild animals, the Brambell Report
illustrates how both zoos and some animal rights activists can view animal well-
being from the perspective of the welfare criterion.

The Authenticity Criterion

Some (though not all) critics of zoos judge the actions of zoos according to an
altogether different standard. The authenticity criterion is more difficult to state
precisely, but it is implicit, for example, in Jamieson's arguments for a presumption that wild animals should remain free (Jamieson 1985, this volume). This
argument suggests there is inherent value in the state of wildness itself; no amount
of caring for an animal can compensate for the freedom and authenticity of
experience lost when the animal is removed from its habitat, which is fraught
with both danger and opportunity, and in which the animal succeeds or fails on
its own strengths or weaknesses. The authenticity criterion of well-being might
be stated as follows: A wild animal achieves a state of authentic well-being when
it survives and reproduces offspring, based on its own genetic abilities and behavioral adaptations, in a truly natural (as opposed to naturalistic) environment.

This criterion should not be interpreted to require that animals in the wild should encounter no human impacts—wild animals differ from domesticated ones not in being free of human impacts but in being free to adapt and evolve to such impacts through their own behavior, rather than being adapted to have characteristics useful or cherished by humans. In terms of our obligations to animals, captive animals are more analogous to domesticated animals than wild animals—by domesticating them, we have entered an implicit contract with domestic animals. We owe them protection as reciprocation for the benefits they bestow on us in a community of intermingled species. We have limited their free behavior patterns by breeding and conditioning; we thereby accept a responsibility toward their protection (Norton 1991, 242).

To illustrate the authenticity criterion, we introduce a literary analogy. Kurt Vonnegut, in *Slaughterhouse-Five,* suggests the importance of authentic autonomy (with regard to humans) as he describes a book, *The Big Board,* that the protagonist had read:

[The book] was about an Earthling man and woman who were kidnapped by extraterrestrials. They were put on display in a zoo on a planet called Zircon-212.

These fictitious people in the zoo had a big board supposedly showing stock market quotations and commodity prices along one wall of their habitat, and a news ticker, and a telephone that was supposedly connected to a brokerage on Earth. The creatures on Zircon-212 told their captives that they had invested a million dollars for them back on Earth, and that it was up to the captives to manage it so that they would be fabulously wealthy when they were returned to Earth.

The telephone and the big board and the ticker were all fakes, of course. They were simply stimulants to make the Earthlings perform vividly for the crowds at the zoo—to make them jump up and down and cheer, or gloat, or sulk, or tear their hair, to be scared shitless or to feel as contented as babies in their mothers' arms.

The Earthlings did very well on paper. That was part of the rigging of course. And religion got mixed up in it, too. The news ticker reminded them that the President of the United States had declared National Prayer Week, and that everybody should pray. The Earthlings had had a bad week on the market before that. They had lost a small fortune in olive oil futures. So they gave praying a whirl.

It worked. Olive oil went up. (Vonnegut 1969, 174)

Although the characters freely chose to exhibit behaviors that might be called typical of the human species, it would seem ludicrous to call the kind of autonomy experienced by these fictitious captives authentic or real. We think many critics of zoos would insist that an authentic relationship between behavior and outcomes—what was lacking in the passage quoted—is a component of animal, as well as human, well-being (e.g., Heim 1971, 150; Singer 1975, 47; Jamieson, this volume). As Maple et al. point out, Snowdon found human authenticity of experience relevant to his discussion of psychological well-being for non-human primates (Snowdon 1991). Because the passage from Vonnegut's novel deals with humans instead of zoo animals, it provides no proof, but it does serve to illustrate just how divergent are the two criteria. At the very least it shows that with regard to the species *Homo sapiens,* stimuli that evoke species-typical behavior do not necessarily entail real autonomy or authenticity of experience, and attempting to justify captivity on the grounds that behavior in captivity and behavior out of captivity are roughly indistinguishable falls short of the authenticity criterion. Simply because humans tend to sleep in the same place every night does not mean we would be just as well off sleeping in prison. Similarly, the fact that some migratory birds elect to return to the same spot every year or that other animals stake out rather limited territories does not, from the perspective of the authenticity criterion, serve to justify their confinement.

While Maple et al. do recognize that "good zoos provide opportunities for animals to express their autonomy," the difference between their recognition and the perspective of the authenticity criterion is that the latter does not believe it is *possible* for zoos to provide such opportunities. The welfare criterion sees some conditions of captivity as acceptable and other conditions unacceptable. The authenticity criterion finds captivity itself to be unacceptable.

There are probably more than two criteria employed by zoo critics and the public at large as yardsticks for the measurement of animal well-being; nevertheless, we believe the pair presented here captures one important divergence in these yardsticks. Certainly not all rights advocates adhere to the authenticity criterion. But so long as some animal rights and welfare advocates understand animal experience as diminished if it falls short of authenticity (as illustrated in the passage from Vonnegut), it is difficult to see how zoos can fulfill the expectations of those animal advocates short of setting animals free. As long as some critics of zoos employ the authenticity criterion of animal well-being, zoos will remain on the defensive regarding the *fact* of captivity, not just regarding the *conditions* of captivity. We therefore question whether it is possible for zoos to get ahead of those critics who embrace the authenticity criterion.

We do not believe however that the intractability of the debate over criteria for well-being undermines the basic direction proposed by Maple, McManamon, and Stevens. Even if consensus is not achievable, a proactive policy is likely to expand the base of support in the community and forestall interference from government and regulatory agencies. If zoos have a well-defined, well-articulated, and scientifically defensible policy regarding the treatment of animals in captive breeding programs, and if they fund, implement, and enforce the policy meticulously, they will have more success in defending their actions. If zoos do not examine their policies and improve them when necessary, they will be forced to adopt a defensive posture.

THE ROLE OF SCIENCE IN PUBLIC CARE-GIVING

Maple et al. suggest that the welfare of captive animals can be enhanced not only by the adoption of proactive policies aimed at consensus with the animal rights activist community but also by a marked increase in scientific activity in zoos. We agree. But we are concerned that science and community-based zoo management, without the guarantee of an explicit, articulated, and continued commitment to animal welfare, could wind up raising more questions than answers. Science is indeed vital to determining what types of stimuli can best increase the well-being of captive animals, but it cannot, without the help of softer disciplines, decide what well-being actually is, or to what extent it should be sought at the expense of other goals. While Maple et al. do propose a framework for prioritizing the four directives of the AZA, they do not consider the possibility that the four directives themselves could be fraught with inherent conflict—that sometimes difficult choices must be made between incommensurable aspects of otherwise complementary goals.

Maple et al. call on science as a solution to the welfare issue: "Scientific measurements of stress in both captive and wild animals would aid in addressing this dilemma" of whether to try to duplicate "normal stressful behavior" in captivity. Science, we agree, would no doubt aid in addressing the dilemma. But suppose we had at our disposal all the scientific information about both captive and wild stressors we could possibly accumulate. Would we then know what level of stressful behavior ought to be facilitated in zoos? The answer to that question has more to do with public sentiment, ethics, and policy than science. Increased scientific knowledge is indeed helpful and should be furthered. But science cannot fully address the problem at hand, because the question of how much stress we impose on captive animals is as much a normative as a scientific issue. It may depend not just on the animal's response but also on contextual questions such as whether the animal is being prepared for reintroduction. Increased university-zoo relations, therefore, should not focus solely on contact with the more scientific departments when confronting the issue of appropriate stress levels (and other zoo issues as well).

There is a reasonable tendency for zoo managers to want to claim that their animals are at least as well off or even better off than their wild counterparts. Indeed, this is a worthy aspiration, if the existence of animals in captivity is a given (and it probably is). But what follows if zoos are able to prove this? What if, through science and other disciplines, it is unequivocally determined that animals are better off in captivity? Are we then obligated (if committed to acting in the best interest of animal welfare) to attempt to take into captivity all the wild animals of the world? What if the converse is true? Are we then irresponsible if we do not set free every animal in every collection worldwide? "If there is only one principle governing our treatment of other animals, and [if] that principle is that we should do all in our power to reduce their pain, then we should remove wild animals from their natural habitats where they are threatened daily by pain and death and place them in zoos where they can be fed and protected" (Norton 1991, 241–242; see also Sagoff 1984, Callicott 1989).

But the fact of the matter is that there is more than one principle governing our treatment of animals, often held by the same people under various circumstances and on different scales. The criteria of welfare and authenticity are but two of them. Thus we need ethics and policy to sort out these competing principles and the associated tactical positions of interested parties, and to facilitate decision-making by presenting them to zoos and the public in a manner that places scientific knowledge in its larger social and moral context. As Norton explains, "Contextual thinking encourages us to value actions differently depending on the context in which they are analyzed" (1991, 242). To encourage contextual thinking in the face of new scientific discoveries is the necessary role of the nonscientific or social-scientific community.

Maple et al. remind us that "the well-being of captives is a subject that will always be susceptible to the imposition of human values." This is no shortcoming. Though less rigid than scientific data, human values are nonetheless a necessary and useful component of policy decisions for zoo managers. If we are to ensure that the data accumulated from scientific inquiries are used responsibly and with foresight, then the call for science must be supplemented with a call for more scientists with training in ethics and other humanistic disciplines, or at least for the consideration of human values in the process of scientific inquiry. New scientific discoveries lead inevitably to new questions. Scientific inquiry is only the beginning of a difficult process that involves the implementation of policies always in the dim light of imperfect information (see Hutchins et al., this volume).

Attempting to provide animals not just with physical well-being and stimuli that evoke species-typical behavior but also with *authentic* experiences may reduce conflicts with proponents of the authenticity criterion. But it will also lead zoos into difficult and largely uncharted areas of animal care, as would be the case if it is decided that predators must be provided authentic hunting opportunities, for example. Here is the dilemma—increased science and public input may only serve to generate new controversies. Some of these new difficulties are illustrated by Maple et al. in their discussion of stress levels appropriate for captive animals.

STRESS LEVELS IN CAPTIVITY AND IN THE WILD

While it may not be clear whether animals are generally better off in the wild or in captivity, everyone can at least agree that living conditions in the two environments are markedly different. It may be possible to agree on descriptions of the differences, but not on the moral valuation of such observations. This is due in no small part to the problem of trying to act in the interests of a group that cannot represent its own position to us in a way we can understand, and we are left with no other faculty than our own anthropomorphically biased opinions in attempting to gauge the meaning of well-being for nonhumans.

If, as Maple et al. assert, "the exhibition of species-typical behavior should be the goal of all zoo husbandry," then it follows that every reasonable step should be taken to include every possible stressful stimulus encountered by a wild animal in the daily regimen of captive life, even if this means introducing stimuli that evoke predator-avoidance behavior. After all, predator-avoidance behavior often *is* species-typical behavior. In the enclosures of threatened or endangered species, shadows of predatory birds or scents of predatory mammals could be introduced. In areas where surplused or less-than-threatened species are kept—especially those being prepared for release into the wild—the actual introduction of live predators could add jarring authenticity (see Beck, this volume). Maple et al. recognize the dilemma posed by the facilitation of normal stressful behavior: should it be encouraged because it is natural, or discouraged because it is stressful? It would seem that the fact that stressful behavior is distressful is not an adequate rebuttal. The facilitation of normal stressful behavior may liken zoo professionals to the zookeepers of Zircon-212, but if the existence of zoos is a given, this is as close as we can get to satisfying proponents of the authenticity criterion.

Provided the captives subjected to such introductions were given the spatial or structural elements necessary to avoid their assailants, and to the extent that the introduction of a live predator and its prey into a confined area is an authentic experience for either animal, such introductions may allow zoos to at least approach levels of authentic autonomy that satisfy the authenticity criterion (bearing in mind that captivity is still captivity). In fact, this has been done on several occasions. North American river otters have been exhibited in the same pools as their surrogate prey, the African mouth brooder (Markowitz 1982); and maned wolves have been separated by only a mesh-wire fence from a collection of anteaters, capybaras, storks, screamers, geese, and wild turkeys (Bartmann 1980).

Whether the exhibition of species-typical behavior is the most appropriate goal of zoo husbandry requires an examination of values and moral priorities as well as scientific understanding. As to whether animals are any happier in their new naturalistic habitats, we cannot be certain (Siebert 1991; Maple et al., this volume). But we are not relieved of having to ask ourselves a similar question: Is zoo husbandry's goal of evoking species-typical behavior intended to benefit individual animals directly by increasing their well-being, or to "make ourselves happier about keeping animals in captivity" (Siebert 1991), or to prepare animals for reintroduction into the wild (Beck, this volume; Hancocks, this volume)? Unless it is predicated on the goal of increasing individual captive animal welfare, the facilitation of species-typical behavior may not be the best goal of zoo husbandry after all.

WHEN GOALS CONFLICT: SOME PRACTICAL CONSIDERATIONS

The issue of facilitating normal stressful behavior in captive animals is but one question susceptible to the variations in criteria understood as animal welfare. More generally, the criteria of welfare and authenticity rest on such differing value positions that full resolution may never occur, and zoos are in no better a position than anyone else to force the sides to consensus on positions, much less to unify conceptually these two divergent approaches. A realistic assessment of the prognosis for achieving consensus suggests there will always be conflict. Providing an open forum for discussion and debate, and reasonable procedures for conflict resolution when discussion fails to achieve consensus, may be the most that can be accomplished. We turn to an examination of some practical situations in which conflicts about goals and priorities remain.

Interests of Captives versus Interests of Other Clients

Maple et al. explain that the decision to treat or not to treat a domesticated animal's condition traditionally rests solely with the owner, while in the case of zoos, "management generally preselects the outcome without proactive consultation with the financial provider—the community or clientele." But in the case of zoos, the zoo manager must take into consideration more than just the financial provider. The interests of the individual animal, the reputation of the zoo itself, the zoogoer, the animal rights activist, those who fund zoos, those who legitimate and regulate zoos, and the species that the animal represents must all be considered. Often these crucial decisions must be made not by an individual but through committee meetings and consensus-building. The person giving the care may have had nothing to do with the decision to give it.

The individual captive animal is but one of a wide array of clients the zoo must accommodate, and the sheer number of clients' interests the zoo manager must consider makes it likely that the interests of the animals in captivity will not always be consistent with the interests of all the other clients. Indeed, this book is predicated on the recognition that sometimes conservation goals will require us to sacrifice the interests of some animals. Consequently, *a zoo cannot always give top priority to the welfare of its captives.* Sometimes the best way for a zoo manager to consider the welfare of captives is indirectly: a zoo manager may have to limit funding of programs that focus on captive well-being in favor of some project that exclusively serves the public's interest, such as advertising or direct marketing, in the hope that this diversion will eventually result in more funding in general, trickling down to the animals in captivity. But sometimes the best interests of another client may be in direct conflict with the best interests of the individual animals in a collection. Often, for example, animals are kept in less comfortable holding pens during hours when zoos are closed, showing that zoos sometimes choose to balance the well-being of animals against economic costs. It is unlikely that any solution exists that will eliminate altogether such inherent conflicts. But ethics and policy, when coupled with science and community input, can aid in the determination of when one goal should overpower another (for a discussion of contextualism, see Norton 1991, 242).

Soliciting versus Changing the Public's Opinion

Maple et al. assert, "The motivational zoo realizes that its mission is to conserve wildlife and natural habitats through *changing* the attitudes of its visitors" (emphasis added). Yet at another point they say, "Rather than seeing the community, including animal welfare activists, as outsiders, the good zoo should *solicit* its input in establishing what the institution stands for" (emphasis added). These two ends are practically antithetical, yet we do not deny the importance of either one. Attempting to both solicit and change public opinion requires an extraordinary skill. Some may call it a gentle deception. We see this delicate balance as a spiraling process of give and take, in which the two goals are played off one another in such a way that the zoogoer can tell the difference between being solicited and educated but is willing to accept both because it leads to a learning experience.

Defining the Public

If we can agree that zoos should seek to both solicit and change public opinion, the next problem will be to define what is meant by "the public." Do we mean zoogoers, nonzoogoers, activists, or some combination of these groups? Do we mean to solicit the opinion of animal rights activists while educating the zoogoer, or do we mean to try to solicit and change the opinion of both groups? Who shall represent the positions of these groups to the zoo? Should input be sought at the level of individual zoos, the AZA, or some other level? How can we expect to discern the views of the general public by any means other than costly phone or mail surveys? How much weight should be given the results of such surveys?

The variety of public opinions will only exacerbate problems such as conflicts represented by those animal rights activists who favor authenticity. If zoos set out to solicit public input as a means to reduce conflict, they may find that they have only generated a greater number of more heated conflicts, both among advocacy groups and between those groups and zoos. The goal, then, should not be eventually to become such a leader in animal welfare that conflicts with animal rights activists cease to arise but rather to establish stable and permanent avenues for conflict resolution between zoos and activists. In an earlier chapter Jamieson reminds us there will always be someone who has an interest in protesting something. Zoos should concentrate on articulating, funding, implementing, and enforcing a rational and achievable agenda that is compatible with a high level of welfare, meanwhile forging new, open, and permanent pathways for communication with advocacy groups and other clients.

As suggested by Maple et al., the public's needs or expectations for zoos in a rural community may vary widely from those of an urban one. The same may also be true for other demographic factors affecting zoo clientele, such as mean income, age, ethnic origin, or gender. In Part One Hancocks has argued persuasively for an emphasis on local species and habitats. Given the potential for great disparity in the composition of and perspectives on zoo management and priorities of an individual zoo's clientele, some zoos in some geographic areas may wind up serving one client better than the others. This could be good or bad. But simply because the demographic distribution of clients in an individual zoo's area may favor a certain program or action is not reason enough to implement it. The zoo manager must balance public input—from both local and national forums—against the demands of all the zoo's clients in every decision made and, in close calls, should err on the side of individual animal welfare. We do not argue for reducing public input—only for a careful balancing of it with the needs of other clients, especially those clients that are nonhuman.

CONCLUSION: CALL FOR A NEW DIRECTIVE

Let us consider conservation, education, science, and recreation in the context of who benefits from these directives and at whose expense. Conservation has as its end the benefit of populations of wild animals, which sometimes conflicts with the interests of individuals in captivity. Education attempts to impact the zoogoer (possibly at the expense of captive animals, if we embrace the authenticity criterion). Science strives toward the betterment of all animals and all humans at the expense of captive animals. Recreation benefits the zoogoer at the same expense. None of these aims (with the possible exception of science) directly benefits the welfare of the individual captive animal.

This is not to say that the care, maintenance, and welfare of captives is not presently of grave concern to zookeepers and zoo professionals everywhere. Maple et al. seem to imply that the directive for captive animal welfare is implicit in the other four. We disagree. Although science can be employed in the task of improving the well-being of captive animals, there is nothing in the nature of science itself that lends itself to captive welfare. The same is true of education, recreation, and conservation. They all *can* be employed in the betterment of captive welfare, but does such an implicit and potential commitment ensure that our obligations to captive animals will be met? Granted, the AZA has not gone without contributing explicit requirements in the area of captive animal welfare. AZA has had a strong hand in formulating standards in such federal statutes as the Animal Welfare Act and the Marine Mammal Protection Act, which regulate conditions of facilities, standards of care, and research in zoos (Schmitt 1988). But individual animal welfare should be added as a fifth directive as a positive, proactive goal toward which zoo professionals can aspire, to complement these more regulatory guidelines aimed at ensuring compliance with minimum standards. A fifth directive would remind us that none of the other four has any inherent safeguards that relieve us of the duty to further the welfare of captives. The scientific directive, or some other one, may in the eyes of zoo managers already fill the gap that a fifth directive would provide. But a directive does more than remind just zoo managers of the goals toward which they should strive. It also sends that same message to the zoo's many clients, including activists, regulators, and the general public.

REFERENCES

Bartmann, W. 1980. Keeping and breeding a mixed group of large South American mammals at Dortmund Zoo. *International Zoo Yearbook* 20:271–274.

Brambell, F. W. R., chair. 1965. *Report of the Technical Committee to Enquire into the Welfare of Animals Kept under Intensive Livestock Husbandry Systems.* Command Paper 2836. London: Her Majesty's Stationery Office.

Callicott, J. B. 1989. *In Defense of the Land Ethic: Essays in Environmental Philosophy.* Albany: State University of New York Press.

Carpenter, S., and W. J. D. Kennedy. 1988. *Managing Public Disputes.* San Francisco: Jossey-Bass.

Clarke, A. S., C. J. Juno, and T. L. Maple. 1982. Behavioral effects of a change in the physical environment: A pilot study of captive chimpanzees. *Zoo Biology* 1:371–380.

Donnelley, S. 1990. The troubled middle *in medias res.* In Animals, science, and ethics, ed. S. Donnelley, and K. Nolan, 2–4. *Hastings Center Report* 20, No. 3 (May–June): Special supplement, 1–32.

Heim, A. 1971. *Intelligence and Personality.* Baltimore, Md.: Penguin Books.

Jamieson, D. 1985. Against zoos. In *In Defense of Animals,* ed. P. Singer, 108–117. New York: Harper & Row.

Kellert, S. R. 1989. Perceptions of animals in America. In *Perceptions of Animals in American Culture,* ed. R. J. Hoage, 5–24. Washington, D.C.: Smithsonian Institution Press.

Markowitz, H. 1982. *Behavioral Enrichment in the Zoo.* New York: Van Nostrand Reinhold.

Norton, B. G. 1991. *Toward Unity among Environmentalists.* New York: Oxford University Press.

Novak, M. A., and K. Drewson. 1989. Enriching the lives of captive primates: Issues and problems. In *Housing, Care, and Psychological Well-Being of Captive and Laboratory Primates,* ed. E. F. Segal. Park Ridge, N.J.: Noyes Publications.

Sagoff, M. 1984. Animal liberation and environmental ethics: Bad marriage, quick divorce. *Osgoode Hall Law Journal* 22:306–332.

Schmitt, E. C. 1988. Effects of conservation legislation on the professional development of zoos. *International Zoo Yearbook* 27:3–9.

Siebert, C. 1991. Where have all the animals gone? The lamentable extinction of zoos. *Harper's,* May, 49–58.

Singer, P. 1975. *Animal Liberation: A New Ethics for Our Treatment of Animals.* New York: Avon Books.

Snowdon, C. T. 1991. Naturalistic environments and psychological well-being. In *Through the Looking Glass: Issues of Psychological Well-Being in Captive Nonhuman Primates,* ed. M. A. Novak and A. J. Petts. Washington, D.C.: American Psychological Association.

Vonnegut, Kurt, Jr. 1969. *Slaughterhouse-Five.* New York: Dell Publishing.

THE USE OF ANIMALS IN RESEARCH

Conservation cannot occur in the absence of knowledge. Consequently, much research is needed to develop effective methods of captive propagation and animal care, as well as to preserve animals in nature. In this exchange, Michael Hutchins, Betsy Dresser, and Chris Wemmer argue that captive studies have made significant contributions to our knowledge of wildlife biology and management. Furthermore, they reject the idea that individual animal welfare should be our primary and paramount concern when the survival of whole species and ecosystems lies in the balance. Roger Fouts agrees that zoo and aquarium science is important but cautions zoo biologists to approach their work with humility rather than arrogance. All of the authors agree that zoos and aquariums should have well-designed research protocols and programs in place to help ensure that the highest standards of animal welfare and scientific method are maintained.

ETHICAL CONSIDERATIONS IN ZOO AND AQUARIUM RESEARCH

Michael Hutchins, Betsy Dresser, and Chris Wemmer

lthough it is generally thought that public recreation was the primary impetus behind the initial development of zoos and aquariums, scientific investigation was an explicit objective of many early institutions. For example, the ancient zoo in Alexandria was built to satisfy the scientific curiosity of Emperor Ptolemy II (Mullin and Marvin 1987). The menagerie of the Museum National de Historie Naturelle in Paris was established primarily for zoological investigation in 1793, making it one of the first modern zoos (Mullin and Marvin 1987). As early as 1817, Stamford Raffles, the founder of Singapore, had suggested the "need for a collection of animals for scientific purposes as well as the general interest," and this led ultimately to the creation of the Zoological Society of London. The society's charter aspired to create "a collection of living animals such as never yet existed in ancient or modern times . . . to be applied to some useful purpose, or as objects of scientific research, not of vulgar admiration" (Olney 1980).

In time, however, zoos and aquariums in industrialized countries became motivated more by the need for public recreation and civic pride than by the lofty ideals of these early zoos (Olney 1980). In the 1950s and 1960s comparatively little research was being conducted in North American zoological institutions. However, research is again being identified as one of the primary goals of zoos and aquariums, along with conservation and public education (Conway 1969, Hutchins 1988). In 1983, Finlay and Maple found that 70 percent of North American zoos and aquariums (out of a sample of 120) were conducting research of some kind, and 46 percent intended to expand their programs (1986). These efforts have continued to grow, but the need for and appropriateness of zoo and aquarium research, and of animal research in general, are being questioned by animal rights and welfare advocates (e.g., Regan 1983, Jamieson 1985, Fox 1990).

Our objectives are threefold: (1) to examine the types of animal-oriented research that takes place in zoos and aquariums today, (2) to discuss ethical issues related to research on captive animals, and (3) to outline ways of insuring that high standards of animal welfare are maintained during the pursuit of institutional and societal goals, including those of science and conservation.

AN OVERVIEW OF RESEARCH IN MODERN ZOOS AND AQUARIUMS

In the last decade, zoos and aquariums have been cast into a new role as conservators of wildlife (Conway 1969). In fact, many species, such as the Asian wild horse, Pere David deer, Arabian oryx, California condor, and black-footed ferret, owe their existence to the cooperative efforts of modern zoos (Tudge 1991). Along with this heavy responsibility has come a realization that little is known about the basic biology of many wild animals or about their behavioral and environmental requirements in captivity. Zoo biologists are tackling these challenges with a vengeance, and the quality of animal husbandry has rapidly improved. Underpinning all captive animal management and breeding efforts are essential scientific studies. Ongoing research in animal behavior, nutrition, reproduction, genetics, pathology, and clinical veterinary medicine are conducted by zoo and aquarium scientific staff and by collaborating scientists from local colleges and universities.

The investment of modern, professionally managed zoos and aquariums in conservation science is vastly underappreciated. In 1992–1993 alone, the AZA and its 164 accredited institutions initiated or supported nearly 1,100 scientific and conservation projects in more than sixty countries worldwide (Wiese et al. 1993). From 1990 to 1993, they also produced more than 1,300 technical and mainstream popular articles on wildlife biology, conservation, and captive animal management, including contributions in respected scientific journals such as *Science, Primates, Endocrinology, Biology of Reproduction, Journal of Mammalogy, Journal of Reproduction and Fertility, Conservation Biology, Condor, Animal Behaviour, American Naturalist,* and *Heredity* (Hutchins et al. 1991a; Wiese et al. 1992, 1993).

Scientific staff and collaborators at 48, 46, and 50 AZA member institutions reported publishing at least 1 paper during 1990–1991, 1991–1992, and 1992–1993, respectively, and the range was from 1 to 83 papers per institution. Since 1981, the journal *Zoo Biology,* produced by Wiley-Liss and Sons in affiliation with

AZA, has published twelve volumes containing more than 350 articles, all focused on zoo and aquarium research. Additional papers have been published in other international zoo publications, such as the *International Zoo Yearbook* and *Dodo: The Journal of the Jersey Wildlife Preservation Trust.* The AZA recently initiated a symposium series of its own. The first volume, *Biotelemetry Applications to Captive Animal Care and Research,* was produced in 1991 (Asa 1991). In addition, the association is cooperating with the Smithsonian Institution Press to produce a series of books focusing on zoo biology and conservation, of which this volume is the first.

Some larger zoos and aquariums, such as the National Zoological Park, Bronx Zoo (Wildlife Conservation Park), Chicago Zoological Park, New England Aquarium, Vancouver Public Aquarium, Sea World, Cincinnati Zoo and Botanical Garden, London Zoo, Frankfurt Zoo, and San Diego Zoo, have, over many years, made substantial investments in conservation and science. However, even medium-sized to smaller institutions—such as the Roger Williams Park Zoo in Providence, Rhode Island; Zoo Atlanta; Washington Park Zoo in Portland, Oregon; Woodland Park Zoo in Seattle, Washington; Dallas Zoo; Omaha's Henry Doorly Zoo; and the Jersey Zoo—now have full-time staff scientists and growing research programs.

Efforts to coordinate the research activities of individual zoos and aquariums are improving. The AZA recently established a research coordinator's committee (RCC) and seven scientific advisory groups. The RCC was established to address broad-based questions concerning the conduct and administration of research programs in AZA institutions, including ethical and organizational matters. Membership in the RCC is open to all individuals who are responsible for coordinating or administering the scientific programs of their respective institutions (Hutchins and Wiese 1991). More than forty individuals now serve on this committee, and membership continues to grow.

AZA's seven existing scientific advisory groups include those for contraception, animal nutrition, behavior and husbandry, genome banking, reintroduction, small population management (genetics and demography), and veterinary science. The function of such groups is (1) to provide a network for zoo biologists and collaborating university scientists working on projects of interest to the zoo and aquarium community; (2) to advise the AZA Board, Wildlife Conservation and Management Committee, director of conservation and science, and various other committee chairs on technical issues related to their respective disciplines; (3) to coordinate certain collaborative studies, particularly when they involve several AZA institutions; and (4) to create liaisons with appropriate scientific societies, IUCN Species Survival Commission (SCC), Birdlife International (formerly the International Council on Bird Preservation), and taxonomic specialist groups as needed (e.g., Animal Behavior Society, IUCN/SSC Reintroduction Specialist Group).

ETHICAL ISSUES IN ZOO AND AQUARIUM RESEARCH

Animals are used extensively in both basic and applied research, but the appropriateness of their use is being challenged by animal rights activists (Moss 1984). Proponents of animal rights have traditionally opposed the use of animals in biomedical and toxicological studies, especially when animals are caused to suffer pain or when their lives are sacrificed in order to benefit humans (e.g., Singer 1975, Regan 1983, Ryder 1985, Barnes 1986). In *The Case for Animal Rights* Regan states that "animals are not to be treated as mere receptacles or as renewable resources. Thus the practice of scientific research on animals violates their rights. Thus it ought to cease, according to the rights view" (1983, 385).

There are differing opinions on animal use, which have resulted in a wide variety of ethical questions for scientists and society. As Donnelley states, "Particular uses of animals demand the careful consideration of various ethical claims, which will always require the developed art and wisdom of ethical judgment and will never be reduced to a systematic science" (1990, 3). This does not, however, preclude the use of animals in science, nor does it diminish our responsibility toward them (Allen and Blackshaw 1986). In fact, there is a clear trend in public attitudes that favors giving more attention to the welfare of research animals (Moss 1984). We agree with Driscoll and Bateson, who noted that stubborn and reactionary opposition to this trend will reinforce "a growing public distaste for science" (1988, 1569). It is therefore important for zoo and aquarium biologists, and indeed all scientists who conduct research on animals, to take ethical considerations seriously (Hutchins 1988, Hutchins and Fascione 1991, Rowan 1991).

Some animal protectionists do not oppose the use of animals in research but argue that every effort must be made to minimize animal suffering and loss of life (Fox 1986). When a particular research project is deemed necessary, the potential benefits obtained from the research must outweigh the ethical costs as measured in terms of animal suffering (see Bateson 1986, Orlans 1987, Driscoll and Bateson 1988). It is important here to draw a distinction between the terms "animal rights" and "animal welfare" (Allen and Blackshaw 1986). Indeed, Regan and Francione point out that there are "fundamental and profound differences between the philosophy of animal welfare and that of animal rights." Whereas animal rights advocates recognize the "moral inviolability of the individual," animal welfare advocates are "committed to the pursuit of 'gentle usage' " (1992, 40). One implication of the latter view is that it is morally acceptable to use animals in research but that scientists should attempt to minimize any suffering and loss of life.

In this context, how should we view zoo and aquarium research? Studies conducted at zoos and aquariums generally fall under the umbrella of a relatively new scientific discipline known as conservation biology (Soulé and Wilcox 1980; Soulé 1985, 1986). The goal of this applied science is to preserve naturally occurring biological diversity. Conservation biology is one of the most interdisciplinary of sciences, encompassing not only the biological sciences but also economics and the social sciences. Conservation biology is the scientific foundation of the conservation ethic, the goal of which is much broader than either the animal rights or animal welfare ethics (Norton 1987). According to Ehrenfeld, the broad goal of conservation is "to ensure that nothing in the existing natural order is permitted to become permanently lost as the result of man's activities except in the most unusual and carefully examined circumstances" (1972, 7). It is therefore directly concerned with the rights of wild species to continue to exist in natural ecological communities.

It should be noted that unlike biomedical research, most zoo and aquarium research tends to be opportunistic, noninvasive, and nonterminal and does not involve large numbers of subjects. Often the subjects are endangered species. Rather than having a human-focused goal, it is aimed at improving conditions and ensuring a future for individual animals, populations, species, and even ecosystems (Hutchins 1988). Like any research conducted on animals, however, it does have the potential to violate the rights of individuals as defined by Regan (1983). In fact, the rights view would preclude all practices that "cause intentional harm." Regan states, "This objective will not be accomplished merely by ensuring that test animals are anesthetized, or given postoperative drugs to ease their suffering, or kept in clean cages with ample food and water, and so forth. For it is not only the suffering that matters—though it certainly matters—but it is the harm that is done to animals, including the diminished welfare opportunities they endure" (1983, 387). But what are the consequences of this view?

If our interpretation of Regan's concept of intentional harm is correct, the animal rights ethic could preclude many practices commonly used by zoo and aquarium researchers and other conservation biologists, including but not limited to minor surgery to conduct reproductive studies (e.g., those related to experimental in vitro fertilization, embryo transfer, or artificial insemination procedures); any anesthesia or physical restraint solely to collect biological materials, such as tissue, blood, and semen; any testing of immobilizing drugs; and some forms of marking for individual identification (e.g., ear tagging, tattooing, and freeze branding).

Despite the obvious implications for disciplines such as zoology, ecology, ethology, veterinary medicine, and conservation biology, Regan claims that the animal rights view is not antiscientific in that it "calls upon scientists to do science as they redirect the traditional practice of several disciplines away from reliance on 'animal models' toward the development and use of non-animal alternatives." He further states that "all the rights view prohibits is science that violates individual rights. If that means that there are some things that we cannot learn, then so be it" (1983, 388).

But how realistic is it for zoologists, ecologists, ethologists, and conservation biologists to search for alternatives to the use of animals in their respective sciences and, more specifically, in conservation science? In fact, it would be virtually impossible to substitute computer models, tissue cultures, or other alternatives for every living organism on earth, as well as for their interactions with nonliving nature in functioning ecological systems (see Smyth 1978, Balls 1983). Despite many recent advances, modern biology and ecology are still in their infancies, especially when it comes to an understanding of the inner workings of biotic systems or of the taxonomy and basic biology of the vast majority of extant species (Wilson 1989). Thus, we feel that it would be impossible for conservation biologists to cease conducting research on captive or wild animals in the foreseeable future.

Scientific research is one means by which humans gain an understanding of the natural world, and we believe that such an understanding is critical to wildlife conservation efforts. It is impossible to evaluate the ethics of zoo and aquarium research without placing it in this larger context. In fact, Poole and Trefethen have stated, "Knowledge is the essential prerequisite to making a management decision respecting a species, population or group of wildlife. A decision made in the absence of information about a species or population, depending on the result, is at worst, an act of ignorance, or, at best a stroke of good fortune" (1978, 344). Given the overriding goal of the conservation ethic (i.e., to conserve biological diversity), we cannot agree that all research on captive or wild animals should cease. In fact, in order to save endangered species, it is essential that such efforts be expanded as rapidly as possible (see Hutchins 1988, Roberts 1988, Mlot 1989, Soulé and Kohm 1989).

The development of innovative technologies and the attainment of new knowledge are critical for the future of endangered species conservation. It is estimated that as many as one million species of animals and plants could be lost in the next few decades, primarily through habitat destruction (Ehrlich and

Ehrlich 1981). The implications of this impending loss are enormous, not only for natural ecosystems but also for the quality of human life and for the lives of millions of individual animals (Ehrlich and Ehrlich 1981, Regenstein 1985). What will it take to save wildlife in the context of growing human populations, habitat alteration and fragmentation, pollution, or even more pervasive ecological changes, such as global warming and holes in the ozone layer? Animal rights advocates believe that they have the answer. According to Regan, "with regard to wild animals the general policy recommended by the rights view is: let them be" (1983, 361). But what would be the consequences of such inaction?

Ecologists have noted that even the largest of national parks will lose much of their biological diversity in the absence of careful management (Soulé et al. 1979). For example, fragmented islands of habitat, no matter how large, can prevent normal gene flow from occurring, resulting in a rapid loss of genetic diversity (Shaffer 1978). This factor alone can lead to population and species extinctions. Small, isolated populations, however, are susceptible to a variety of other risks, including disease and various natural catastrophes (e.g., hurricanes, fires, or volcanic eruptions). Even if protection is successful, additional problems can develop. For example, local populations can become overabundant when predators or other natural checks and balances are removed. When this occurs, population pressures or social conflict may force animals to leave protected areas and therefore come into conflict with humans (Sukumar 1991). Densely populated animals can also alter their habitats to such an extent that they threaten their own existence or the existence of other species (Caughly 1981). Thus, the laissez-faire concept of wildlife conservation championed by Regan is unlikely to succeed, at least not in view of the political, economic, and biological realities under which conservation must occur (Hutchins and Wemmer 1987, Howard 1990).

Most conservationists, including those who work at modern zoos and aquariums, agree that habitat preservation should be our highest priority (see Hutchins and Wemmer 1991, Hutchins and Wiese 1991). Like it or not, however, the survival of many species, especially the larger vertebrates, is going to require unprecedented levels of human intervention (Hutchins and Fascione 1993). Intensive management actions, such as population culling, habitat modification and restoration, translocation, captive breeding for reintroduction, supplemental feeding, field veterinary care, and control or elimination of introduced or exotic species, will become increasingly necessary (Duffy and Watt 1971, Merton 1977, Wilbur 1977, Foose 1983, Younghusband and Myers 1986, Hutchins and Wemmer 1987, Cairns 1988, Conway 1989, Flesness and Foose 1990, Hutchins et al. 1991b, Diamond 1992, Packer 1992). Consequently, evolving technologies such

as biotelemetry, reintroduction, embryo transfer, artificial insemination, antibiotics, immunization, contraception, sterilization, and chemical immobilization
are likely to become important tools of wildlife conservationists (Cade 1988,
Dresser 1988, Conway 1989, Hutchins et al. 1991b, Hutchins and Fascione 1993).

Some large African national parks have been completely fenced to protect both
wildlife and humans, and have essentially become megazoos (Younghusband and
Myers 1986). Thus, the technologies being developed by modern, professionally
managed zoos and aquariums are directly relevant to field conservation efforts
(Conway 1989, Hutchins and Wemmer 1991, Hutchins and Wiese 1991). A
failure to conduct the needed studies and to take widespread and immediate
action will almost certainly result in numerous species extinctions.

To our knowledge, only one previous attempt has been made to evaluate zoo
and aquarium science from an ethical perspective. In an essay titled "Against
Zoos," Jamieson considers and subsequently rejects the notion that research is a
cogent argument for the existence of zoos. He suggests that "there is a moral
presumption against keeping animals in captivity" and that "this presumption can
be overcome only by demonstrating that there are important benefits that must
be obtained in this way if they are to be obtained at all" (1985, 114). Based on
his analysis of the current state of zoo research, as well as other factors, Jamieson
concludes that "both humans and animals will be better off when they [zoos] are
abolished" (117).

We believe that it is entirely appropriate for the public to ask zoological parks
and other government-supported institutions to justify their existence (Hutchins
and Fascione 1991). We will not discuss all of the reasons for the existence of
modern, professionally managed zoos and aquariums here, but they include conservation, education, and research (Conway 1969). It is important to note, however, that Jamieson's views are based on some highly questionable assumptions.
For example, Jamieson states that research conducted in zoos can "be divided
into two categories: Studies in behavior and studies in anatomy and pathology"
(1985, 112). This in no way reflects the actual diversity of research that is
conducted by modern zoos and aquariums, nor does it show any appreciation for
the benefits that have been derived from it. As summarized above, modern zoos
and aquariums conduct research on a variety of topics, including ecology, reproductive biology, genetics, nutrition, behavior, disease, anatomy, physiology,
and clinical veterinary medicine, both in captivity and in nature.

Jamieson also states that "very few zoos support any real scientific research.
Fewer still have scientists with full-time research appointments. Among those
that do, it is common for their scientists to study animals in the wild rather than
those in zoo collections" (1985, 112). While we agree that zoos and aquariums

need to continue to strengthen their commitment to science (Hutchins 1988, Kleiman 1992, Thompson 1993), we still question the accuracy of these statements. They do not, for example, take into account the fact that many zoos now employ curatorial staff with advanced scientific training. In addition, a number of zoos, although they may not employ large scientific staffs or have extensive laboratory facilities, collaborate frequently with major universities, professional scientific societies, and larger zoos that do employ full-time scientists. For example, in 1991, AZA held a joint symposium on the captive breeding and conservation of reptiles and amphibians with the Society for the Study of Amphibians and Reptiles (SSAR) and the Herpetologists' League. SSAR, which publishes the *Journal of Herpetology,* has long recognized the importance of zoo research and has an official zoo liaison (Wiese and Hutchins 1994).

Jamieson used his prejudicial view of zoo research to argue that zoos should be abolished. He states that "the fact that zoo research contributes to improving conditions in zoos is not a reason for having them. If there were no zoos, there would be no reason to improve them" (1985, 113). Yet, improving captive animal management and care is only one aspect of zoo and aquarium research. Indeed, many results of zoo- and aquarium-based studies have contributed directly to field conservation efforts. For example, zoo biologists were responsible for verifying the deleterious effects of inbreeding on wildlife populations (Ralls and Ballou 1983)—an extremely important insight for conservationists (Lande 1988).

Zoo biologists have also developed or refined various technological advances that have the potential to make major contributions to wildlife conservation (Benirschke 1983). For example, reproductive technologies, such as cryopreservation, in vitro fertilization, embryo transfer, and artificial insemination will soon allow conservationists to move genetic material between isolated populations in captivity and in the wild and thus avoid the deleterious effects of inbreeding (Dresser 1988, Wildt 1989). Such methods may also allow zoo biologists to accelerate rates of reproduction or to stimulate reproduction in especially recalcitrant species (Dresser 1988, Wildt 1989, Cohn 1991). Disease is one of the most significant threats to the future of small, isolated wildlife populations. Advances in zoo veterinary medicine have helped to diagnose and treat various ailments in both wild and captive animal populations (Hutchins et al. 1991b). For example, veterinarians with zoo experience recently treated wild mountain gorillas *(Gorilla gorilla berengi)* for injuries and vaccinated them against disease, a practice that itself has stimulated numerous ethical questions (see Hutchins and Wemmer 1987).

Jamieson also states that behavioral research conducted on zoo animals has been "very controversial. Some have argued that nothing can be learned by studying animals that are kept in the unnatural conditions that obtain in most zoos" (1985, 112). Although living in a zoo environment can influence animal behavior to a certain extent, it is widely recognized that much can be learned through captive studies, especially now that animals are being maintained in larger, more naturalistic enclosures and social groups (Rumbaugh 1972, Hutchins et al. 1984, Kleiman 1992). Studies of birth, parental care, and various social interactions have been particularly enlightening. Indeed, much of what we know about the behavior of secretive, arboreal, nocturnal, and aquatic animals has come from zoo and laboratory studies (Hutchins 1988). In many cases, zoo and laboratory studies complement field studies and provide a more comprehensive picture of an animal's behavioral repertoire. Zoo and laboratory studies of animal behavior not only contribute to the welfare of captive animals (e.g., Maple and Finlay 1989, Shepherdson et al. 1993) but also help in formulating more effective methods of captive breeding and reintroduction (Eisenberg and Kleiman 1975; Kleiman 1980, 1992; Hutchins et al. in press). For example, behavioral studies figured prominently in the successful reintroduction of captive-bred golden lion tamarins to their native habitat in Brazil (Beck et al. 1988, Kleiman 1989).

It is also important to note that many kinds of research—for example, those that involve the routine collection of biological materials, such as blood, urine, milk, feces, semen, or tissues, or body measurements—can be conducted much more humanely in zoos and aquariums than they can in nature. Habituated animals are less likely to experience stress in the presence of humans (e.g., Frank et al. 1986). In addition, some captive animals can be easily trained to submit themselves voluntarily for sample collection, whereas the capture of wild animals can be an extremely stressful and risky procedure (Fowler 1978). For example, killer whales *(Orcinus orca)* in oceanariums have been habituated to people and taught to present their tail flukes for routine blood collection. Development of this method has allowed biologists to determine, for the first time, the characteristics of the female's reproductive cycle (Robeck et al. 1993). Because they involve the regular collection of samples over a period of time, similar studies would be extremely difficult, or even impossible, to conduct in nature.

In some cases, zoo biologists have used common domestic or laboratory animals as models to develop techniques for application to endangered species. The World Conservation Union's policy on research involving species at risk of extinction encourages basic and applied research that contributes to the survival of threatened species, but opposes research that might directly or indirectly impair their survival (IUCN 1989). By inference, this encourages the use of common or domestic species as models for more endangered varieties. It also assumes, however, that endangered animals are more valuable than common ones, and that individuals can be sacrificed or caused to suffer for the greater good (i.e., to preserve a population, species, or ecosystem).

Animal rights advocates consider this view to be speciesist. They consider all sentient animals to be worthy of equal moral consideration and thus do not make distinctions based on the rarity of the species. Regan is very clear about this point. He states that the "rights view is about the moral rights of individuals. Species are not individuals, and the rights view does not recognize the moral rights of species to anything, including survival. What it recognizes is the prima facie right of individuals not to be harmed, and the prima facie right of individuals not to be killed. That an individual animal is among the last remaining members of a species confers no further right on that animal, and its right not to be harmed must be weighed equitably with the rights of any others who have this right" (1983, 359). One implication of Regan's viewpoint is that when a conflict of interest exists, humans should allow a rare species to go extinct rather than violate the rights of even a single individual of a common species.

Regan has labeled any attempt to subordinate the rights of individual animals to the species or ecosystem as "environmental fascism" (1983). From the perspective of the conservation ethic, however, the rights of individual animals must be viewed as secondary to those of the population, species, or ecosystem as a whole (Rodman 1977, Callicott 1980, Gunn 1980, Hutchins et al. 1982, Sagoff 1984, Hutchins and Wemmer 1987, Norton 1987). In fact, without the latter, there is no way that the former could even exist. As Soulé and Wilcox have pointed out, "Death is one thing—an end to birth is something else" (1980, 8). Proponents of the conservation ethic argue that endangered species should be given special status solely because of their scarcity and because their loss is irreversible (Gunn 1980, Rolston 1985, Hutchins and Wemmer 1987, Norton 1987). The underlying rationale of this view is that naturally occurring biological diversity is intrinsically good and that every effort should be made to conserve it. It also recognizes that species in an ecological community are often interdependent, so that the loss of one species may have detrimental effects on many others. Here the focus is on the population, species, or ecosystem as a whole, rather than on individual organisms (Rodman 1977, Callicott 1980, Sagoff 1984, Hutchins and Wemmer 1987, Norton 1987).

Fox has recognized the apparent weakness of the animal rights philosophy, noting that "the ecological imperative of responsible stewardship concerns our relationship with all creation, both sentient and non-sentient" (1979, 54). In this respect, conservation biologists appear to be less speciesist than animal rights advocates. Indeed, Peter Brussard recently said, "A conservation biologist is blind to species: The butterfly is worth as much as an elk" (Gibbons 1992, 21).

While we agree that a recognition of the value of wild animals and their habitats may be important to both the animal rights advocate's and conservationist's goals (see Leopold 1949, Stone 1974, Gunn 1980, Regenstein 1985, Rolston 1985, Norton 1987), we also stress that responsible stewardship can involve difficult decisions (Hutchins et al. 1982, Howard 1986, Hutchins and Wemmer 1987, Rolston 1992). In some cases, our actions may result in the death or suffering of other sentient beings. Of course, this does not imply that animals can be treated without care and respect. For example, when the need to study or manage a population of animals, whether in captivity or in the wild, has been identified, it should be accomplished in the most humane manner possible. However, when the purpose of such research or management is to conserve natural ecosystems or to protect endangered species of animals or plants, it should not be perceived as fascist or inhumane.

We conclude that the conservation ethic is generally compatible with the animal welfare ethic but largely incompatible with the animal rights ethic, except under very specific circumstances (Gunn 1980, Hutchins et al. 1982, Hutchins and Wemmer 1987, Norton 1987). As Norton says, "When one justifies differential treatment of animals because of the status of their species, one treats individuals as a means to the preservation of species and denies that they are ends-unto-themselves. This is surely a damaging conclusion for anyone . . . who hopes to base protectionist policies on claims of intrinsic value of individuals" (1987, 165). He further states that "concern for welfare of individuals seems, in general, to be the wrong direction to look if the goal is to preserve species" (167).

Our intent in this section has been to show that the use of animals in zoo and aquarium science, and in conservation science in general, is a moral imperative if many species are to be saved from extinction. We do not wish to imply, however, that researchers have carte blanche to conduct science without taking ethical considerations, including the welfare of their subjects, into account. In the next section, we show how the design of zoo and aquarium research programs can help ensure that animal welfare and other relevant issues are taken into consideration and that various research projects will produce significant and meaningful results.

ANIMAL WELFARE AND THE DESIGN OF ZOO AND AQUARIUM RESEARCH PROGRAMS

How should zoological institutions structure their research programs to help ensure that animal welfare issues are taken into account? How should they evaluate proposed research projects to help ensure that they are of the highest quality and thus increase the probability that useful and original information will be generated?

Any zoological facility that has an active research program should also have a mechanism for making decisions regarding which projects it will and will not support (Kleiman 1985, Hutchins 1988). Typically this is accomplished by a zoological research committee composed of staff scientists, university scientists, veterinarians, zoo administrators, and animal curators. The functions of such committees are to (1) determine the institution's policies with regard to research; (2) review proposals made by staff or university affiliates to conduct research on the animal collection; (3) review manuscripts produced as a result of this research; (4) monitor compliance with the institution's research protocols and regulations, including its animal welfare guidelines; and (5) decide if funds will be used to support specific research projects. The New York Zoological Society (now NYZS—The Wildlife Conservation Society) has such a committee and has developed an effective model for the design of zoo and aquarium research programs (Hutchins 1988, 1990a, 1990b).

With regard to animal welfare, it is critical that individual zoological institutions develop written policies on the care and use of animals in research (Kleiman 1985, Hutchins 1988). United States institutions that receive federal funding for research are required to have an institutional animal care and use committee (IACUC). Such committees, which consist of both institutional and outside representatives, monitor researchers' compliance with established legal guidelines such as the Animal Welfare Act of 1970. However, this legislation also recognizes that professionally managed zoos and aquariums are already heavily regulated and committed to quality animal care. It is therefore recommended that zoos and aquariums establish formal IACUCs only if they are conducting research that has the potential to be controversial. It makes no sense to create additional bureaucracy if an institution's research program is based solely on observational research or on the opportunistic collection of biological materials that occurs during routine animal management or veterinary care procedures.

When developing a policy on the use of animals in research, zoos and aquariums can adopt guidelines prepared by other organizations, or portions thereof. A number of guidelines currently exist, and these focus on the care and handling of both captive and free-ranging animals. For example, the American Society of Mammalogists has produced a document titled "Acceptable Field Methods in Mammalogy" (American Society of Mammalogists 1987). Similar guidelines have been developed for other vertebrates, including birds (American Ornithologists' Union 1988), reptiles, amphibians, and fish (Schaeffer et al. 1992). Excellent guidelines for the care and use of laboratory animals have been produced by many professional and governmental organizations, including the National Institutes of Health (1985), the Sigma Xi Committee on Science and Society (Sigma Xi 1992), and the Association for the Study of Animal Behaviour and the Animal Behavior Society (ASAB 1986). Some taxonomic groups may require special attention. For example, recent (1985) amendments to the Animal Welfare Act now contain some minimum standards designed to promote the "psychological well-being" of captive primates (Novak and Suomi 1988).

It should be noted, however, that zoos and aquariums differ from many other research institutions, especially biomedical laboratories. For example, animals exhibited in modern, professionally managed zoological institutions are usually maintained under conditions that far exceed the minimum standards for laboratory animals. In addition, the number of individual animals used in research is typically much smaller and the number and diversity of species studied are considerably greater than in biomedical or university laboratories. The diversity of animals maintained and studied by zoos and aquariums is not a trivial issue, with each species having its own specific requirements for care and maintenance. In addition, much of the research conducted in zoos and aquariums is focused on endangered or threatened species, and this can result in additional concerns. For example, research conducted on endangered animals should not threaten their ability to reproduce, particularly if such animals are involved in organized cooperative breeding or reintroduction programs, such as those coordinated under the AZA's Species Survival Plan (Foose 1983). The World Conservation Union has developed a policy on research involving species at risk of extinction to which all zoological institutions should conform (IUCN 1989).

The issue of what kinds of research should be allowed in zoos and aquariums is complicated. Research in zoological institutions should, of course, conform to all legal guidelines related to animal welfare (Kleiman 1985, Hutchins 1988). However, the question of whether certain kinds of research should be conducted must be left up to individual institutions. In short, each proposed project must be considered on a case-by-case basis, taking into account not only the potential benefits to be derived but also the costs to the individual subjects. In each case, every effort must be made to consider alternative approaches and to reduce, as much as possible, any pain, discomfort, stress, or loss of life.

Kleiman suggests that zoological research committees ask themselves the following questions when evaluating research proposals (1985): Are all federal, state, and local regulations being followed? Is the species being used the most appropriate for this study? Is the smallest number of animals required (for statistical analysis) being used? Are there alternative procedures available that might achieve the same goals? Are animals being maintained so that their species-typical and individual needs are being met? Does the scientific gain outweigh the cost to the individual animals or species in terms of unavoidable stress or discomfort? Has unavoidable stress or discomfort been minimized to the best possible extent?

In the case of laboratory studies, several decision-analysis models have been developed to assist review committees in their deliberations (Bateson 1986, Orlans 1987), and a similar approach might prove useful for zoos and aquariums. The most obvious projects to proceed with are those in which the amount of animal suffering is negligible, the quality of the research is high, and the benefit is certain. Conversely, proposals should be rejected when the potential for suffering is great and the quality of the work and the benefit are uncertain (Driscoll and Bateson 1988). We agree with Kleiman's statement that "all research supported by a zoo, whether using animals in the collection or in the field, must be conducted in as humane a manner as possible, in line with the zoo's mission to preserve and respect life" (Kleiman 1985, 97). We also note, however, that when the fate of populations, species, or ecosystems lies in the balance, the scales may be tipped in favor of a given project, regardless of its implications for individual animals.

Disputes will probably arise as to where this line should be drawn. For example, Eaton studied the development of predatory behavior in captive-reared lions *(Panthera leo)* and cheetahs *(Acinonyx jubatus)*. To observe predation, he released live domestic goats, which were subsequently killed and eaten by the cats (1972a, 1972b). The rights view certainly would not condone such experiments, yet despite the consequences for individual goats, this work appears to be compatible with the more holistic conservation ethic. Many carnivorous species, including the large cats, have been forced to the brink of extinction by humans (Seidensticker and Lumpkin 1991). One method by which conservationists hope to save some of these species is through captive breeding for reintroduction. However, reintroducing captive-bred animals into their natural habitats poses many difficult problems, including the possible inability of the animals to obtain their own food (Kleiman 1989). Although young felids come equipped with an instinctive sequence of predatory behaviors, practice is necessary to increase killing efficiency (Leyhausen 1973). Live prey can be both unpredictable and dangerous, and efficiency is important. Thus, a knowledge of how captive-bred predators learn to recognize their prey and how predatory behavior improves with practice will be essential to any serious reintroduction effort (Bogue and Ferrari 1976).

One could argue that such experiments are inhumane or unethical in that they involve a high degree of human intervention (see Huntingford 1984 for a discussion of ethical issues involved in studies of predatory behavior). One could also argue that domestic ungulates offer little challenge for a large predator, even if the predator has been reared in captivity. However, both wild and domestic ungulates are frequently killed by predators in nature. Is the human intervention factor enough justification to preclude such studies, especially given their potential significance to conservation? If not, should we move to eliminate all predation in nature (see Sagoff 1984)? This offers an excellent example of the complexity of the ethical issues facing conservationists, animal welfare advocates, and society today.

Because of the nature of scientific inquiry, assessing the potential benefit of a project is not an easy task (Will 1986). Although a scientist typically starts with a working hypothesis or question, he or she does not know a priori whether the results of any project will actually be beneficial. Much of the research conducted by zoo and aquarium biologists is applied (i.e., aimed at finding answers to practical questions regarding animal care, breeding, and conservation). However, basic, or pure, research (i.e., aimed at gaining an understanding of a particular theory or testing a specific hypothesis) is also conducted, and this is critical to the future of conservation biology. As Thompson has pointed out, "Basic research should be encouraged, as it is virtually impossible to presume that such research has, or will have, no applied value: often the distinction between basic and applied research is simply one of context and interpretation" (1993, 158). Thus, there is a danger in rejecting projects simply because of a perceived lack of applicability.

Assessing the quality of a given research project is somewhat easier, although it involves having access to the relevant expertise. Projects should not be repetitive, although some replication might be necessary to confirm certain findings (Will 1986, Dressler 1989). The proposal should demonstrate the applicant's knowledge of previous studies conducted on the species or topic of interest. In addition, the methods to be used to collect and analyze data should be appropriate and likely to lead to valid results. In the case of invasive or potentially stressful research, the number of subjects used should be the minimum required for statistical validity (Still 1982). This makes it critical that zoos with active research programs but no trained scientists on staff have access to individuals with advanced scientific training (Hutchins 1988).

While we recognize the necessity of some regulation, we are concerned that it might drown legitimate conservation organizations, including professionally managed zoos and aquariums, in a sea of paperwork. At present, federal permit regulations make it extremely difficult to conduct basic studies of protected species, and this could hamper long-term conservation efforts (Ralls and Brownell 1989). Similarly, the continual imposition of more stringent regulations on the use of laboratory and zoo animals may result in more paperwork for administrators and scientists, while making little, if any, difference in the care and treatment of animals (see Briefing 1992). We agree with Rowan, who suggested that the greatest advances might be made by training and sensitizing researchers and animal caretakers to welfare issues (1991).

Finally, the importance of communicating the role and methods of conservation science, including the role of zoo- and aquarium-based research, to the general public cannot be overemphasized (see Birke 1990). The Cincinnati Zoo and Botanical Garden, for example, recently opened a working research facility to zoo visitors. The public can tour the laboratories and see zoo scientists working on various projects involving both animals and plants. They also learn about the essential role that zoo studies play in wildlife conservation efforts. It is also critical that extensive efforts be made to educate the public about pressing environmental issues and about the impending loss of biological diversity. We feel that very few animal activists and other laypersons currently understand the nature or scope of the problem or what it is going to take to find solutions. With more than 105 million visitors a year, AZA-accredited zoos and aquariums are in an excellent position to undertake this task.

Perhaps most important, it is critical that the public sees conservation biologists as they are—not as cold, calculating scientists but as sensitive, caring people, people who value the lives of individual animals and worry about the future of life on this planet (Ratcliffe 1976, Rowan 1991). As responsible stewards, we recognize that some difficult decisions will have to be made, but we feel we should also strive to be sensitive to people's emotional reactions and not condemn them for well-meaning intentions. In this context, and despite some fundamental philosophical differences, there may be much room for compromise and cooperation between conservationists and animal welfare and rights advocates (see Kellert 1982, Hutchins and Wemmer 1987, King 1988, Ehrenfeld 1991). We also feel, however, that the conservation issue presents a severe challenge for ethical theory and agree with E. O. Wilson, who wrote, "In ecological and evolutionary time, good does not automatically flow from good or evil from evil. To choose what is best for the near future is easy. To choose what is best for the distant future is also easy. But to choose what is best for the near future and distant future is a hard task, often internally contradictory, and requiring ethical codes yet to be formulated" (1984, 123).

CONCLUSIONS AND RECOMMENDATIONS

Animal protectionists have underestimated the role that zoo and aquarium research will play and, indeed, is playing in wildlife conservation efforts. Rather than being curtailed, these efforts must grow exponentially if extinctions are to be prevented and natural habitats preserved.

The conservation ethic, on which much of zoo and aquarium research is based, is generally compatible with the animal welfare ethic but largely incompatible with the animal rights ethic as defined by Regan (1983). If carried to its logical conclusion, the rights view would place heavy restrictions on the nature of all biological research, including that which might be called conservation biology. One result would be that essential information could not be collected, thus increasing the probability of species extinctions.

It is virtually impossible for conservation biologists to adopt alternatives to animal models in their research, as has been suggested for biomedical studies. There is simply no way that computer models, tissue cultures, or other alternatives can be substituted for the direct study of thousands of unique species and their interaction with nonliving nature in functioning ecological systems. Zoo and aquarium studies often complement field studies by offering opportunities that would be difficult or even impossible to replicate in nature.

Conservation of species will require unprecedented levels of human intervention. It should also be recognized that responsible stewardship can involve difficult decisions. This does not imply, however, that animals can be treated without care and respect. When the need to study or manage a population of animals, whether in captivity or in the wild, has been identified, it should be accomplished in the most humane manner possible. When the purpose of such research or management is to conserve or restore natural ecosystems or to protect endangered species of animals or plants, it should not be perceived as inhumane.

Like all biologists, zoo and aquarium biologists need to be fully aware of animal welfare issues. Zoological research committees should be formed to evaluate research proposals and to monitor adherence to institutional protocols regarding animal care and use. Institutions that are conducting potentially controversial research should establish formal IACUCs to aid in and document the decision-making process. Such committees should have public representation. It is equally important that zoological institutions increase their efforts to educate the public about the role and scope of zoo and aquarium studies and especially about how they contribute to wildlife conservation efforts.

Despite some fundamental philosophical differences, there is much room for compromise between conservationists and animal welfare and rights advocates. Any serious compromise, however, must take biological realities into account.

ACKNOWLEDGMENTS

The authors would like to thank N. Fascione, E. F. Stevens, and B. Norton for reading and commenting on an earlier version of this manuscript.

REFERENCES

Allen, D. J., and J. K. Blackshaw. 1986. Ethics, welfare, and laboratory animal management. In *Advances in Animal Welfare Science 1986–1987,* ed. M. W. Fox and L. D. Mickley, 1–8. Washington, D.C.: Humane Society of the United States.

American Ornithologists' Union. 1988. Guidelines for use of wild birds in research: Report of the American Ornithologists' Union, Cooper Ornithological Society, and Wilson Ornithological Society, Ad Hoc Committee on the Use of Wild Birds in Research. *Auk* (Supplement) 105:1A–41A.

American Society of Mammalogists. 1987. Acceptable field methods in mammalogy: Preliminary guidelines approved by the American Society of Mammalogists, Ad Hoc Committee for Animal Care Guidelines. *Journal of Mammalogy* (Supplement) 68(4): 1–18.

Asa, C. F., ed. 1991. *Biotelemetry Applications to Captive Animal Care and Research.* AAZPA Symposium 1. Wheeling, W.Va.: AAZPA.

ASAB. 1986. Guidelines for the use of animals in research, Association for the Study of Animal Behaviour and Animal Behavior Society. *Animal Behaviour* 34:315–318.

Balls, M. 1983. Alternatives to experimental animals. *Veterinary Record* 113(7): 398–401.

Barnes, D. J. 1986. The case against the use of animals in science. In *Advances in Animal Welfare Science 1986–1987,* ed. M. W. Fox and L. D. Mickley, 215–225. Washington, D.C.: Humane Society of the United States.

Bateson, P. 1986. When to experiment on animals. *New Scientist,* 20 February: 30–32.

Beck, B. B., I. Castro, D. G. Kleiman, J. M. Dietz, and B. Rettberg-Beck. 1988. Preparing captive-born primates for reintroduction. *International Journal of Primatology* 8:426.

Benirschke, K. 1983. The impact of research on the propagation of endangered species in zoos. In *Genetics and Conservation: A Reference for Managing Wild Animal and Plant Populations,* ed. C. M. Schonewald-Cox, S. M. Chambers, B. McBryde, and W. L. Thomas, 402–413. Menlo Park, Calif.: Benjamin/Cummings.

Birke, L. 1990. Selling science to the public. *New Scientist,* 18 August: 40–44.

Bogue, G., and M. Ferrari. 1976. On the predatory "training" of captive reared pumas. *Carnivore* 3(1): 36–45.

Briefing. 1992. Rat and mouse care. *Science* 255:539.

Cade, T. J. 1988. Using science and technology to reestablish species lost in nature. In *Biodiversity*, ed. E. O. Wilson, 279–288. Washington, D.C.: National Academy Press.

Cairns, J. 1988. Increasing diversity by restoring damaged ecosystems. In *Biodiversity*, ed. E. O. Wilson, 333–343. Washington, D.C.: National Academy Press.

Callicott, J. B. 1980. Animal liberation: A triangular affair. *Environmental Ethics* 2:311–338.

Caughly, G. 1981. Overpopulation. In *Problems in Management of Locally Abundant Wild Mammals*, ed. P. A. Jewell and S. Holt, 7–19. New York: Academic Press.

Cohn, J. P. 1991. Reproductive biotechnology. *Bioscience* 41(9): 595–598.

Conway, W. G. 1969. Zoos: Their changing roles. *Science* 163:48–52.

———. 1989. The prospects for sustaining species and their evolution. In *Conservation for the Twenty-First Century*, ed. D. Western and M. C. Pearl, 199–209. New York: Oxford University Press.

Diamond, J. 1992. Must we shoot deer to save nature? *Natural History* 8:2–8.

Donnelley, S. 1990. The troubled middle *in medias res*. In Animals, science, and ethics, ed. S. Donnelley, and K. Nolan, 2–4. *Hastings Center Report* 20, No. 3 (May–June): Special supplement, 1–32.

Dresser, B. L. 1988. Cryobiology, embryo transfer, and artificial insemination in *ex situ* animal conservation programs. In *Biodiversity*, ed. E. O. Wilson, 296–308. Washington, D.C.: National Academy Press.

Dressler, R. 1989. Measuring merit in animal research. *Theoretical Medicine* 10(1): 21–34.

Driscoll, J., and P. Bateson. 1988. Animals in behavioural research. *Animal Behaviour* 36:1569–1574.

Duffy, E., and A. S. Watt, eds. 1971. *The Scientific Management of Plant and Animal Communities for Conservation*. Oxford: Blackwell Scientific Publications.

Eaton, R. L. 1972a. An experimental study of predatory and feeding behavior in the cheetah *(Acinonyx jubatus)*. *Zeitschrift für Tierpsychologie* 31:270–280.

———. 1972b. Predatory and feeding behavior in adult lions: The deprivation experiment. *Zeitschrift für Tierpsychologie* 31:461–473.

Ehrenfeld, D. 1972. *Conserving Life on Earth*. New York: Oxford University Press.

———. 1991. Conservation and the rights of animals. *Conservation Biology* 5(1): 1–3.

Ehrlich, P., and A. Ehrlich. 1981. *Extinction: The Causes and Consequences of the Disappearance of Species*. New York: Random House.

Eisenberg, J. F., and D. G. Kleiman. 1975. The usefulness of behaviour studies in developing captive breeding programmes for mammals. *International Zoo Yearbook* 17:81–88.

Finlay, T. W., and T. L. Maple. 1986. A survey of research in American zoos and aquariums. *Zoo Biology* 5:261–268.

Flesness, N., and T. Foose. 1990. The role of captive breeding in the conservation of species. In *1990 IUCN Red List of Threatened Animals*, pp. xi–vv. Gland, Switzerland, and Cambridge, U.K.: World Conservation Union.

Foose, T. 1983. The relevance of captive populations to the conservation of biotic diversity. In *Genetics and Conservation: A Reference for Managing Wild Animal and Plant Populations*, ed. C. M. Schonewald-Cox, S. M. Chambers, B. McBryde, and W. L. Thomas, 374–401. Menlo Park, Calif.: Benjamin/Cummings.

Fowler, M. E. 1978. *Restraint and Handling of Wild and Domestic Animals*. Ames: Iowa State University Press.

Fox, M. W. 1979. Animal rights and nature liberation. In *Animal Rights: A Symposium,* ed. D. Paterson and R. D. Ryder, 48–92. Sussex, U.K.: Centaur Press.

———. 1986. *Laboratory Animal Husbandry: Ethology, Welfare, and Experimental Variables.* Albany: State University of New York Press.

———. 1990. *Inhumane Society.* New York: St. Martin's Press.

Frank, H., L. M. Hasselbach, and D. M. Littleton. 1986. Socialized vs. unsocialized wolves *(Canis lupus)* in experimental research. In *Advances in Animal Welfare Science 1986–1987,* ed. M. W. Fox and L. D. Mickley, 33–49. Washington, D.C.: Humane Society of the United States.

Gibbons, A. 1992. Conservation biology in the fast lane. *Science* 255:20–22.

Gunn, A. S. 1980. Why should we care about rare species? *Environmental Ethics* 2:17–37.

Howard, W. E. 1986. *Nature and Animal Welfare: Both Are Misunderstood.* Pompano Beach, Fla.: Exposition Press of Florida.

———. 1990. *Animal Rights vs. Nature.* Davis, Calif.: Published by the author.

Huntingford, F. 1984. Some ethical issues raised by studies of predation and aggression. *Animal Behaviour* 32:210–215.

Hutchins, M. 1988. On the design of zoo research programs. *International Zoo Yearbook* 27:9–19.

———. 1990a. Serving science and conservation: The biological materials request protocol of the New York Zoological Society. *Zoo Biology* 9:447–460.

———. 1990b. *New York Zoological Society Zoo and Aquarium Research Manual.* New York: New York Zoological Society.

Hutchins, M., and N. Fascione. 1991. Ethical issues facing modern zoos. *Proceedings of the American Association of Zoo Veterinarians:* 56–64.

———. 1993. What is it going to take to save wildlife? In *Proceedings: American Association of Zoological Parks and Aquariums Regional Conferences,* 5–15. Wheeling, W.Va.: AAZPA.

Hutchins, M., and C. Wemmer. 1987. Wildlife conservation and animal rights: Are they compatible? In *Advances in Animal Welfare Science 1986–1987,* ed. M. W. Fox and L. D. Mickley, 111–137. Washington, D.C.: Humane Society of the United States.

———. 1991. Response: In defense of captive breeding. *Endangered Species Update* 8:5–6.

Hutchins, M., and R. J. Wiese. 1991. Beyond genetic and demographic management: The future of the Species Survival Plan and related AAZPA conservation efforts. *Zoo Biology* 10:285–292.

Hutchins, M., V. Stevens, and N. Atkins. 1982. Introduced species and the issue of animal welfare. *International Journal for the Study of Animal Problems* 3(4): 318-336.

Hutchins, M., D. Hancocks, and C. Crockett. 1984. Naturalistic solutions to the behavioral problems of captive animals. *Zoologische Garten* 54:28–42.

Hutchins, M., R. J. Wiese, K. Willis, and S. Becker, eds. 1991a. *AAZPA Annual Report on Conservation and Science, 1990–1991.* Bethesda, Md.: AAZPA.

Hutchins, M., T. Foose, and U. S. Seal. 1991b. The role of veterinary medicine in endangered species conservation. *Journal of Zoo and Wildlife Medicine* 22(3): 277–281.

Hutchins, M., C. Sheppard, A. Lyles, and G. Casedi. In press. Behavioral considerations in the captive management, propagation, and reintroduction of endangered birds. In *Captive Conservation of Endangered Species,* ed. J. Demarest, B. Durrant, and E. Gibbons. Stony Brook: State University of New York Press.

IUCN. 1989. Research involving species at risk of extinction, policy statement. Gland, Switzerland: World Conservation Union.

Jamieson, D. 1985. Against zoos. In *In Defense of Animals,* ed. P. Singer, 108–117. New York: Harper & Row.

Kellert, S. 1982. Striving for common ground: Humane and scientific considerations in contemporary wildlife management. *International Journal for the Study of Animal Problems* 3(2): 137–140.

King, W. 1988. Animal rights: A growing moral dilemma. *Animal Kingdom* 91(1): 33–35.

Kleiman, D. G. 1980. The sociobiology of captive propagation. In *Conservation Biology: An Evolutionary-Ecological Approach,* ed. M. E. Soulé and B. A. Wilcox, 243–261. Sunderland, Mass.: Sinauer Associates.

———. 1985. Criteria for the evaluation of zoo research projects. *Zoo Biology* 4:93–98.

———. 1989. Reintroduction of captive mammals for conservation. *Bioscience* 39(3): 152–161.

———. 1992. Behavior research in zoos: Past, present, and future. *Zoo Biology* 11:301–312.

Lande, R. 1988. Genetics and demography in biological conservation. *Science* 241:1455–1460.

Leopold, A. 1949. *A Sand County Almanac.* New York: Oxford University Press.

Leyhausen, P. 1973. On the function of the relative hierarchy of moods (as exemplified by the phylogenetic and ontogenetic development of prey-catching in carnivores). In *Motivation of Human and Animal Behavior: An Ethological View,* ed. K. Lorenz and P. Leyhausen, 133–247. New York: Van Nostrand Reinhold.

Maple, T., and T. Finlay. 1989. Applied primatology in the modern zoo. *Zoo Biology* (Supplement) 1:101–116.

Merton, D. V. 1977. Controlling introduced predators and competitors on islands. In *Endangered Birds: Management Techniques for Saving Threatened Species,* ed. S. A. Temple, 121–134. Madison: University of Wisconsin Press.

Mlot, C. 1989. The science of saving endangered species. *Bioscience* 39(2): 68–70.

Moss, T. H. 1984. The modern politics of laboratory animal use. *Bioscience* 34:621–625.

Mullin, B., and G. Marvin. 1987. *Zoo Culture.* London: Weidenfeld & Nicholson.

National Institutes of Health. 1985. *Guide for the Care and Use of Laboratory Animals.* Bethesda, Md.: U.S. Department of Health and Human Services, Public Health Service, National Institutes of Health.

Norton, B. G. 1987. *Why Preserve Natural Variety?* Princeton, N.J.: Princeton University Press.

Novak, M. A., and S. J. Suomi. 1988. Psychological well-being of primates in captivity. *American Psychologist* 43(10): 765–773.

Olney, P. J. S. 1980. London Zoo. In *Great Zoos of the World,* ed. S. Zuckerman, 35–48. London: Weidenfeld & Nicholson.

Orlans, F. B. 1987. Review of experimental protocols: Classifying animal harm and applying "refinements." *Laboratory Animal Science* (special issue) January: 50–56.

Packer, C. 1992. Captives in the wild. *National Geographic* 181:122–136.

Poole, D. A., and J. B. Trefethen. 1978. The maintenance of wildlife populations. In *Wildlife and America,* ed. H. Brokaw, 339–349. Washington, D.C.: Council on Environmental Quality.

Ralls, K., and J. Ballou. 1983. Extinction: Lessons from zoos. In *Genetics and Conservation: A Reference for Managing Wild Animal and Plant Populations,* ed. C. M. Schonewald-Cox, S. M. Chambers, B. McBryde, and W. L. Thomas, 164–184. Menlo Park, Calif.: Benjamin/Cummings.

Ralls, K., and R. L. Brownell. 1989. Protected species: Research permits and the value of basic research. *Bioscience* 39(6): 394–396.

Ratcliffe, D. A. 1976. Thoughts towards a philosophy of nature conservation. *Biological Conservation* 9:45–53.

Regan, T. 1983. *The Case for Animal Rights.* Berkeley: University of California Press.

Regan, T., and G. Francione. 1992. The animal "welfare" vs. "rights" debate. *Animals' Agenda* 12(1): 45.

Regenstein, L. 1985. Animal rights, endangered species, and human survival. In *In Defense of Animals,* ed. P. Singer, 118–132. New York: Harper & Row.

Robeck, T. R., A. L. Schneyer, J. F. McBain, L. M. Dalton, M. T. Walsh, N. M. Czekala, and D. C. Kraemer. 1993. Analysis of urinary immunoreactive steroid metabolites and gonadotropins for characterization of the estrous cycle, breeding period, and seasonal estrous activity of captive killer whales *(Orcinus orca). Zoo Biology* 12:173–187.

Roberts, L. 1988. Beyond Noah's ark: What do we need to know? *Science* 242:1247.

Rodman, J. 1977. The liberation of nature? *Inquiry* 20:83–131.

Rolston, H. III. 1985. Duties to endangered species. *Bioscience* 35(11): 718–726.

———. 1992. Ethical responsibilities toward wildlife. *Journal of the American Veterinary Medical Association* 200:618–622.

Rowan, A. 1991. Animal experimentation and society: A case of an uneasy interaction. In *Bioscience/Society,* ed. D. J. Roy, B. E. Wynne, and R. W. Old, 261–282. New York: John Wiley & Sons.

Rumbaugh, D. M. 1972. Zoos: Valuable adjuncts for instruction and research in primate behavior. *Bioscience* 22:26–29.

Ryder, R. D. 1985. Speciesism in the laboratory. In *In Defense of Animals,* ed. P. Singer, 77–88. New York: Harper & Row.

Sagoff, M. 1984. Animal liberation and environmental ethics: Bad marriage, quick divorce. *Report from the Center for Philosophy and Public Policy* (University of Maryland) 4(2): 6–8.

Schaeffer, D. O., K. M. Kleinow, and L. Krulisch, eds. 1992. *The Care and Use of Amphibians, Reptiles, and Fish in Research.* Bethesda, Md.: Scientist's Center for Animal Welfare.

Seidensticker, J., and S. Lumpkin, eds. 1991. *The Great Cats.* Emmaus, Pa.: Rodale Press.

Shaffer, M. L. 1978. Minimum viable population sizes for species conservation. *Bioscience* 31:131–134.

Shepherdson, D., K. Calstead, J. Mellen, and J. Seidensticker. 1993. The influence of food presentation on the behavior of small cats in confined environments. *Zoo Biology* 12:203–216.

Sigma Xi. 1992. Sigma Xi statement on the use of animals in research. *American Scientist* January: 73–76.

Singer, P. 1975. *Animal Liberation.* New York: Random House.

Smyth, D. H. 1978. *Alternatives to Animal Experiments.* London: Scholar Press.

Soulé, M. E. 1985. What is conservation biology? *Bioscience* 35:727–734.

————, ed. 1986. *Conservation Biology: The Science of Scarcity and Diversity*. Sunderland, Mass.: Sinauer Associates.

Soulé, M. E., and K. A. Kohm. 1989. *Research Priorities for Conservation Biology*. Washington, D.C.: Island Press.

Soulé, M. E., and B. A. Wilcox, eds. 1980. *Conservation Biology: An Evolutionary-Ecological Perspective*. Sunderland, Mass.: Sinauer Associates.

Soulé, M. E., B. A. Wilcox, and C. Holtby. 1979. Benign neglect: A model of faunal collapse in the game reserves of East Africa. *Biological Conservation* 15:259–272.

Still, A. W. 1982. On the numbers of subjects used in animal behaviour experiments. *Animal Behaviour* 30:873–880.

Stone, C. D. 1974. *Should Trees Have Standing?* New York: Discus, Avon Books.

Sukumar, R. 1991. The management of large mammals in relation to male strategies and conflict with people. *Biological Conservation* 55:93–102.

Thompson, S. 1993. Zoo research and conservation: Beyond sperm and eggs toward the science of animal management. *Zoo Biology* 12:155–159.

Tudge, C. 1991. *Last Animals at the Zoo: How Mass Extinction Can Be Stopped*. London: Hutchinson Radius.

Wiese, R., and M. Hutchins. 1994. The role of zoos and aquariums in amphibian and reptilian conservation. In *Captive Management and Conservation of Reptiles and Amphibians*, ed. J. B. Murphy, J. T. Collins, and K. Adler, 37–45. Ithaca: Cornell University Press.

Wiese, R., M. Hutchins, K. Willis, and S. Becker, eds. 1992. *AAZPA Annual Report on Conservation and Science, 1991–1992*. Bethesda, Md.: AAZPA.

Wiese, R., K. Willis, J. Bowdoin, and M. Hutchins, eds. 1993. *AAZPA Annual Report on Conservation and Science, 1992–1993*. Bethesda, Md.: AAZPA.

Wilbur, S. R. 1977. Supplemental feeding of California condors. In *Endangered Birds: Management Techniques for Saving Threatened Species*, ed. S. A. Temple, 135–140. Madison: University of Wisconsin Press.

Wildt, D. E. 1989. Reproductive research in conservation biology: Priorities and avenues for support. *Journal of Zoo and Wildlife Medicine* 20(4): 391–395.

Will, J. A. 1986. The case for the use of animals in science. In *Advances in Animal Welfare Science 1986–1987*, ed. M. W. Fox and L. D. Mickley, 205–213. Washington, D.C.: Humane Society of the United States.

Wilson, E. O. 1984. *Biophilia: The Human Bond with Other Species*. Cambridge, Mass.: Harvard University Press.

————. 1989. Threats to biodiversity. *Scientific American* 261(3): 108–116.

Younghusband, P., and N. Myers. 1986. Playing God with nature. *International Wildlife* 16(4): 4–13.

SCIENCE IN ZOOS

Arrogance of Knowledge versus Humility of Ignorance

Roger Fouts

While this chapter commends the practice of research science in zoos, it does warn against some of the more human dangers that science can cause. I find little in the arguments of Hutchins, Dresser, and Wemmer in the previous chapter with which I disagree. I echo their sentiments with regard to making sure that researchers take ethical and welfare issues into account, the concept of "gentle usage," the recognition that a zoo's mission is to preserve and respect life, and that zoo researchers and caretakers must be trained and sensitized to animal welfare issues. If zoos are going to play an active role in science, however, then they should be aware that there are grave dangers and pitfalls that await scientists, and great care must be taken to avoid them. The pitfalls are seductive, and if the scientist is caught in one it is usually the animal that suffers, as well as science.

The following points are made: The pursuit of knowledge is worthwhile only if it is sought with an appreciation of our ignorance and with a strong dose of humility. In our scientific approach we must use a method that takes the animal on its terms by being willing to learn from it rather than merely proving our pet theories. So we must take care not to overstructure the experimental situation such that the animal can answer only our question or essentially nothing else at all. Aristotle's *scala naturae* is appealing to us but incorrect. It puts man on top and has traditionally been used as a justification for exploitation. We are really stewards for our fellow animals, and we must guard against becoming pretentious kings. Finally, the two sides of anthropomorphism are examined. Anthropomorphism must be used to understand the animal in question and not as an excuse to deindividualize and exploit someone who is different from us.

KNOWLEDGE PUFFS UP

Hutchins et al. begin the previous chapter by advocating the notion that knowledge is the essential aim and goal of science. This is appealing to scientists—many of us fancy ourselves as quite knowledgeable. However, if scientists focus on knowledge they may be led astray by its appeal to their vanity. In turn this vanity tends to replace a prerequisite for doing science, which is something quite opposite of knowledge, namely, the humility that comes with the acceptance of our ignorance. In other words, thinking of oneself as knowledgeable too often leads to arrogance. This idea is not something new, but is really quite ancient. For example, Paul states in First Corinthians 8:1, "Knowledge puffs up, but love builds up." This puffing up is arrogance, and as is well known, it can lead to other unworthy pretensions and eventually to misuse and even abuse. At a more contemporary level, Robertson Davies reflects a similar concern in his novel *What's Bred in the Bone* when a character, a professor, states, "Science is the theology of our time. . . . The new priest in his whitish lab-coat gives you nothing at all except a constantly changing vocabulary . . . and you are expected to trust him implicitly because he knows what you are too dumb to comprehend. It's the most overweening, pompous priesthood mankind has ever endured in all its recorded history" (Davies 1985, 16). In other words, just because something is called science is no guarantee that it is necessarily the road to truth, even though at times some of its more pompous practitioners give the impression that they have exclusive access to truth. Too often science leads its practitioners off onto the road of greed and arrogance. Such arrogance can make a scientist act as if his or her endeavors are above question; this turns scientific theory into scientific dogma. Kurt Vonnegut, Jr., reflects the danger of this arrogance with a warning in *Cat's Cradle*, "Beware of the man who works hard to learn something, learns it, and finds himself no wiser than before. . . . He is full of murderous resentment of people who are ignorant without having come by their ignorance the hard way" (Vonnegut 1963, 187). Just as science or the seeking of knowledge may be a noble quest, it can also be a shameful deed when the human element is added.

A prerequisite to seeking knowledge is that we should first embrace our ignorance. This is necessary because when we embrace our ignorance we can approach our subject matter with humility and as a result a whole world of discovery awaits us. On the other hand, if we justify our activities as the only road to knowledge and forget that we must always start and continue on that journey in ignorance, then sometimes scientific arrogance can lead a scientist to the position of using the ends to justify the means, and the animals always lose when this is done.

This message of embracing one's ignorance is not a popular one because it points to a reality that many scientists see as a weakness, namely, that we are ignorant. There is a tremendous amount of information in nature about which we are terribly ignorant. I am absolutely sure that we shall never know it all. Perhaps it is not even possible to begin to approach that position. As Vonnegut points out, this is not a popular message among people who have spent most of their lives trying to remove or deny ignorance's presence. If we embrace our ignorance it helps us to approach our study with humility and openness. In this way we can actually learn something from our fellow animals, rather than requiring them to prove a pet theory or maintain the particular dogma that might be in vogue.

TWO SCIENTIFIC APPROACHES: ONE ARROGANT, ONE HUMBLE

I come from two traditions in science. This is because I studied with two major professors, one an experimental psychologist and the other an ethologist. The way they blended the two different approaches resulted in a unique way of doing science, one that has both rigor and humility, objectivity and love, which is all overlaid with an awe of discovery.

The traditional approach of experimental psychology is to emphasize the importance of experimental design and rigor. This is certainly a worthy approach, but sometimes we have forgotten the reason we have the controls, and it works against us. Too often an experiment will be designed in such a manner that the organism being studied can answer only the question posed by the investigator or essentially nothing else at all. In such cases we are finding out more about the mental capacity of the experimenters than we are about the animals they claim to be studying. At the very worst the experimenter sets up the experiment so that the animal can only prove a pet theory. Pertinent to this point, Wolfgang Kohler made the following points in 1921:

Lack of ambiguity in the experimental setup in the sense of an either-or has, to be sure, unfavorable as well as favorable consequences. The decisive explanations for the understanding of apes frequently arise from unforeseen kinds of behavior, for example the use of tools by the animals in ways very different from human beings. If we arrange all conditions in such a way that, so far as possible, the ape can only show the kinds of behavior in which we are interested in advance, or else nothing essential at all, then it will be less likely that the animal does the unexpected and thus teaches the observer something. (Kohler 1971, 215)

In short, methodology must be carefully designed so as not to exclude information. One way to avoid this is to embrace our ignorance and be humble enough to learn from our fellow animals.

Ethology represents the other extreme from experimental psychology. Whereas experimental psychology runs the risk of arrogance, ethology runs of the risk of being too humble. Ethology evinces its humility in that it truly is receptive to the animals. Rather than asking them if this theory is right or wrong, it simply records their behavior and learns from them. This is fine if we have no question to ask and are willing to accept only the information that the animals may provide us when we systematically observe them. Sometimes, however, we are in situations where we must ask a specific question. The first thing that the experimenter must do here is make sure that the question is relevant to the species, which requires that we have a thorough knowledge of the biology of the animal as well as the species-preserving function of the behavior we wish to study (Eibl-Eibesfeldt 1970, 25). In this way, by being humble enough to learn carefully about the animal we wish to study, we hope to avoid asking stupid questions, to be able to understand the animal's answer to our question, and as a result to learn something new from the animal. Put another way, the experimental design must be set up so that it adjusts to the particular nature of the animal being studied, rather than tradition. Running an antelope in a maze would be quite stupid, but a maze would make a great deal of sense to study a pocket gopher. So both rigor and humility are needed for proper research. Having too much rigor without the humility to take the organism being studied on its own terms often leads to meaningless results. On the other hand, sometimes we must structure the experimental situation to answer a specific question, but this always requires humility, a willingness to learn from and adjust to the organism being studied. The message here is that you should know your critters and adjust your design and questions to them, regardless of whether they are chimpanzees, white rats, or college freshmen.

THE PHYLOGENETIC SCALE VERSUS ARISTOTLE'S SCALA NATURAE

An implicit patronizing attitude toward animals is expressed in some of the chapters of this book. As far as I can determine by the way some people have referred to zoo animals ("lower" or "less complex" and so on), they seem to assume that Aristotle's *scala naturae* is the operative model for zoos and science alike. This is a popular approach, and it is well received because it implicitly tells the human using it, "You are special and certainly much above the lower animals whom you must care for in your zoo." Likewise, the use of euphemisms like "ambassadors" to hide the unwilling confinement to which we subject our fellow animals is one more manifestation of this trait. This is comparable to youths' trying to persuade someone who is quite homely to run for homecoming king or queen. In our children this behavior is seen as cruel.

The patronizing point of view can be traced to Aristotle and Descartes. It basically says that mankind is special and above the defective have-nots who make up the rest of the animal kingdom, such as the apes, women, dogs, and horses. Its appeal to male vanity and arrogance is why this model of nature has been so popular and even hangs on today. After all, Darwin's theory has been around for only a little more than one hundred years, whereas the Aristotelian view in Western thought is several thousand years old.

In Aristotle's *scala naturae,* man is on top, followed by elephants, dolphins, and women, in that order. The reason I point this out is that it clearly demonstrates that if you were not a member of the dominant class, you were lacking. Women know only too well the result of such an outlook on nature. It has been only a little more than one hundred years since the rule-of-thumb law was taken off the books in England. That law stated that it was illegal for a man to beat his wife with a stick any thicker than his thumb. This Aristotelian view, later supported by Descartes in the early modern period, basically held that creatures without a rational soul were inferior. In reality it meant that if you were not like the people in charge, then those differences, whatever they may have been, were defects. This noted difference turned into a justification for exploitation and abuse.

With Darwin's theory this Aristotelian view, which was seen as obvious scientific truth, was discredited. Darwin turned a vertical model into a horizontal model. Now all animals are different in degree rather than kind, and there is no top or bottom of a scale. This robbed man of his high position; however, it seems that many men are not ready to give it up. The implication of this model change is that we must carefully examine the relationship we wish to have with our fellow animals. Those of us who are responsible for their care might wish to examine ourselves carefully to make sure that we do not maintain the false arrogance with which Aristotle and Descartes have seduced our culture. In my own situation, we make sure this attitude is absent by insisting that the people who work with and around the chimpanzees do not view themselves as superiors, deigning to clean up after and feed the chimpanzees. Instead we insist that they assume the role of domestic—that of the trusted butler, always there when needed for anything; the creative cook, who knows the tastes or preferences of his or her clients; or the ever reliable maid, who carefully cleans and offers fresh bedding. This also means that we do not feed the chimpanzees; we *serve* them breakfast, lunch, dinner, and snacks. Thus, in line with our not-so-special status, we are humble as well.

THE STEWARDSHIP MODEL

If we fancy ourselves care-givers and stewards, it has a profound effect on the way we deal with our fellow animals. We would view animals quite differently from the way we would see them if we fancied ourselves predators or parasites. Too often predators try to pass themselves off as care-givers. This behavior occurs in nature, as is evidenced by the mimicry of the coloration of a cleaner fish by a predator fish. The cleaner fish cleans the fins of larger fish; the mimic predator eats the fins of the unsuspecting fish. We must be aware that Cartesians or Aristotelians may claim to be stewards when they actually behave much more like predators and parasites. This is not to demean the important role of predators and parasites; it is just to say that in our species we hold honesty to be a valuable trait and that this honesty goes beyond our own species to our treatment of our fellow animals as well.

Perhaps the problem lies in what David Ehrenfeld refers to as the "snare of stewardship" (1981). To illustrate his point Ehrenfeld quotes from J. R. R. Tolkien's *Two Towers*. Boromir, the elder son of the Steward of Gondor, inquires of his father, how long must a steward wait before he becomes king, if the king does not return? "His father answers, 'Few years, maybe, in other places of less royalty. . . . In Gondor ten thousand years would not suffice' " (p. xi). The snare of stewardship, Ehrenfeld points out, is that too often the steward forgets that he or she is not the king or queen.

In a captive situation it is too easy to forget our stewardship role and to be seduced by the king role. It goes without saying that it is a much bigger boost to our egos to play the role of prison warden than the humble role of maid and cook. When we are in the roles of scientist and healer the snare of stewardship is even more likely to capture us, because we tend to believe the pretensions that society gives us in regard to our special nature, because we alone are supposed to possess special knowledge. Once again, humility is the antidote to this illness.

ANTHROPOMORPHISM AND RUBBER RULERS

Another issue that has come up in a few places is the notion of anthropomorphism. It is an argument that stems from scientific arrogance because it invokes objectivity as its main support. Anthropomorphism warns us against ascribing human characteristics to nonhumans. We must be objective and not attribute anything that we might have to other animals. This has led members of the zoo community to argue against giving animals names. They think that this is anthropomorphic, yet when we consider that our names really do nothing more than point out our individuality, why is it that animals may not have a name? Are they not individuals as well? Too often this argument against naming an animal is nothing more than an attempt to deindividualize the animal. If that is achieved, then ethics and responsibility usually go out the window. Our country did this during the Vietnam War. By turning individuals into "gooks," we made the napalming seem less horrid.

More often this criticism is used against those of us who might ascribe some allegedly human trait to nonhumans. One of the favorites is thought. If a person thinks that nonhuman animals might think, then that person is thought to be anthropomorphizing. It does not seem to occur to these critics that it is just as anthropomorphic to say that a nonhuman animal thinks as it is to say that they do not think. Indeed, the person leveling these criticisms is using a rubber ruler, one that gives different measures when it measures different animals.

Several years ago I had the occasion to meet one of the primary proponents of the war against anthropomorphism. I was concerned about this because I often found anthropomorphism to be a useful device for understanding some chimpanzee behavior (which is not surprising when one considers that they are anthropoids). The person was Dr. Heini Hediger. In the course of our discussion he did admit finally that there was good anthropomorphism and bad anthropomorphism. The good anthropomorphism is really better named zoomorphism. The point here is that if you know the animal you are studying well enough, you can infer motivating states and predict possible behaviors. Some of these states and behavior may also be shared by our species. This point is very important—just because our species might evince a certain behavior, it does not mean that we have exclusive rights to that behavior.

Hediger gave me an example of bad anthropomorphism. He told me a story of a couple who lived in Africa and raised a lion cub that grew to an unmanageable size. Realizing that they could not put their pet back in the wild, they arranged to have her shipped to a European zoo. The surrogate human mother was disturbed to see her pet placed in a shipping crate that looked terribly uncomfortable. So perhaps to ease her own guilt but under the guise of making the lion's trip more comfortable, she placed a pillow in the crate so the lion could rest her head on it. When the crate arrived at the European zoo, there was no pillow, but the sole occupant was one very constipated lion.

Hediger also gave me an example of good anthropomorphism, or zoomorphism, as I like to call it. He stated that if a tiger escaped at his zoo, he would say to himself, "Now where would I go if I were a tiger?" and then act accordingly. This brings up the point of humility in regard to knowing your organism. By being humble enough to take the time to learn everything you can from a fellow animal, you can begin to anticipate what it might do. For example, if I lost a *Rattus norvegicus* in my office, I would look for it among the piles of books on the floor, in my bicycle shoes under the bookcase, or behind the wastepaper basket in the corner. However, if it was a *Rattus rattus* that had escaped in my office, I would look for it on top of the bookcase, on the shelf on top of the chalkboard, and so on. I would do so because I have taken the time to take these two different species of rat on their own terms and learn from them about themselves. In Europe *Rattus norvegicus* is also known as the cellar rat and will run down when frightened. On the other hand, *Rattus rattus* is known as the roof rat and lives in trees and attics and will run up when frightened.

One final point about the criticism of anthropomorphism—it often embodies something that is inherently bad science. The assumption in such criticism is that because we humans do something, the have-nots do not do this thing. So the critics might accuse me of anthropomorphizing if I say that a chimpanzee is grieving. They would say that chimpanzees do not grieve; only we special humans have that capacity. Such statements are based on ignorance, or the absence of evidence. Many people assume that the absence of evidence is evidence of absence. This is bad scientific thinking. The most that scientists can do when they have no evidence, or when they are ignorant of any evidence, is to withhold judgment. They must admit that they do not know, for example, whether chimpanzees are capable of grief or not. Unfortunately, as mentioned earlier, the presumptuousness of some of our scientific colleagues prevents them from admitting any ignorance.

DISCOVERY AWAITS THE IGNORANT

In summary, I am advocating that we scientists embrace our ignorance and approach the study of our fellow animals with humility and reverence. As commanded in the physician's creed, we should "First do no harm," and add to that a warning to make darn sure we are going to do some good for the animal we are about to study. This attitude creates a symbiotic relationship between us human scientists and our fellow animals. By treating them with respect and reverence we can light a small lamp of knowledge and in so doing drive back a tiny bit of the darkness of ignorance, which will reveal to us the exciting world of discovery.

REFERENCES

Aristotle. *De Anima II,* 3, 414a(28)–415a(10).

Davies, R. 1985. *What's Bred in the Bone.* New York: Elisabeth Sifton Books, Viking.

Ehrenfeld, D. 1981. *The Arrogance of Humanism.* 2d ed. New York: Oxford University Press.

Eibl-Eibesfeldt, I. 1970. *Ethology: The Biology of Behavior.* New York: Holt Rinehart & Winston.

Kohler, W. 1971. Methods of psychological research with apes. In *The Selected Papers of Wolfgang Kohler,* ed. M. Henle, 197–223. New York: Liveright.

Vonnegut, K., Jr. 1963. *Cat's Cradle.* New York: Dell Publishing.

PUBLIC RELATIONS, FUND-RAISING, AND DISCLOSURE

Zoos are public institutions and, as such, have a responsibility to taxpayers and the visiting public. Public attitudes toward animals, however, vary depending on a number of factors. In this complex social environment, Karen Allen argues that zoos need to consider their public relations and development initiatives carefully. She also notes that achieving consensus may be impossible and that controversial decisions are often necessary in conservation. The author cautions that should zoos and animal welfare advocates continue their "righteousness wars," it is wildlife that may lose. Roger Caras believes zoos should be as open as possible in disclosure of their activities but recommends avoiding zealots in favor of those who are open to discussion, reason, and compromise. He suggests a new term for animal rights, which may help to avoid polarization among zoos and activists.

PUTTING THE SPIN ON ANIMAL ETHICS

Ethical Parameters for Marketing and Public Relations

Karen Allen

The zoological profession has experienced profound changes in recent years; what, in many cases, were menageries have become modern conservation way stations. With those changes have come new considerations regarding conservation and animal welfare ethics. Because of its virtual overnight revolution, the zoological profession has not had adequate time to address the myriad of ethical conundrums with which it is faced. Thus, it has opened itself to criticism from animal rights and animal welfare critics who want immediate solutions. That criticism has the potential to threaten the zoological community's credibility seriously, unless it establishes ethical guidelines that will further engender public trust. In exploring the need for an expanded ethical base, it becomes evident immediately that conservation ethics and animal rights or animal welfare ethics often follow divergent tracks. While there is some potential for establishing common ground, it appears that the two camps are going to have to agree to disagree; otherwise, a righteousness war will be perpetuated and its greatest casualties will be the animals everyone claims to care about.

A friend I would characterize as a nonanimal type—a well-read, middle-class Washington bureaucrat with no particular interest in animal activism—has this to say about individuals dedicated to zoological and animal welfare endeavors: "People in the business world compete for power and money. The game is brutal, but everyone knows the rules. But the way I see the animal people, they are competing for righteousness. One group is trying to one-up the other in terms of who cares more about animals. And to make matters worse, nobody knows the rules, because there are not any. You are never going to have any winners, just a bunch of battles. It seems to me the animals get the short end of the deal" (Asis 1990).

289

A notable skirmish in the righteousness war occurred in Tokyo in February 1992 when two representatives of People for the Ethical Treatment of Animals (PETA) stripped down to their underwear and paraded through a shopping district in 46°F weather during an international fur trade show. Placards covering strategic parts of their anatomy read, "We'd rather go naked than wear fur." One of the demonstrators, Dan Mathews, explained, "By showing some of our skin, we hope to save some animal skins" (Reuter News Service 1992). This was not unlike an earlier publicity stunt in which two paid PETA staff members flew from Washington, D.C., to Iowa to throw a pie in the face of the newly crowned, teenaged Iowa Pork Queen (Eide 1991). A conservative estimate for the cost of the Tokyo and Iowa outings (if the Tokyo duo immediately put on their clothes for a return flight to D.C.) is $4,500—a high price to pay when one considers that probably no mink, fox, nutria, or pig lives were spared as a direct result of these media events. Yet that same $4,500 could have covered the cost of one month's operations—for eighty park staff—at the Ujung Kulon National Park in Indonesia, 284.9 square kilometers (110 square miles) of habitat that is home to the remaining fifty to sixty Javan rhinoceroses left in the world (R. Tilson, personal communication).

We zoo and aquarium people have managed to keep our clothes on in our skirmishes in the righteousness war, but historically, our critics claim to have caught us with our pants down. Admittedly, we (professionals who run zoos and aquariums) have made some mistakes and some regrettable judgment calls—not from ignorance or a lack of compassion for animals but, in some cases, a lack of maturity. Twenty years ago, most zoos struggled to get their animals to breed; surplus animals were not a problem and probably would have been a novel idea. Sophisticated education and conservation programs were the stuff of science fiction. As opposed to today's trend of nonprofit institutions, most facilities were municipally funded operations attracting directors with a park mentality instead of animal management experience. The people who ran zoos were not the bad guys; they were operating within their limits of scientific and technological knowledge, as well as the ethical limits deemed appropriate by the public at the time.

Today's zoological managers still operate with good intentions, but times have changed. Any meaningful social, political, or economic change is always preceded by a change in values (Kidder 1992). Environmental and animal welfare values have changed radically in the last decade. Those changes, along with the breakneck speed of scientific and technological advancements, have created a revolution within the zoological profession. As with all revolutions, there is a period of transition during which a new order must be established. The zoological profession is at that crucial stage in its development. Zoo critics, however, appear to have no patience for this evolutionary progress. They want instant answers for the problems that evolved from our virtual overnight conversion from menageries to conservation way stations—they might as well ask the Commonwealth of Independent States to have its economic and social programs in perfect order within the next year. Initially, we should have put more effort into evaluating the ethical consequences of our successes as they occurred, particularly in reproductive physiology. But we have learned a valuable lesson; if we want to maintain the highest level of values, we must look far into the future (Kidder 1992). The deeper we delve to identify distant—and not-so-distant—issues, the better our chances of predicting problems and finding solutions before problems occur. We must also anticipate ramifications of the decisions and standards we adopt today.

When I say "we," I am not only referring to husbandry and science professionals; I am also talking about those who determine marketing policies for zoological facilities. We look to these individuals to use their zoo or aquarium as a revenue builder whose money supports not just operating expenses but also wildlife conservation, research, and education programs (Ann 1988). But beyond that, marketing professionals, especially those in public relations, have the potential to play an important role in helping to form the public's perceptions of wildlife conservation through captive management. In other words, public support for the work of zoos and aquariums hinges on marketing efforts. If zoological marketers mobilize forces—something the animal rights community has mastered—their potential for significantly molding public opinion is tremendous. But with that power comes new ethical considerations.

Being a dove, I am not keen on this righteousness war business; I would rather work in harmony with the animal activist community. Nevertheless, in certain areas, battle lines have been drawn, and we are going to have to fight our critics in the interest of self-preservation.

Zoos and aquariums need to fight smart, not dirty. First and foremost, we need to correct weaknesses and tighten our ranks. Someone once said a kick in the butt is a step forward. Since the animal activist community has been aiming its well-placed foot at our posteriors for the last few years, I would like to think we have taken more than one step forward in terms of correcting problems. We have become more aggressive in dealing with accredited institutions whose conduct defies the American Zoo and Aquarium Association's (AZA) Code of Professional Ethics. Wrongdoers have the potential to give everyone a bad name, and our critics gleefully use them as examples to generalize about problems in the zoological profession. The critics need find only minor infractions to do major damage to all our reputations and to undermine the public's trust in us.

I suggest we go even a step further and actively crusade to put substandard roadside zoos out of business, which may mean more government regulation in terms of animal welfare standards. Jay Leno once joked on the *Tonight Show*, "President Bush says in his new health care program he will encourage people to avoid drugs, drink less, and practice safe sex. Does he really think Congress is going to pass something that they themselves aren't eligible for?" In the same vein, are we willing to propose legislation that would put substandard facilities out of business but could also interfere with some of our own operations through potential overregulation? Perhaps government intervention is not the best answer, but we had better find a solution soon, because those who do not make the grade or follow the ethical straight and narrow are dragging the profession's image through the mud.

Novelist Channing Pollock once quipped that "a critic is a legless man who teaches running"—like animal rights people trying to teach zoos about wildlife conservation. We are systematically chastised by the animal rights community for not embracing its philosophy, a philosophy that appears to say the only way to save animals is to be nice to them (Budiansky 1992). Herein lies a monumental ethics conundrum for the zoological marketing and public-relations community. We believe that although well intentioned, the animal rights "save them with kindness" philosophy is ethically and biologically flawed. The reality is that in order to save species, we must abide by Mother Nature's code of ethics, a force much larger and harsher than the morality the animal rights movement has tried to impose on the natural world. To work in harmony with the dictates of nature to preserve species, we will have to make some difficult animal management decisions—for example, to cull overrepresented populations—decisions that will offend the morality of animal activists. We say this with conviction, yet we wince every time the activists lob epithets against us that suggest we are ruthless "species fascists" because of our commitment to consider the good of a species over an individual animal. As a result of this pressure, we often try to present ourselves not as we really are but as what we think will be most palatable to the public, a public whose ideals have been influenced by animal activism.

To a certain extent, good business dictates that we market ourselves to the desires of the public. After all, one of the basic tenets of marketing is "Give the customer what he wants." However, just how far will we go to please the public? When do we decide that what the public wants is not necessarily best for our institutions or, more important, for wildlife conservation? So far, we have been able to answer these concerns through education and positioning. In other words, we design the market. We have been effectively telling the public what their expectations for zoos and aquariums should be (humane care of animals, natural habitats, sanctuaries for endangered animals, education centers); we then aspire to fulfill those desires. I think we are to be commended for what we have accomplished so far, but we have made it too easy for ourselves. It is time for us—not the animal activists—to change public expectations for zoos and aquariums even further. Beyond that, it is time for us to stop the animal activists from setting our agenda.

When are we going to get the gumption to dispel the notion that all zoo births are blessed events and explain how some of those babies grow to become surplus animals and may even have to be culled? If what we are doing is ethically correct, then give the public a little more credit and explain our position to them without the sugar coating. As a public affairs professional, I should qualify this statement by saying we need to approach the problem with honesty but also with sensitivity. It will not be easy or necessarily popular. A Jacksonville State University study titled "Understanding Your Visitors" found that the presence of an infant animal can increase the holding power of an exhibit by at least 100 percent (Bitgood et al. 1986). The best way to deal with this misplaced public affection is to quit rhapsodizing about baby animals. Instead, let us promote the significant births and aggressively market the value of our contraceptive programs. Let us also reevaluate the ethics of publicly naming animals, thus portraying them as individual personalities. This is not a problem if institution directors are willing to deal with public reactions when well-known, well-loved animals are moved to another facility. It is unrealistic, however, to expect the public to turn its emotions on and off to suit the agenda of the zoological facility.

This is not meant to suggest we should ignore the emotions of the public. In fact, the public needs to see more of our human side—rolled up sleeves, a loosened tie, an occasional tear in the eye, and an absence of scientific euphemisms (Thomas 1991). In spite of appearing human, we can still communicate the missions and goals of zoos and aquariums to the public with dignity and without apology. Being candid about our real missions and goals is bound to alleviate another dilemma, the ethics of disclosure. Just how much information do we owe the public? I maintain we should give them only what they are able to assimilate, which means we need to give them enough of an education for them to make informed judgments. Yet, if we continue to explain our actions with euphemisms and allow ourselves to be intimidated by the name-calling of animal activists, we will effectively keep the public in the dark. When the public is then unable to process information, we will feel justified in not wanting to release critical data. Not only is this grossly unethical, but it will also eventually undermine the public's trust and support of our programs.

If we owe the public as much information as possible about our operations, I believe we should confine our debates with activists to venues such as this, not forums controlled by the news media. When we talk about our contributions to wildlife conservation, what we are trying to achieve cannot be encapsulated into a thirty-second sound bite. But the problem goes beyond brevity and depth. When we allow ourselves to be put in a debate situation with animal rights adversaries, we are forced into a defensive position that allows us little time to speak positively about our accomplishments. The activists' agenda is not to discuss their contributions to wildlife conservation, since, in most cases, they have none; their discourse is centered totally around discrediting us in order to bring credibility to themselves. Ergo, the combination of the two factions yields a disparate, counterproductive spectacle that serves only to confuse the public further.

Not only is the public getting conflicting messages about zoos and aquariums from animal activist groups, but it is also getting contradictory impressions from the zoological community itself. Institutional policies vary widely in regard to the off-site use of animals, animal rides, animal demonstrations, and the portrayal of animals in promotional materials. This is an observation, not necessarily a criticism. If we homogenize ourselves to the extreme, we will become nothing more than a chain of McZoos. Where do we draw the line on standardization of policies and ethical philosophies? Ray Kroc, the founder of McDonalds, insisted that his burgers, no matter where they were purchased, measure 9.8 centimeters (3.875 inches) in diameter and weigh 45.4 grams (1.6 ounces), not including 7.1 grams (0.25 ounce) of onions (Boas and Chain 1987). Many of us have probably stayed in so many Holiday Inns that we know the way to the bathroom in the middle of the night without turning on the light. If we were to succumb to such extreme standardization, the future of zoos and aquariums may be doomed; however, we can also learn from the success of the Golden Arches and Holiday Inns. They have managed to survive because of quality control, which brings us back to the issue of strengthening our standards and working to close down substandard facilities. If we take those issues seriously, we will be better able to validate our role as animal welfare advocates. Public criticism of zoos and aquariums by animal activists often creates an us-versus-them situation that undermines our credibility as animal welfare advocates. By taking the offensive position and thus portraying themselves as the guys in the white hats, the activists are able to justify the presumption that they can and should set our ethical agenda.

One of our greatest ethical imperatives is to deliver on our claim, our promise, to be conservationists, to back up rhetoric with actions; in short, there must be substance before image. The importance of maintaining conservation and research programs in our institutions speaks for itself. Yet how we are judged as conservationists—and animal welfare advocates—also depends on where we seek financial support. Is it ethical to accept money—money that is critical to both operational and conservation efforts—from individuals or organizations whose actions are labeled unethical or questionable by the public? Regardless of its source, the money spends just as well, but does the end justify the means? Will the public perceive us as conservationists with integrity or as foolishly principled purists if we turn down major sources of support? The time has come for us to address these questions formally and come to a consensus. I feel we should judge potential sponsors by where they are going, not by where they have come from. Such a protocol may elicit criticism, but we are in the business of helping to maintain natural diversity, not sitting in judgment of big business. If our standards are exceedingly restrictive, we will undermine our own efforts. It should also be noted that there are two sides to this issue; responsible businesses should also have the prerogative to judge our performance records and standards before they align themselves with us.

Concerns about where and how zoos and aquariums derive their income go beyond corporate sponsors. Traveling animal exhibits, white tigers, and all manner of baby fuzzy wuzzies are crowd pleasers and therefore revenue producers. Critics call this exploitation. Some zoo and aquarium managers say it pays the way for legitimate and much-needed programs; others feel that some of these revenue enhancers should be audited for their long-term value to the goals and missions of professionally run zoological facilities and, ultimately, to wildlife conservation. I agree that we need to put more thought into the long-term ethical implications of these programs. However, I am a realist and acknowledge that there is a fine line between maintaining only the purest of principles and ensuring financial solvency. In our effort to avoid financial deficits, we have become leaders in devising fund-raising schemes, particularly special events. But even they draw criticism. A zoo enthusiast from Baltimore recently wrote, "I am not a zoo-goer who will be found at a 'kangaroo hop' or a summer solstice dance under the stars. These public relations activities are surely necessary for fund-raising, but in some ways they also reinforce unhealthy notions of the zoo as a fantasy Disney World, a cartoon for our enjoyment" (Turbyville 1992). God forbid that people should enjoy themselves at a zoo or aquarium!

I wholeheartedly support critics who are offended if the dignity of nature, particularly wildlife, has been violated through some of our promotional or special events endeavors. Beyond that, what is inappropriate should be judged by common sense and public trust, not Puritan ethics. As English statesman Thomas Macaulay put it, "The Puritan hated bear-baiting not because it gave pain to the bear, but because it gave pleasure to the spectators." Ingrid Newkirk, national director of PETA, tells a story about how she asked a vegetarian community to petition the government to maintain its requirement that cows and steers be given water during their long wait to be butchered at the slaughterhouse, a practice the cattle lobby apparently opposed. Members of the vegetarian community sent their regrets to Newkirk with a note that read, "We are ethically opposed to the slaughter of animals for food, therefore we cannot get involved" (Newkirk 1992). In order to maintain their stand against slaughtering, this segment of the animal rights movement was willing to allow the cows to suffer without water. Ecologist David Ehrenfeld writes, "Life is or should be a tightrope walk between the realm of rights and the realm of obligations. Tilt too far to either side and the results will be unpleasant" (1991). I urge everyone not to let our perceived ethics get in the way of common sense and humanity, because if that happens, the righteousness war becomes an ethical holocaust.

REFERENCES

Ann, M. 1988. Entertainment and economics: Why zoos? *Universities Federation for Animal Welfare Courier* (England) 24 July: 9–14.

Asis, K. 1990. Animal activism: Who are the players? In *Proceedings: American Association of Zoological Parks and Aquariums Annual Conference*, 302–307. Wheeling, W.Va.: AAZPA.

Bitgood, S., D. Patterson, and A. Benefield. 1986. *Understanding Your Visitors: Ten Factors That Influence Visitor Behavior.* Technical Report 86-60. Jacksonville, Fla.: Psychology Institute, Jacksonville State University.

Boas, M., and S. Chain. 1987. *The Sacred Cow and the Abominable Pig.* New York: Touchstone, Simon & Schuster.

Budiansky, S. 1992. *The Covenant of the Wild.* New York: William Morrow & Co.

Ehrenfeld, D. 1991. Raritan letter. *Orion* Summer: 5–8.

Eide, V. 1991. Of pigs and pie. *Animals' Agenda* November: 3–7.

Kidder, R. 1992. Ethics: A matter of survival. *Futurist* March-April: 10–12.

Newkirk, I. 1992. Total victory, like checkmate, cannot be achieved in one move. *Animals' Agenda* January-February: 43–45.

Reuter News Service. 1992. U.S. anti-fur activists strip for walk in Tokyo. 18 February.

Thomas, E. M. 1991. A manner of speaking. *Orion* Summer: 3.

Turbyville, L. 1992. Where the wild things are. *City Paper* (Baltimore), 21–27 February, 8–19.

A View from the ASPCA

Roger Caras

H ow much should we reveal to the public? My lawyer of thirty years said to me (and I have not been in jail once): "Never do anything operating on the premise that it will never be known. In that direction lies madness. Always assume that everything you do will become common knowledge. You're always going to have a disgruntled employee, no secret ever gets kept for very long, and you end up with an incredibly messy chin—egg all over it." Determining (or thinking we can determine) what the public can handle and what they cannot handle at the very least smacks of elitism. We cannot predict what the public can handle. Let me say that I believe the ASPCA and zoos face many of the same problems in this area. I feel that if we hold things back, keep secrets, have locked back rooms beyond those necessary for security, we are really looking for trouble. We are public institutions, and if we do not want to serve the public, then we should close the doors and call ourselves a laboratory and seek our sustenance elsewhere. If we are public institutions, then we must be public.

The exhibiting of baby animals—the idea of teddy bears—is here to stay, and so is Bambi and so is cuteness, and no zoo on this earth can ban, outlaw, or effectively restrict this. Objectivity on the part of zoogoers simply cannot be mandated. Indeed, if it could be, you would see your position dwindle. Each zoogoer expects to view animals on an individual basis, because everyone who comes through the gate is an individual. We must cater to each one. Perhaps we must be all things to all people, but I cannot buy the mechanistic approach. Please the public.

I remember a speech Bill Conway made many years ago. He said, "At one time, the one branch of conservation you simply could not get any support for was whale protection." They were simply too remote; they were too far removed. "And then two things happened," Bill said. "They trapped a killer whale in a fenced-in area of Seattle Harbor. At the same time, an unusual television show appeared—*Flipper*." All of a sudden there were two cetaceans with names, and from that day to this, whale conservation is one of the easiest fields in which to raise funds—and it happened just in time for those animals remaining. In a community, the zoo animals belong to the people, and if zoo animals cease to be the community's pets, people will stop coming to the zoo. They will stop supporting the zoo, and they will lose their interest in wildlife. We must hook the public with cuteness. Go with cuteness. That is why your gift shop does so well. I do not believe in the mechanistic approach. Higher scientific minds come in on freebies; they have a pass. The lower minds pay, do they not?

I constantly hear about the issue of returning animals to nature. This seems to indicate that outside, there is an absolute, constant habitat—fixed, like a painting from the Hudson River School. Inside here we have animals, and that is a mistake, and we ought to put them back out there. Nature is dynamic, not constant. The chainsaw is not the cause, time is. The chainsaw is a passing element in time. A short time ago it did not exist. In a few years it will be replaced, probably by lasers. Storing animals as intact organisms or as genes eventually might take us to a world we cannot even imagine. Long Island may have gone. San Francisco and Los Angeles may have gone to sea. They are going to go eventually. All of the barrier islands along the coast of the continents of the world may be gobbled up if the world is warming. If Antarctica warms only slightly, then the coastline of the United States will be somewhere in central Pennsylvania. You would have to take a boat trip out to see the Statue of Liberty's torch. Nature is a dynamo. We do not know what will happen. No one really knows if global warming is in fact occurring. All I know is that winter has not yet come this year, and it's getting a little late for it. Maybe it will come in July. Do not think of nature as being some constant, wonderful thing that is out there just waiting for you. It is changing rapidly and constantly. Sometimes we see the change, and sometimes a few grains of sand wash off a beach on Long Island, and we do not see the change.

Animal welfare, animal protection, even animal rights can be thorns in the side of the zoo professional. Indeed, that is the tactic. They can be constructive, and they can be obstructive. Thorns cause pain, but the purpose of pain is to tell us when and where something is wrong. Michael Hutchins spoke of cruelty, suggesting that animals in accredited zoos are not treated cruelly. But I would add that animals both in the wild and in captivity are frequently afraid. That is a survival mechanism. Some practices in zoos can unknowingly, and with the best of intentions, impose fear and anxiety on animals. It may be projection on my part, but excessive, unwarranted, unrelenting apprehension to me is a cause of pain, an element of cruelty in the extreme. Therefore, you can expect to hear from the ASPCA and other humane groups from time to time. That should neither surprise nor alarm you: we all have our job to do.

Terry Maple asked, "What is humane?" Opinions are generally based more on the emotional needs of the determiner than on the actual treatment of the animal. My son is, among other things, a psychiatrist. Because I deal with obsession and illusion in my profession, I asked him to explain these concepts to me. He said, "Obsession with illusion is, by textbook definition, a system of thought where the patient resists any evidence to the contrary." You come to me and say, "I am Napoleon." I say, "Well, here's your birth certificate and here's Napoleon's death certificate." You say, "Yes, but I'm still Napoleon." Do not try to deal with people who have an obsession or people who are deluded. You will never move them an inch. This reasoning leads one to a principle that has served me well: seek out those—even those with differing opinions—who are open to discussion, reason, and compromise, and avoid the zealots.

Someone suggested that a new author always speaks a new idea, but I do not quite agree. I am the eighteenth president of the ASPCA. The founder and first president was Henry Bergh. He got the idea for the ASPCA after seeing a man beating a horse in the streets of St. Petersburg. Bergh went out into the middle of the street and took the stick away from the man and beat him instead. He became the avenging angel that is represented today on our seal—a man beating a horse and an angel standing over him with a horn. In the original seal, which we found in our files, the angel was a man. In 1866 Bergh founded the society to alleviate (not terminate, because that is not yet possible) pain, fear, and suffering in animals. Our goal is to engage in an open discourse with the zoo community and the public, and to gain as much of a consensus as possible regarding the acceptable treatment of animals. Our contribution to this goal, I hope, is a constant message in favor of reducing unnecessary suffering.

Today we are trying to do exactly the same thing. I propose a phrase that may replace "animal rights." "Animal rights" is polarizing and is in itself antagonizing. "Appanage" is a word derived from medieval French, referring to provisions made by royalty for daughters and the younger son. The oldest son got the crown. The daughters and younger son had to be provided for, and so there was appanage. This concept expanded and came to mean anything awarded by custom or tradition in recognition of natural necessity. If that is not what the zoo profession and the humane community are all about, then I cannot imagine what they are about.

Consensus-Searching Efforts

DESIGN OF THE CONFERENCE

The following introduction and guidelines (reproduced with a few slight refinements) were distributed to those who were to participate in the conference Animal Welfare and Conservation: Ethical Paradoxes in Modern Zoos and Aquariums.

When we gather in Atlanta on 19 March 1992, we will examine the serious ethical issues that inevitably arise when we decide to breed in captivity species that are threatened by extinction as their wild habitats disappear. Our goal for the conference is to focus narrowly on ethical issues associated with captive breeding programs, especially Species Survival Plans, but to do so while recognizing that these ethical issues can be fully understood only in their broader conservation and social contexts. While we need, for example, to explore concerns regarding the welfare of animals maintained in captivity for breeding programs, we must also recognize that a comprehensive understanding of those ethical dilemmas will require a broader exploration of the role of captive breeding programs in conservation, and also the entire role of zoos and aquariums in modern society. The purpose of this background paper will be to divide the basic quandary into what we hope are manageable units and to pose questions that seem to us central in each of six more specialized areas.

To this end, we have chosen three broad themes and six issue areas. We will have three plenary sessions on the broad themes. In the opening keynote session, Dr. William Conway and Dr. George Rabb will state the challenge of protecting biological diversity. We will also have two panel discussions on broad social and philosophical issues: (1) the role of zoos in the twenty-first century; and (2) conservation targets—what is the object of protectionism?

The questions to be examined in the panel discussions are as follows:

1. Are animal rights and welfare philosophies compatible with wildlife conservation? When are they compatible? When are they not?

2. What is it going to take to save wildlife, given the diversity and severity of threats facing it today?
3. Will intensive management of all wildlife be necessary, including captive breeding, invasive research, culling, veterinary care, translocation, and habitat modification, in order to stave off mass extinctions? Are such actions compatible with animal rights and welfare philosophy? If not, what are the implications?
4. Should we take no action, let species go extinct to preserve individual rights, or reduce pain and suffering? Is this environmentalism or is it just good stewardship based on biological, ecological, and political realities?
5. Are extreme ethical approaches, such as those espoused by animal rightists, hurting or helping wildlife conservation?
6. What will be the role of zoos and aquariums in the twenty-first century?
7. Should that role be mainly educational, or should zoos become more active in conservation in the wild?

We will also explore in more detail six issue areas. A plenary session will be held on each issue area, and smaller working groups will be formed for discussion. These smaller issue clusters will be addressed by working groups, which will be charged to bring forward consensus recommendations for policies and also to identify and describe clearly areas where consensus could not be reached. We will have six point-counterpoint presentations to stimulate discussion in the subsequent working-group meetings.

For convenience, we have arranged these topical plenaries into two sessions, as listed below, and each topic will be discussed in a point-counterpoint exchange.

The Captive-Wild Interaction

1. Procurement of animals for captive breeding programs
2. Reintroduction, training, and welfare of released animals
3. Captive breeding, surplus animals, and population regulation

Good Stewardship

1. Captive care, maintenance, and welfare
2. The use of animals in research
3. Public relations, fund-raising, and disclosure

OVERVIEW OF ISSUES FOR WORKING-GROUP DISCUSSIONS

What follows are the brief presentations and lists of thought-provoking questions that we gave to working groups in each of the six issue areas. We admitted that the opinions reflected in the point-counterpoint summaries represent extreme viewpoints, but we explained that they were intended to be a starting point for further discussion.

Issue I: Procurement of Animals for Captive Breeding Programs

Point

Zoos, proponents say, contribute to wildlife conservation in many ways, including educating an urban public that has become largely divorced from nature; propagating and
sometimes reintroducing endangered wildlife; conducting reproductive, nutritional, behavioral, genetic, and veterinary research; developing technologies that can be applied to
field conservation; training of zoo and wildlife professionals; direct support of field conservation, etc. Many species, such as the California condor, black-footed ferret, Arabian
oryx, Pere David deer, and Asian wild horse would be extinct today if it were not for the
efforts of zoos. As human population growth and habitat destruction increase, the problem
of endangered species is going to get worse, making such institutions and their expertise
even more critical.

Counterpoint

Zoos have recently come under attack from conservation organizations and animal rights
and welfare groups who claim they are exploiters of wildlife—taking them from the wild
for entertainment and profit. This, opponents say, is threatening the survival of species in
the wild by creating a market. Opponents say captive breeding has only a limited ability
to deal with the endangered species problem and is extremely costly. After all, zoos focus
only on the large, so-called charismatic vertebrates and ignore the more numerous invertebrates. Zoos should therefore be closed and the money used to save wildlife and their
habitats in situ.

Questions

1. Under what conditions should endangered wild animals be procured for captive breeding programs? Who should make these decisions and determine priorities? Does extreme exigency (such as severe danger of extinction) justify removing specimens from
   the wild to strengthen captive breeding stocks?
2. Are zoos focusing only on large mammals? Is this strategy detrimental to conservation
   or is it beneficial?
3. What is the current and (perhaps more important) future relationship between captive
   animal populations and wild animal populations (between in situ and ex situ conservation efforts)? Do zoos currently serve wildlife conservation aims? If not, then how
   can these resources be utilized better? Is it realistic to suggest that all zoos should be
   closed? Would the money saved be spent on conservation, or is this just rhetoric used
   by zoo opponents?
4. Under what economic and political realities will wildlife conservation efforts have to
   take place? Are animal rights and animal welfare philosophies compatible with actions
   that will need to be taken? Are they flexible enough to allow the kinds of actions that
   will be necessary for conservation to occur in the face of growing human populations,
   habitat destruction, etc.?

5. Do zoos create a market for endangered wildlife during their procurement efforts? How can zoos avoid creating a market for wildlife and potentially being part of the problem rather than part of the solution? What should zoos do about orphaned endangered animals? Is zoo collecting contributing to the demise of species in the wild, and are current laws adequate to protect wild populations from depletion?

6. Should the focus of zoo breeding efforts be on furthering the welfare of individual animals or on maintaining viable populations of threatened species as a hedge against extinction? What is the highest imperative from a conservation, evolutionary, ecological viewpoint? Is compromise possible? How are the problems faced by zoos similar to or different from those faced by wildlife managers?

7. Should a distinction be made between private breeders, roadside zoos, and institutions accredited by AZA? Should a distinction be made between institutions that support formal captive breeding programs, such as the SSP, and those that do not?

8. What kinds of institutions should future zoos be? What should the emphasis of such institutions be? How could such institutions play a larger role in conservation efforts?

9. Should a species be saved in zoos even if its habitat is completely destroyed? Is it justified to take the last remaining animals of a population out of the wild to save them through captive breeding? Should species be in captive breeding programs only if there is an opportunity to save them in the wild?

10. Does a wild animal have a right to stay wild? If not, how would we characterize the apparently felt obligation that, other things being equal, we ought to respect an animal's right to noninterference?

Issue 2: Reintroduction, Training, and Welfare of Released Animals

Point

Reintroduction efforts have been successful in reestablishing species that were once lost in nature. The Arabian oryx, Puerto Rican crested toad, and black-footed ferret are a few examples. Though the science of reintroduction is still in its infancy, great strides are being made. As more and more species become threatened with extinction, reintroduction techniques will assume greater importance. Though cost is currently high, improvements in technology will likely result in more cost-effectiveness. Loss of individuals is expected during the reestablishment process. Nature is not a Garden of Eden, and wild animals are subjected to parasites, diseases, predators, severe weather conditions, etc. When appropriate, captive-bred animals should receive training and preparation to help increase their chances of survival. However, the need for training will differ from species to species.

Counterpoint

Reintroduction will be able to save only a few species at best and is extremely costly. We should give up on such species, close zoos, and use the money to save animals in their natural habitats. Reintroduction efforts have not been very successful, and captive-bred animals simply do not have the tools to make it in the wild. It is cruel to subject these compromised individuals to the rigors of nature since they are obviously ill prepared. There is no way that animals raised in impoverished zoo environments will be able to survive. The losses of individuals experienced during experimental reintroductions cannot be tolerated.

Questions

1. What responsibility do zoos have to return animals to their natural habitats? Is that the primary purpose of organized captive breeding programs for endangered species (e.g., SSP programs)? Is it a realistic goal?
2. What are the characteristics of a successful reintroduction? Should captive-bred animals be trained before their release to increase the probability of success? Should translocated animals receive constant veterinary care or be allowed to make it or break it on their own? What kinds of training should animals receive? Are there species differences in this regard?
3. Should losses and suffering of animals be expected in cases of reintroduction? If so, what level is acceptable? Again, how should welfare of individuals be weighed against population, species, or ecosystem survival?
4. Under what conditions should reintroductions be attempted? Do conditions for release have to be perfect to ensure welfare of individuals, or is individual welfare a lesser consideration than the future survival of the species in the wild? What is the higher moral imperative? What are the practical considerations?
5. For how long and to what degree should the zoos monitor the reintroduced animals in terms of veterinary care and political intervention?

Issue 3: Captive Breeding, Surplus Animals, and Population Regulation

Point

Surplus is inevitable, even in intensively managed breeding programs. Zoos should be allowed to manage (i.e., control) captive animal populations, including SSP animals, through sales, trades, and loans to appropriate facilities; contraception, sterilization, and related techniques; and culling (humane euthanasia by veterinarians following established and legally accepted methods). Without such flexibility it will be impossible to use captive space efficiently and maintain genetically viable populations of many endangered species. The result will be extinctions. Species survival must take higher priority than individual welfare. Maintaining individual animals into perpetuity will take limited resources away from conservation and make it difficult, if not impossible, to maintain viable populations. This is especially true because captive animals live much longer than their wild counterparts (as a result of nutritionally balanced diets and access to veterinary care, etc.). If populations cannot be controlled, then fewer species can be allowed aboard the zoo ark. Such intensive management will also be necessary to maintain viable populations of wild animals. The demands of animal rights and welfare groups have been based purely on emotion and are unrealistic and anticonservationist. Stewardship means making difficult decisions, and these issues cannot be ignored.

Counterpoint

Surplus is not inevitable and could be completely controlled through contraceptive techniques. Zoos breed to have baby animals and attract more visitors so that they can make more money through the sale of animals to dealers and hunting ranches. They are exploiting animals for monetary gain and cause animal suffering in the process. Zoos should be responsible into perpetuity for all animals they produce. All animals under their care, no matter what their age, physical condition, or genetic makeup, must be given lifetime care under acceptable conditions. Zoo and wild animals should simply be left alone, free from human interference.

Questions

1. What is the definition of "surplus"? Is there a distinction between legitimate surplus (SSP animals that have made their genetic contribution to the population) and non-legitimate surplus (animals that are bred indiscriminately outside the SSP)?
2. Is the production of surplus inevitable? Are contraception or other intensive management protocols the answer? Will such techniques eliminate all surplus?
3. Is there an inherent conflict between the maintenance of genetically and demographically viable captive populations as a hedge against extinction and the welfare of individual animals? If there is a conflict of interest, then which should take precedence? When should the emphasis be on populations, species, and ecosystems, rather than on individual animals? Can an individual animal be respected even if its well-being is considered subservient to a larger goal such as perpetuation of its species?
4. When should various methods of population control be implemented (contraception, sale, euthanasia, culling)? Is predation in the zoo an option? Which should take precedence in various cases? In the case of males contributing to the majority of a surplus problem, should male fetuses be aborted? Should male neonates be euthanized?
5. Under what conditions is euthanasia or culling acceptable (genetic defects, ill health, social incompatibility, behavioral and physical abnormalities, etc.)? Is euthanasia of healthy animals acceptable if it frees up limited captive space (i.e., if it allows more efficient use of space and allows viable populations of more endangered species to be maintained)? Is it acceptable to euthanize healthy animals if such individuals have met their genetic goals under the SSP master plan?
6. Human individuals often risk their lives for a perceived benefit to a social system or the human species; here, some form of consent is usually considered an adequate justification for altruism. Since animals are incapable of deliberation and consent, does it follow that no animals can justifiably be sacrificed for a larger good? Are there substitute arguments that could explain and justify animal altruism?
7. Should zoos be responsible into perpetuity for maintaining animals they breed? What are the long-term financial and conservation implications of such a policy?
8. Should infant animals be bred to draw visitors even if their disposition is difficult? Should all captive populations be managed? What are the financial and conservation implications?
9. What responsibility do zoos have to animals when they leave their collections? Is

euthanasia a better alternative than sending the animals to inadequate facilities, hunt-ing ranches, laboratories, etc.? If so, then how can zoos proceed (realistically) in the face of criticism from animal rights and welfare organizations and the general public?

10. Is it ever appropriate for endangered species to be bought or sold, and if so when? Are there differences between for-profit and nonprofit institutions in this regard (e.g., does the money profit individuals or the institution and its programs)?

11. Are the AZA's surplus guidelines sufficient? How should they be improved? What should be done about roadside zoos, private breeders, or AZA members that do not conform to accreditation standards and ethical codes? What can be done from a practical viewpoint?

Issue 4: Captive Care, Maintenance, and Welfare

Point

A better understanding of the behavior and ecology of animals in the wild is making it possible for zoos to construct better exhibits and holding facilities for captive wildlife. Indeed, exhibit technology has improved considerably in the last decade, resulting in larger and more naturalistic zoo exhibits. Such exhibits are not only better for the animals but also more educational because they make a connection between an animal and its habitat. In addition, AZA-accredited zoos have worked with the federal government and within the profession to improve minimum housing standards for exotic animals. Behavioral enrichment is considered for many species. Some zoos still have antiquated exhibits, but the transition to new facilities takes time and money (costly exhibits cannot be built for every species that is maintained). Much research is aimed at developing better methods of animal care, including improved nutrition. Zoo animals also have access to veterinary care, and many species tend to live considerably longer than their wild counterparts.

Counterpoint

Zoos are essentially animal prisons, which have no place in modern society. Some zoos have built costly, large exhibits, but this is the exception rather than the rule. There is no way that the complexity of the zoo environment can match that of the wild. Animals should be left in nature to live in the habitats in which their specific adaptations have evolved. Zoo exhibits tend to be sterile, boring environments that cause animals to suffer psychologically. Animals in captivity also suffer from diseases to which animals in the wild are not subjected; despite veterinary care, their lives are often shorter and of lower quality than they are in the wild. The public learns nothing by visiting a zoo—at least nothing that cannot be learned by watching a television documentary.

Questions

1. Are accredited zoos providing adequate environments and care standards for animals? Do these requirements differ on a species-by-species basis? Who should determine if a given enclosure is adequate, and on what basis should this determination be made?

2. Are zoos animal prisons or are they sanctuaries? Do health and longevity take precedence over living in a naturalistic environment (with all of its inherent risks)? Is there any compromise possible between zoo professionals and animal rights and welfare advocates? Are all zoos painted with the same broad brush by animal welfare and rights organizations (i.e., accredited versus nonaccredited)?

3. Do animals have a right to freedom? Is the concept of freedom even applicable to animals?

4. What kinds of enrichment should zoos be required to provide for captive animals in their daily care routines? How will these differ between species? What is an acceptable cost of such efforts?

Issue 5: The Use of Animals in Research

Point

Conservation cannot occur in the absence of knowledge. Much research is needed to develop effective methods of captive propagation and animal care. To this end, zoos have established research programs. Zoos should be allowed to use common or domestic models to study problems that affect endangered species. (Risky or invasive research on endangered animals should be undertaken only as a last resort.) Endangered animals are given priority because of their rarity and because they could be irreplaceably lost, perhaps threatening the existence of other species and ecosystems as well. Many topics that can be studied in captivity cannot be studied in the wild, and captive studies have made significant contributions to our knowledge. All research should conform to established legal and ethical standards. In each case, animal welfare considerations should be weighed (i.e., Is the cost to individual animals justified by the benefits to species conservation?). If the research will contribute to species conservation, and the procedures conform to accepted standards, then the work should be approved. Preservation of species has the highest moral imperative.

Counterpoint

All research on animals is cruel because they have no choice in the matter. Sacrificing common or domestic animals to save another species is utilitarian and speciesist and should be considered environmental fascism. Invasive research should not be conducted because it causes pain and suffering to individual animals. Terminal research violates their right to life. Sentient rare species should be considered under the same moral yardstick as common ones. The labeling of an animal as "rare" is based on human values and references. Research on zoo animals is invalid because the animals are in captivity; such studies tell us nothing about the lives or biology of animals in the wild.

Questions

1. What is the role of zoo research? Is it contributing to conservation efforts in situ? If not, does it have the potential to do so?

2. What kinds of research are acceptable on captive endangered wildlife? What levels of invasiveness, risk, stress, and pain are acceptable? Is this true across the board, or can infertile or aged animals be used for higher-risk procedures?
3. Should animal husbandry be compromised to facilitate research? Should research collections be housed differently from exhibit animals (i.e., are smaller cages justifiable for research collections)?
4. Is it acceptable to sacrifice individuals of a common species to save an endangered variety (e.g., when common species are used as research models to develop techniques and knowledge that can be used for conservation purposes)?
5. How should zoos make decisions about research priorities and weigh animal welfare considerations? Are the current policies of zoos adequate to meet animal welfare considerations? What are the legal implications?

Issue 6: Public Relations, Fund-Raising, and Disclosure

Point

Zoos are public institutions and as such have a responsibility to taxpayers and the visiting public. However, some practices by zoos that may be necessary for conservation (e.g., euthanasia) may also be difficult to explain and may open an institution to unrealistic and emotional criticism from emotional and inflexible opponents. In such cases, coordinated programs that provide information that the public is able to assimilate are essential. However, it is not necessary to disclose completely all information.

Funds for captive breeding, research, and reintroduction programs are difficult to obtain. Accepting funds from major polluters or exploiters of wildlife, or selling the animals themselves, could ultimately result in a benefit. However, each potential donor should be evaluated on a case-by-case basis. Captive animals should be considered ambassadors for their species and, as such, can be used to raise money for conservation in the wild.

Counterpoint

As public institutions, zoos should open all their records and practices to public scrutiny. If the public does not support the practice or policy, it should be banned. The use of live animals to raise money (e.g., through rides, public handling, and demonstrations) is stressful and should not be allowed, even if it does contribute to species preservation in the wild.

Questions

1. Where does the money raised by zoos go? Is it appropriate to use animal collections to raise money to support zoo programs? If so, in what ways should they be used? Is the use of animals for fund-raising exploitation? When is the off-site use of animals to solicit donations acceptable?

2. How much secrecy, if any, can be justified in maintaining a captive breeding and reintroduction program, the exigencies of which would require treatment of individual animals (e.g., euthanasia) that could be unpopular with a vocal minority of the public?
3. Are captive breeding programs and other high-visibility programs justified by their public relations and educational value, or must they be justified purely on their contributions to conservation?
4. Should zoos name animals and promote them as individual personalities?
5. Should zoos promote baby animals, regardless of their importance to conservation? Do zoos overemphasize animals that are appealing to the public? What are the implications of such a policy?
6. What involvement should zoo public relations professionals have in zoo management decisions? Are zoos sugar-coating their activities to minimize public criticism?

OUTCOMES OF THE SEARCH FOR CONSENSUS

Conference Organizers' Reflections on the Working-Group Reports

The conference Animal Welfare and Conservation: Ethical Paradoxes in Modern Zoos and Aquariums had two perhaps not fully compatible goals. First, we intended to provide an open forum for the discussion of diverse viewpoints, to be expressed by both theoreticians and practitioners, on the ethical problems arising in captive breeding programs. But we also intended from the start to state as much consensus as possible and to make conflict-reducing policy recommendations to the AZA and to directors of captive breeding programs. For the first goal, the wide-ranging expression of diverse opinions was essential; but for the latter, more practical objectives led us also to press toward consensus. We hoped to fulfill both goals by publishing in book form the diverse opinions expressed at the meetings (as we have done in the main text of this book), as well as a set of recommendations that the conference participants who were in the working groups developed through discussion and consensus-searching (as we have done below).

The groups, which were organized around the six issues introduced in the plenary sessions, met to discuss general principles and to address the lists of more specific questions that conference organizers proposed for each topic (see "Overview of Issues for Working-Group Discussions"). They were specifically instructed to state areas of agreement and disagreement as clearly as possible and to make specific policy recommendations wherever possible. In organizing the conference, we tried to ensure diversity of opinion in each group, although it was impossible to have every viewpoint represented on every issue, especially because the groups were small. Participants were chosen to achieve a balance in each group between technical experts in captive breeding and conservation ecology and representatives of humanistic and animal welfare activist viewpoints.

The committees agreed on a great number of important principles and were able to propose many concrete policy suggestions. Often these were based on an intellectual, moral, and pragmatic consensus. In other cases, significant intellectual and practical differences remained intractable. This was especially true when the issue was how to balance ethical obligations to respect the well-being of individual animals against the pressing needs of conservation action driven by the worldwide biodiversity crisis. It was impossible to reach intellectual unanimity regarding the comparative importance of individual animal welfare and of conservation goals when these came into unavoidable conflict. Indeed, some of the deepest disagreements among participants arose regarding whether there is a strong moral presumption in favor of leaving wild animals free; if there is such a case, captive breeding programs are justified only under some limited conditions. Many remaining disagreements centered, in one way or another, on whether participants endorsed such a strong presumption.

Not surprisingly, some groups achieved more consensus than others. While in most working groups significant recommendations were possible based on the full range of ethical viewpoints presented, less progress was made on the difficult problems of population regulation and the problem of surplus animals. It is in this area that the theoretical and moral disagreements between moral individualists, conservation ecologists, and captive breeders are most in conflict. It is clear that more discussion, both scientific and ethical, will be required in this area.

One attitude shared by all participants was a sense of urgency. If we do not act now to save many animal species and populations in this generation, it will be forever too late for many of the magnificent species and for many of the less noticeable but essential species that share the planet with us. Further, there was broad agreement that humans are responsible for the biodiversity crisis and that there exists a strong moral obligation to respond—to reverse if possible and to mitigate if not—the impacts of the extraordinary expansion of human populations into hitherto wild habitats. It was also argued by representatives of all moral viewpoints expressed at the conference that zoos, aquariums, and other institutions with captive breeding programs should accept important and special obligations to protect and respect the animals they have removed from the wild for conservation purposes. At the very least, consensus stated that managers of captive breeding programs must provide for and protect the welfare of individual animals in their programs, doing everything in their power to avoid unnecessary pain and discomfort and also to provide opportunities for animals to express natural behaviors to the extent practicable in captivity.

We respectfully submit the following discussions and recommendations, which were prepared by our working groups at the conference. We believe the recommendations, taken as a whole, imply a clear recognition that, despite deep moral concerns both for the welfare of individual captive animals and for the protection of wild populations in their native habitats, zoos can and must undertake captive breeding programs in many situations. Also implied in the recommendations is that activities of these captive-breeding programs—including activities that are fraught with ethical concerns, such as procurement of founder stock from wild populations and population regulation—are necessary and justified, provided they are (1) carried out in a way that respects individual animal welfare, (2) justifiable as part of a broader conservation program designed to perpetuate the species in the wild, and (3) based on some reasonable hope that the captive breeding program will lead to augmentation of wild stocks or to reintroduction into the wild at some future time.

Working-Group Membership

Working groups each consisted of 4–10 members, as listed below.

Group 1: Procurement of Animals for Captive Breeding Programs
Moderator: Dennis Merritt
Recorder: Carolyn Harings
Other members:
  Richard Block
  William Conway
  Ardith Eudey
  Valerius Geist
  Bryan Norton
  George Rabb
  Geza Teleki
  Robert Vrijenhoek

Group 2: Reintroduction, Training, and Welfare of Released Animals
Moderator: David Hancocks
Recorder: Craig Piper
Other members:
  Benjamin Beck
  Robert Loftin

Group 3: Captive Breeding, Surplus Animals, and Population Regulation
Moderator: Eugene Hargrove
Recorder: Lori Perkins
Other members:
  Nancy Blaney
  Dale Jamieson
  Robert Lacy
  Donald Lindburg
  Frederic Wagner
  Peregrine Wolff

Group 4: Captive Care, Maintenance, and Welfare
Moderator: Fred Koontz
Recorder: Elizabeth Stevens
Other members:
  Terry Maple
  Rita McManamon
  Bruce Read
  Tom Regan
  Les Schobert

Group 5: The Use of Animals in Research
Moderator: Peter Jaszi
Recorder: Sene Sorrow
Other members:
  Betsy Dresser

Roger Fouts
Michael Hutchins
Chris Wemmer
Group 6: Public Relations, Fund-Raising, and Disclosure
Moderator: Douglas Myers
Recorder: Gail Bruner Lash
Other members:
Karen Allen
Roger Caras
Bob Jenkins
Mike Kaufman
Rob Reece

Instructions for Working-Group Moderators and Recorders

In preparation for the group discussions, the moderators and recorders were given the following instructions.

The purpose of working-group meetings will be to discuss particular problem areas and achieve maximal consensus. This consensus will then be embodied in a set of recommendations that will be discussed in a final plenary session. Each working group will correspond to issues in one of the orientation point-counterpoint sessions.

Each group will have a moderator and a recorder. Responsibilities will be as follows: The moderator will formulate questions and guide discussion. An easel with newsprint will be provided so that the moderator can write down key ideas in a few words, keeping a running account of topics discussed. The recorder should try to record as much of the substantive discussion as possible. The goal will be to produce an executive summary (no more than three pages, single-spaced) for consideration by the entire group. The executive summary should contain (1) a brief general statement of areas of agreement and consensus, (2) a brief general statement of areas in which agreement could not be reached, and (3) recommendations for future policies. These should be brief but as specific as possible. You may furnish a two- or three-sentence rationale for recommendations if you wish. In addition to the executive summary, we hope the recorders will be able to provide attributions of comments to specific participants. This will not be published but will be an essential guide to conference organizers when they prepare recommendations to the AZA Board of Directors.

Participants have been assigned to a specific working group; invited guests may participate in a group of their choice. It is up to moderators to decide whether (and when) invited guests may participate in working-group decisions.

What follows are the reports written by the working groups during and immediately after two days of intense discussions. Because some of the wording was carefully crafted to achieve consensus, these reports have been edited in only superficial ways that did not alter content; this concern for nuances explains the somewhat irregular format and uneven style of the reports.

Report of Group 1: Procurement of Animals for Captive Breeding Programs

General Consensus

In all countries, the conservation of biological diversity is now a major and shared public concern; the urgency of the need for action is no longer under serious debate. Unfortunately, the resources for achieving these goals are not equally distributed among all nations. Decisions concerning these issues must be made collectively.

If we are committed to sustaining biological diversity for many generations, it follows that we must, as Aldo Leopold said, "save all the pieces," and protect the wild processes that they both constitute and sustain. From many ethical viewpoints, preservation of wild species in their natural habitat is the highest priority in biodiversity protection. From the viewpoint of efficient use of conservation resources, in situ conservation of the elements—genes, individuals, and species—is less costly than ex situ conservation. This is no less true of the charismatic vertebrates than of other species, because the rescue of sufficient habitat to support any large, far-ranging species will simultaneously save countless species that exist in smaller-scaled habitats ranged across by these larger animals. The conservation of wild populations is therefore both a moral priority and an economic bargain, provided we are committed to wildlife as a generator of wealth. More can therefore be done to protect the options for future generations of humans by saving wild populations than in any other way.

Since wild animals also strive to perpetuate their genetic lines, concern for wild species, as ongoing entities, no less than concern for future human generations, demands that we save populations and their associated ecological dynamic whenever possible.

Conservation goals cannot and should not be pursued without concern for other social values, however. Especially, they cannot be pursued without compassion for human cultures trapped in vicious spirals of poverty and population growth in developing countries. Those human demographic forces that cannot be reversed—except by natural disaster or unacceptable interventions into human lives—make in situ conservation of natural resources both imperative and in many cases unfortunately impossible.

Furthermore, evidence mounts almost daily that circumstances such as depletion of the ozone shield may render many species' habitats at least temporarily nonviable within this decade. A holocaust of species loss is upon us. If we fail to establish managed breeding populations of many species, there will never be another chance to do so. However attractive the option of saving species in nature, various forces determine that many species will be protected only in captive breeding programs or in closely managed wild populations supplemented with introductions from captive-bred populations.

Even if one recognizes that it is necessary to deprive animals of their freedom by placing them in zoos, this is a lesser consequence than the wide-scale destruction of the diversity of life and loss of potentially sustainable ecosystems that may result if no animals are removed from the wild as buffers against imminent extinction. Responsible, international conservation of biological diversity, whether based on the value of wild species themselves or on the importance of protecting the options of future human generations, will therefore require the removal of some animals from wild populations to establish breeding stock for a captive population. In some cases, these actions can be justified as necessary to allow later reintroduction after habitat has been restored; in other cases the situation has become so dire that animals must be removed from the wild to avoid imminent extinction even if their reintroduction to the wild cannot be foreseen.

Once acquisition actually begins, primary focus shifts from protection of biodiversity to the individual animals, and all activities must be appropriate, given best available knowledge of physiology, psychology, and ethology of the animal. Furthermore, care should be taken to minimize impacts on the population remaining in the wild.

Ethical Criteria for Acquisitions

Three situations may lead to decisions regarding the acquisition of animals from wild populations: (1) when founder stock is needed to establish a coordinated short-term or long-term captive breeding program, such as a Species Survival Plan of AZA; (2) when genetic or demographic immigrants are needed to supplement an already established breeding program; and (3) when the animals are needed solely to serve other compelling conservation goals. While current acquisition projects are occurring on a first-come, first-served basis, the group recommends that we seek a method of assigning priorities in advance of haphazard opportunities. It is suggested that zoos work in conjunction with the SSC of the IUCN to set ideal conservation priorities.

Six criteria must be weighed before embarking on any ethically appropriate procurement project:

1. *The conservation impact on the species and on overall biodiversity.* The positive effects of removing wild animals from the habitat countries on overall conservation efforts must be greater than the negative impacts on the in situ population. It should also be determined that the program does not diminish available resources for in situ conservation efforts. Where possible, animals already in captive programs in country of origin should be used to meet the needs of the program.
2. *The likelihood of establishing a successful captive population.*
3. *The acceptance of long-term commitment and responsibility for managing the captive population.* This includes full participation with wildlife managers in the habitat country.
4. *Communication and documentation.* Prior to acquisition, all interested parties should be fully informed of the proposal and proper legal documents must be obtained. The project should be designed to maximize scientific opportunities to study the species and enhance its well-being in the wild.
5. *Assurance of proper animal welfare.* There must be a commitment to care properly for individual animals and to propagate and to maintain the health of the population.
6. *Risk assessment.* There should be full evaluation of the risks inherent in the project and of the possible steps to minimize such risks.

Although these six criteria can help to decide if specific acquisition projects are ethically justifiable, they do not provide a step-by-step method of initiating such projects.

The criteria listed above should be applied within the following acquisition procedure.

Step 1: Proposal Development

The working group and its resources should be identified.

Initial contact should be made with other interested parties.

Goals of the breeding program should be clearly defined and stated.

Measures of success should be outlined.

The results of the working group's conservation assessment and overall program feasibility study should be tabulated.

A site for the ex situ program must be selected and justified.

A removal strategy must be selected from the various options, including commercial versus noncommercial avenues of acquisition, after an analysis of the impact of local effects in the country of origin.

Animal care and welfare strategies should be designed.

Future ownership of animals must be assigned.

A project schedule must be established.

Step 2: Review of Proposal

Formal comments from interested parties should be solicited.

The proposal must be modified accordingly, if necessary.

Step 3: Obtainment of Permissions and Permits

All necessary local, national, and international permissions and legal permits must be acquired.

Step 4: Capture, Transport, and Quarantine of Animals

Steps must be taken to provide for behavioral and physical welfare of the animals, especially to minimize stress in the containment and conditions during transport.

Reasonable precautions must be taken to prevent and respond to accidents.

Step 5: Program Establishment or Integration of New Animals

The captive breeding program must be properly documented.

Opportunities must be established to maximize research on the animals relevant to their conservation and welfare.

Step 6: Long-Term Captive Care

Ongoing care must be provided for animals and populations.

Updates on program status must be published regularly.

Report of Group 2: Reintroduction, Training, and Welfare of Released Animals

Overall Recommendations

We cannot reintroduce species if there is no habitat—so what can we do to help with restoration and retention of suitable habitat?

1. The ultimate purpose of any reintroduction should be to preserve or reestablish natural ecosystems and processes.

2. Captive breeding must never be undertaken to facilitate the exploitation of a habitat (e.g., the spotted owl should not be taken into captivity so that we can cut down the old-growth forest and get on with our business).
3. Reintroduction should never be used to solve surplus problems—the surplus problem should be dealt with by premanagement of species with cooperation between zoo biologists.
4. Reintroduction must be a professional and scientific exercise.
5. Three factors influence success and failure of habitat preservation, and thus reintroduction: biological, social, and economic. They are integrated factors. Reintroduction involves longer-term commitment.

Biological Component

The biological factors in the issue of reintroduction can be addressed in the following ways.

1. Recognize that zoos are not ultimately or solely responsible for captive breeding—work in cooperation with other agencies and institutions, provide expertise and holding space.
2. Require all reintroduction to be carried out as a scientific exercise, well documented.
3. Develop coordinated national and international plans for reintroduction strategies.
4. Keep individual animals (and species) destined for reintroduction in conditions that prepare them for survival.
5. Breed and maintain animals in environment most relative to natural habitat conditions (e.g., seasonal fluctuations, population sizes, natural habitat components, and food supplies).

Social Component

Zoos can address the social factors of habitat preservation and animal reintroduction in the following ways.

1. Effect public changes in attitudes and life-style.
2. Inculcate respect and concern for wild things—in all zoo programs including gift-shop items, restaurant services, and general ambience.
3. Seek political and business support by showing how environmental well-being makes good sense to politicians and industry.
4. Concentrate on local conservation issues; focus on regional exhibits.
5. Ensure that the selection of species and exhibit construction priorities are driven by an explicit conservation plan, rather than vice versa.
6. Develop exhibits that concentrate on concepts, not just species; on ecosystems, not just animals.
7. Build local support for reintroduction that will engender pride in uniqueness of local environment.
8. Adopt a wilderness habitat.
9. Set up satellite learning centers in areas of reintroduction.

10. Educate board members to support zoo changes and to recognize that they cannot continue as is.

Economic Component

Strategies to address the economic issues include the following.

1. Provide long-term funding to support habitat preservation—matching funds, 10 percent for conservation program.
2. Build endowments for habitat preservation.
3. Become funnels for funding in situ conservation.
4. Seek ways to improve financial well-being of people in reintroduction areas.
5. Develop action plan so that local populations can participate.

Report of Group 3: Captive Breeding, Surplus Animals, and Population Regulation

Agreements

Members of the working group were in agreement on the following points.

1. Although the maintenance, preservation, and restoration of habitat should be the primary means for species preservation, it is possible that ex situ breeding can play a role in that preservation.
2. In most captive breeding programs, surplus animals are a predictable result. It should be noted that breeding programs are not the only cause of surplus animals. The AZA is encouraged to continue its efforts to define and quantify the dimensions of the surplus problem.
3. Off-site breeding centers should be established as alternatives to traditional exhibit-based breeding, when consistent with the objective of maintaining as few animals as possible in captivity. Surplus animals from these programs should go to zoos for education and exhibition.
4. A zoo is responsible for its animals from acquisition to death. Several actions should be taken to discharge this responsibility:
   (a) Zoos should cooperate to bring all species under well-designed management programs to minimize surpluses.
   (b) Efforts should be made in research on, and use of, reproductive inhibition, including both reversible and nonreversible measures, to minimize surpluses.
   (c) Although programs and missions may be revised as conditions necessitate, a zoo must maintain its lifetime commitment to its animals.
   (d) A portion of the spaces now occupied by common species should be redirected to species at greatest risk.
5. The term "euthanasia" should be understood as killing only for the benefit of the individual animal when it is no longer possible to maintain its quality of life.

Disagreements

The working group did not come to a consensus on the following assertions.

1. When zoos breed or otherwise acquire animals, they should make a commitment to maintain them throughout their lives. This implies that the culling of healthy animals is not an acceptable management tool.
2. When a decision has been made to support a surplus population, one mechanism is the establishment of a fund, by donation or surcharge, to support current surplus populations.

Assumptions, Clarifications, and Justifications

This section reflects group discussion but not necessarily agreement, due to lack of time.

1. These recommendations assume that
    (a) zoos may require a new mission beyond the traditional display of a complete range of traditional megafauna;
    (b) to be viable institutions in the twenty-first century, zoos must enhance their educational and conservation roles with off-site and original habitat projects.
2. The display of on-site species representatives may not be a major role of zoos in the twenty-first century.
3. Separation of breeding populations (which are viewed from a species perspective) from display animals (which are viewed from an individualistic perspective) will help avoid conflicts between holistic and atomistic moral intuitions.
4. The best use of a healthy surplus animal is education and display.
5. Off-site breeding centers may be less expensive and more efficient than current programs.
6. Without endangered-species breeding populations, zoos can provide homes for surplus animals and for new activities—for example, integrated conservation education, regional interpretation, and emphasis on minifauna compatible with an enhanced educational and conservation role in the postmodern world.
7. Off-site breeding centers will permit populations at levels that overcome problems with genetic drift, permit increased opportunities for animals to exhibit natural behavior, and reduce the need for reintroduction training.

Report of Group 4: Captive Care, Maintenance, and Welfare

Our recommendations covered five aspects of captive management: management science, exhibit design, enforcement of standards, management of surplus animals, and transport of animals.

Management Science

There should be an established science of captive management. Scientifically based management is what we should be striving for. All decisions regarding captive care and management need to be knowledge-based. This knowledge is acquired through scientific studies. We should be constantly revising our standards with data. Postoccupancy evaluation can be used to measure suffering.

Attempts should be made to evaluate well-being and suffering. This would include using field data for each species as a baseline for what we are aiming for. The field data would consist of an ethogram, activity budgets, and rates of behavior. Our goal is for animals in captivity to exhibit these same levels of behavior.

All facets of the animal's environment, including the visitors, should be continually evaluated and these data incorporated into the management decision-making process. It is important to provide an analysis of how people engage in the zoo experience as well.

A scientific program should be reviewed by outside scientists at all times.

Quality of life for the animals has to be our highest priority. The role of science in captive care, maintenance, and welfare is to promote the highest quality of life in the animals under our care.

Exhibit Design

Exhibit design involves a holistic approach to the entire facility, both exhibit and holding facilities. The design of enclosures needs to be species specific. We have poor exhibit designs for many taxa. Primates probably have the best, most knowledge-based enclosure design standards. The architects should be the animals. Taking the animals' needs into consideration should be the top priority, and the design of these exhibits will work within the limitations of the physical, veterinary, and management constraints. Holding facilities must comply with the highest quality of life possible. The goal is to create an environment that maximizes the opportunities for the animal to express its natural behavior.

A husbandry manual should be developed for every species held in captivity. This would include exhibit design standards and behavioral enrichment criteria.

The quality of life for the animals probably does not need to involve the word "freedom." An animal in captivity can more correctly be said to be autonomous. Well-being is a better concept than happiness in evaluating quality of life.

Mixed-species exhibits should include species that are noncompetitive and are zoo-geographically sensible. These exhibits should make sense from an educational standpoint. We are willing to accept some amount of stress or arousal that is an inherent part of a mixed-species exhibit for the educational benefit. Always address any risks involved prior to the exhibit in a democratic way.

The jury is still out on which predator-prey interactions should be allowed and which should be prevented. Where do you draw the line? Is it okay to feed live fish and not live goats? Cultural attitudes allow for the feeding of live fish. How do you know that your community will embrace this? What is the rationalization for feeding live animals other than crickets and live worms? It is an ethical dilemma. We recommend that there be a meeting to discuss predator-prey relationships in captivity with respect to (1) the feeding of whole animals that have been euthanized or killed, (2) the release of live prey into animal enclosures, and (3) the exhibiting of predator and prey animals in proximity to one another—is it stressful or beneficial to either or both? Federal regulations will have to change along with any decisions pertaining to predator-prey relationships. If we were to propose this in the absence of democratization, we would be subject to criticism.

Enforcement of Standards

We need to have higher and more objective standards for captive care, maintenance, and welfare within the AZA. *We recommend that through the AZA accreditation process we require compliance to these standards.* We recommend working with USDA to raise their standards so that inferior, non-AZA institutions can be weeded out. We recommend formal liaisons with professional societies, animal protection organizations, governmental organizations, and all other related organizations.

The effort to develop standards should include (1) bringing in people from the outside community to get their opinions; and (2) creating technical advisory board standards for collection management and for veterinary necropsy procedures, to facilitate data collection and comparison.

Management of Surplus Animals

The following topics should be considered in managing surplus animals.

1. *Prevention ethics.* Planned breeding through contraception, etc., is mandatory for all species, endangered and nonendangered. There is no choice—if you are not managing your populations, you are either over- or underproducing. You need to know ahead of time what you are going to do with the offspring—preventive ethics. If we planned well, then we could absorb some of the other surplus animals into our collections. We need to manage nonendangered species as well.
2. *Democratization.* Level with the public—we are doing our best and this is the choice we face: either we kill them or you need to help us with the expense of keeping this animal.
3. *Nonlethal options for surplus animals.*
   (a) Veterinary treatment (expensive, though)
   (b) Retirement homes for geriatric animals
4. *Management euthanasia.* Euthanasia for the purpose of management is the last possibility. At a national level, narrow the options taxon by taxon (e.g., euthanasia for great apes will probably never be acceptable). If animals are euthanized because they are genetic surplus, then there will never be resolution between zoos and animal protection societies.
5. *Reality.* We do practice speciesism in zoos. Therefore, we have different values and ethics attached to each species. These issues need to be addressed at a taxonomic level. Do we want to have egalitarianism as a goal? Dealing with the realities of life, we need to address speciesism. Societal morals have contributed to this speciesism (giving higher well-being to some animals over others). If conservation is our number-one goal, then speciesism is in direct conflict.

Transport of Animals

Preparation for shipping, including behavioral and physical preparation, should be accomplished. There should be a list of approved transporters and recipients of animals. Standards for collection management and for veterinary necropsy procedures to facilitate data collection and comparison should be instituted.

Recommendations

The working group offers the following points and recommendations.

Preamble: Quality of life for the animals in our care has to be our highest priority.

1. Science should play a role in captive management. All facets of the animal's environment, including the visitors, should be continually evaluated and these data incorporated into the management decision-making process. The role of science in captive care, maintenance, and welfare is to promote the highest quality of life in the animals under our care.

2. Attempts should be made to evaluate well-being and suffering.

3. The goal of exhibit design should be to create an environment that will maximize the opportunities for the animal to express its natural behavior. Exhibit design involves a holistic approach to the entire facility, both exhibit and holding facilities.

4. We need to develop better exhibit design standards for many taxa. Primates probably have the best, most knowledge-based enclosure design standards.

5. Mixed-species exhibits should include species that are noncompetitive and zoogeographically sensible.

6. A husbandry manual should be developed for every species held in captivity. This would include exhibit design standards and behavioral enrichment criteria.

7. All forms of natural behavior should be encouraged in captivity; however, we have not reached consensus on predator-prey relationships. We recommend further discussion on the following aspects of predator-prey relationships:
   (a) feeding whole animals that have been euthanized or killed to other animals
   (b) releasing live prey into animal enclosures
   (c) exhibiting predator and prey animals in proximity to one another—is the level of stress or arousal acceptable?

8. Higher and more objective standards for captive care, maintenance, and welfare within the AZA need to be developed. Compliance to these standards should be required through the AZA accreditation process.

9. We need to work with the USDA to raise their standards and to enforce current standards rigorously in all institutions.

10. We recommend formal liaisons with professional societies, animal protection organizations, governmental organizations, and all other related organizations.

11. There should be standardization of collection management and veterinary necropsy procedures to facilitate data collection and comparison.

12. The reality is that we do not treat all species equally in relation to veterinary care and euthanasia decisions.

13. There are several possibilities for dealing with surplus animals. Management euthanasia should be the last choice among the possibilities. All possibilities should be openly communicated to the community. If animals are euthanized because they are genetic surplus, there will never be resolution between zoos and animal protection societies.

14. There should be a list of approved transporters and recipients of animals. Preparation for shipping, including behavioral and physical preparation, should be accomplished.

15. We should think about animals as a national treasure, and we should abandon ownership of animals. We should be stewards and not owners.

16. Training is important.

Report of Group 5: The Use of Animals in Research

Recommendations for Review of Research Proposals

Given that the AZA has identified conservation as its highest priority, the working group recognizes the importance of research and the fact that conservation cannot occur in the absence of knowledge. At the same time, the working group recognizes that the traditional mission of zoos and aquariums is to safeguard the welfare and minimize the suffering of animals for which they have responsibility. The working group recognizes that there is a dramatic increase in scope and volume of sponsored research (i.e., any research activity initiated, endorsed, or funded by a zoological institution). Therefore, the working group believes that there must be an increased institutional focus on welfare considerations in zoo-sponsored research. The working group recommends the following specific measures toward that end.

1. The AZA should develop and promulgate policies governing the conduct of zoo-sponsored research at member institutions, including but not limited to considerations of the welfare of the subjects. The journal *Zoo Biology* should include animal welfare guidelines in its instructions to authors.
2. AZA member institutions conducting research should develop and implement specific policies related to the welfare of research subjects. These policies should include both norms and criteria to be observed in conduct of research and mechanisms for ensuring their observance.
3. Institutions that conduct only research involving minimal welfare risks (see below) should consider welfare issues as part of the general procedure for research review but need not establish formal institutional animal care and use committees (IACUCs). All other institutions should constitute independent IACUCs to consider welfare issues raised by research proposals (as is required for institutions receiving federal research support). The IACUC should include representatives of curatorial, veterinary, and research departments, as well as keeper staff. Nonaffiliated individuals from the community served by the institution should be included; wherever possible, the membership should include more than one such member including a responsible individual representing the animal welfare community.
4. Proposals for zoo-sponsored research submitted by potential investigators should specifically address welfare concerns. These concerns include
   (a) the number and kind of animal subjects involved
   (b) the nature of any adverse effects that may be generated by the procedures involved
   (c) measures adopted to minimize animal suffering
   (d) assurances that the research is nonduplicative, that the sample size is no larger than required to achieve statistical validity, and that all alternative measures have been considered
   (e) a statement of the benefits of the research that may justify any irreducible harm to subjects
5. Individuals proposing research should be entitled to appear before the committee to explain their proposals. Additionally, a full proposal should be available to each member of the committee, as well as to appropriate outside reviewers proposed by committee

members. To the maximum extent possible, the committee's records of deliberations and conclusions should be open to inspection by the community.

6. In addition to reviewing proposals for research, the research committee or IACUC should be charged with the duty of following up on the conduct of the research it approves. In particular, the committee should receive and consider complaints or expressions of concern from keepers and members of the public. Researchers should report to the committee any changes in their protocols that may affect the welfare of animal subjects.

7. Research proposals or requests from SSP committees should include specific discussions of welfare implications. IACUCs should expedite review of SSP-generated proposals and requests.

8. Each institution conducting research should identify an individual with overall responsibility for administering the research program, including the proposal review process. AZA should consider employing a North American research coordinator to be based in the Conservation and Science Office.

Levels of Justification

The working group agrees that both basic and applied research may be appropriate under zoo sponsorship so long as that research promises a potential, nontrivial benefit to non-human animals, species, and/or ecosystems. The working group recognizes that in particular cases, these potential benefits must be weighed against potential welfare costs, and that different research proposals therefore will require different levels of justification. The working group recommends that the committee consider guidelines including but not limited to the following.

1. In general, commercial exploitation of animals under the guise of research, or rationalized for scientific research purposes, is unacceptable.

2. In general, higher levels of justification should be required when more invasive or disruptive procedures are proposed to be employed. The highest level of justification should be required for proposals that entail killing, euthanasia, mutilation, use of inhumanely obtained animals or biological materials, interference with reproduction or survival of endangered species, or significant physical and psychological suffering.

3. In considering the level of justification, IACUC reviewers should take into account the quality of the proposal including the significance of the question to be investigated, the appropriateness of the methodology, and the credentials of the investigator. In addition, the certainty and immediacy of the benefit resulting from the research (as specified above) should be considered.

4. In cases involving minimal welfare risks, the role of the reviewers should be limited to verifying that the research proposed falls within one of the following categories:
   (a) nonmanipulative observation studies
   (b) observation incidental to veterinary care
   (c) passive collection (of feces, urine, scent, menstrual blood, and saliva)
   (d) samples taken incident to normal management

5. In any research proposal involving more than minimal welfare risk, the conservation value of the subject should be considered. In general, researchers should attempt to choose their subjects from the lowest possible level of the following conservation value

hierarchy:

> nonsurplus endangered animals (highest)
> surplus endangered animals
> exotic animals from the collection
> exotic animals not from the collection
> domestic animals
> USDA-regulated laboratory animals
> non–USDA-regulated laboratory animals (lowest)

The committee considered and rejected a hierarchy based on criteria of supposed sentience or complexity.

6. In general, animals used in research should be housed according to applicable minimum standards, regardless of whether they are located on or off the premises of the institution. When available, AZA standards apply on site. In the absence of AZA standards, USDA standards apply to covered animals on site. USDA or AZA standards should apply to off-site animals, unless the IACUC specifically grants an exception. In absence of standards, housing and maintenance should be provided on the basis of all available information, including standards for the most closely related species or species that share the ecological, behavioral, or physiological characteristics of the species in question.

Awareness of Welfare Considerations

The working group believes that there is a need to increase awareness of welfare considerations in zoo-sponsored research within AZA member institutions. To this end, the group recommends that steps be taken, including but not limited to the following.

1. Institutional policies on animal welfare in connection with research should be widely disseminated within the institution. In particular, the policy statement should be provided to all potential research investigators and associated animal caretakers and keepers.
2. Animal welfare considerations should be specifically addressed in all training and in-service programs conducted by member institutions.
3. In the case of ongoing research projects, the institution should organize research and management committees, including representatives of the relevant keeper, curatorial, and research staff, to facilitate communication and monitor progress.

The working group believes that AZA institutions should strive to increase public awareness of their commitment to animal welfare in the conduct of scientific programs. Two recommended strategies are given here.

1. Research coordinators and individual researchers should work closely with institutional education departments and the AZA public relations and education departments to inform the public about their scientific activities. The information presented should acknowledge, candidly, the difficult choices that often are involved in the decision-making process.
2. Wherever possible, zoo representatives should reach out to animal welfare and animal protection groups, making themselves available to discuss their scientific programs with the members of those groups.

The members of the working group believe that zoo-sponsored research has a contribution to make in advancing our understanding of animal welfare and animal suffering. It also can lead to the development of techniques to promote welfare and minimize suffering. To attain this goal, the working group recommends that the following specific actions be taken.

1. AZA should convene a meeting bringing together zoo architects, exhibit designers, individuals familiar with the scientific assessment of animal welfare, and experts on environmental enrichment to discuss the relationship between welfare and zoo design.
2. AZA, in cooperation with animal welfare organizations, should produce and distribute a sourcebook compiling existing research on the scientific assessment of animal welfare and suffering. This volume should include reprints of important articles and an annotated bibliography.
3. AZA and its member institutions should define and pursue a research agenda on the topic of scientific assessment of animal welfare and animal suffering. Specific issues to be addressed could include
   (a) keeper-animal interaction
   (b) pre- and postexhibit occupancy studies
   (c) heart-rate telemetry relative to stress
   (d) animal habitat selection studies
   (e) hormonal studies relative to stress (e.g., cortisol metabolites in urine under different conditions)
   (f) feeding strategies, diet, and nutrition (e.g., effects of diet on morphology)
   (g) stress reduction through behavioral training

Report of Group 6: Public Relations, Fund-Raising, and Disclosure

Agreements

The members of the working group were in agreement on the following issues:

1. *Marketing activities.* Special events and promotions should support the mission statement of the institution. Such events should reflect environmental, humane, and conservation ethics, recognizing that all institutional objectives, goals, and missions must be reflected in all aspects of its operation.
2. *Educational role.* The educational role of zoos should work in concert with the conservation role and should be reflected as such in public relations and fund-raising presentations.
3. *Roadside zoos.* It is our ethical responsibility to help improve or eliminate substandard facilities.
4. *Public disclosure.* Every effort should be made to be factual in answering public inquiries; however, institutions must retain the prerogative to disseminate specialized information only to those trained to interpret it properly. Management should communicate with all staff so they can impart facts to the public.
5. *Individual versus species emphasis.* Both individuals and species represent positive public relations and fund-raising values for zoos. Facilities need to acknowledge more readily

that individual animals matter. Public relations and fund-raising efforts should better emphasize contributions of individuals to species survival.

6. *Underwriting.* Zoos must acknowledge their fiduciary responsibility to operate as self-sustaining businesses. The self-sustaining motive is necessary to fund the efficient and humane operation of zoos and aquariums. The methods used to meet these goals should be regularly reviewed to ensure that they do not compromise the institutional mission or the dignity of their collection.

7. *Communications.* Employees should be familiarized with the practices and policies of their institutions. Ongoing open dialogue on key issues relevant to the conservation mission should build consensus and understanding among staff.

8. *Sponsorship and funding.* Support for increased conservation and education efforts will require an expanded field of funding partnerships. Zoos should evaluate potential sponsors by their current policies and future agendas, in addition to their past performance, and in relation to the institution's mission.

9. *Clarification of terms.* Euphemisms, jargon, and esoteric terms have the potential to mislead the public. It is essential that zoos and aquariums communicate practices and policies in terms the public can understand.

Recommendations

From those areas of consensus, the working group drew the following guidelines.

1. Institutions should review special events and promotions for support of overall mission statement.

2. The AZA should reaffirm the equivalency of education and conservation in its mission statement.

3. Individual institutions should give greater priority to interdisciplinary education in several ways:
    (a) by incorporating education at all levels of operation
    (b) by increasing exposure to living animals, which develops empathy, which translates into conservation action
    (c) by attempting to evaluate and optimize educational opportunities

4. Institutions should include social sciences in their research and operating endeavors.

5. Animal welfare groups, zoos, and aquariums should work together to improve humane standards in substandard facilities.
    (a) AZA should review and strengthen its accreditation standards.
    (b) Government regulatory agencies should be pressured to reevaluate and strengthen current standards and enforcement measures.

6. Individual animals should be promoted both as individuals and as a part of larger groups (e.g., population, species, and ecosystem), thus expanding public sympathy from one to many.

7. Attempts should be made to identify consensus areas with animal activist groups on topics such as euthanasia and surplus animals. Once accomplished, those areas should be supported by mutual public endorsement.

8. Eschew obfuscation and cut the "bio-snot" approach.

GLOSSARY

The following definitions have been adopted for the purposes of this book.

American Zoo and Aquarium Association (AZA) A professional, voluntary association, formerly the American Association of Zoological Parks and Aquariums (AAZPA), representing 167 professionally managed zoological parks and aquariums and more than 6,000 individual members in the United States and Canada.

animal rights movement A movement based on the philosophy that each sentient individual, human or nonhuman, has morally inviolable rights. The movement seeks to abolish all actions that violate individual rights.

animal welfare movement A movement based on a commitment to reduce suffering in animals and to minimize negative impacts (especially pain) that might result from human interactions with or use of animals.

aquarium *See* zoo.

biological diversity The variety of life forms, the ecological roles they perform, and the genetic diversity they contain.

captive breeding The propagation of wild animals under controlled conditions. Controlled conditions may involve maintaining animals in anything from relatively small, indoor enclosures to huge outdoor paddocks (e.g., in zoos or even in small, fenced national parks or equivalent reserves).

captive wild animal A nondomesticated animal that is maintained under controlled conditions.

conservation Careful protection of populations, species, reserves, parks, or ecosystems for long-term sustainability; planned management of a natural resource to prevent over-exploitation, destruction, or neglect.

conservation biology A relatively new discipline, comprising both applied and basic science, which seeks to use the power of the scientific method to preserve biological

329

diversity. It is among the most interdisciplinary of sciences, embracing not only biology and ecology but also economics and the social sciences.

contraception Temporary prevention of pregnancy through the use of hormonal implants, oral administration of hormones, or other methods.

culling Regulation or reduction of animal populations (especially surplus animals) through humane and acceptable methods (euthanasia).

reintroduction Release of captive-bred animals into a natural, relatively natural, or restored habitat in order to reestablish a population that has been extirpated or to augment existing but depleted wild populations.

Species Survival Plan A scientifically based, voluntary, cooperative captive breeding program administered by AZA in which threatened or endangered species are managed collectively for genetic variability, demographic stability, behavioral compatibility, and long-term viability.

sterilization Permanent alteration of animals, either surgically or chemically, to eliminate the possibility of future reproduction.

surplus animal This term has been defined in many ways but generally refers to an individual animal that has made its genetic contribution to a managed population and is not essential for future scientific studies or to maintain social-group stability or traditions, that displays abnormal behavior to the extent that it is no longer compatible with its social group or able to reproduce, that has reached an age that makes it no longer compatible with its social group (i.e., the age at which it would normally disperse from its natal group in the wild), that carries a genetically transmittable disease or physical abnormality that interferes with its integration into social groups or with reproduction, or that carries a transmittable disease organism or parasite that places other individuals or the population at serious risk.

wild animal Animal or species of animal that has not been influenced substantially by the effects of domestication or artificial selection. *See also* captive wild animal.

zoo or aquarium A professionally managed zoological institution accredited by the American Zoo and Aquarium Association and having a collection of living animals used for conservation, scientific studies, public education, and public display.